Sensory Evaluation Techniques

3rd Edition

Sensory Evaluation Techniques

3rd Edition

Morten Meilgaard, D.Sc.
Senior Technical Advisor
The Stroh Brewery Company
Detroit, Michigan

Gail Vance Civille, B.S.
President
Sensory Spectrum, Inc.
Chatham, New Jersey

B. Thomas Carr, M.S.
Principal
Carr Consulting
Wilmette, Illinois

CRC Press
Boca Raton London New York Washington, D.C.

Library of Congress Cataloging-in-Publication Data

Meilgaard, Morten.
   Sensory evaluation techniques / Morten Meilgaard, Gail Vance
Civille, B. Thomas Carr. -- 3rd ed.
      p. cm.
   Includes bibliographical references.
   ISBN 0-8493-0276-5 (alk. paper)
   1. Sensory evaluation. I. Civille, Gail Vance. II. Carr, B.
Thomas. III. Title.
TA418.5.M45 1999
664′.07—dc21
99-19553

Visit the CRC Press Web site at www.crcpress.com

No claim to original U.S. Government works
International Standard Book Number 0-8493-0276-5
Library of Congress Card Number 99-19553
Printed in the United States of America   6 7 8 9 0
Printed on acid-free paper

Preface to the Third Edition

How does one plan, execute, complete, analyze, interpret, and report sensory tests? Hopefully, the practices and recommendations in this book cover all of those phases of sensory evaluation. The text is meant as a personal reference volume for food scientists, research and development scientists, cereal chemists, perfumers, and other professionals working in industry, academia, or government, who need to conduct good sensory evaluation. The book should also supply useful background to marketing research, advertising, and legal professionals who need to understand the results of sensory evaluation. It could also give a sophisticated general reader the same understanding.

Because the first edition was used as a textbook at the university and professional level, partly in courses taught by the authors, the second and third editions incorporate a growing number of ideas and improvements arising out of questions from students. The objective of the book is now twofold. First, as a "how to" text for professionals, it aims for a clear and concise presentation of practical solutions, accepted methods, and standard practices. Second, as a textbook for courses at the academic level, it aims to provide just enough theoretical background to enable the student to understand which sensory methods are best suited to particular research problems and situations, and how tests can best be implemented.

The authors do not intend to devote text and readers' time to resolving controversial issues, but a few had to be tackled. We take a fresh look at all statistical methods used for sensory tests, and we hope you like our straightforward approach. The second edition was the first book to provide an adequate solution to the problem of similarity testing. This was adopted and further developed by ISO TC34/SC12 on Sensory Evaluation, resulting in the current "unified" procedure (Chapter 6, Section II, p. 60) in which the user's choice of α- and β-risks defines whether difference or similarity is tested for. Another "first" is the unified treatment of all ranking tests with the Friedman statistic, in preference to Kramer's tables.

Chapter 11 on the Spectrum™ method of descriptive sensory analysis, developed by Civille, has been expanded. The philosophy behind Spectrum is threefold: (1) the test should be tailored to suit the objective of the study (and not to suit a prescribed format); (2) the choice of terminology and reference standards should make use not only of the senses and imagination of the panelists, but also of the accumulated experience of the sensory profession as recorded in the literature; and (3) a set of calibrated intensity scales is provided which permits different panels at different times and locations to obtain comparable and reproducible profiles. The chapter now contains full descriptive lexicons suitable for descriptive analysis of a number of products, e.g., cheese, mayonnaise, spaghetti sauce, white bread, cookies, and toothpaste. Also new is a set of revised flavor intensity scales for attributes such as crispness, juiciness, and some common aromatics, and two training exercises.

The authors wish the book to be cohesive and readable; we have tried to substantiate our directions and organize each section so as to be meaningful. We do not want the book to be a turgid set of tables, lists, and figures. We hope we have provided structure to the methods, reason to the procedures, and coherence to the outcomes. Although our aim is to describe all tests in current use, we want this to be a reference book that can be read for understanding as well as a handbook that can serve to describe all major sensory evaluation practices.

The organization of the chapters and sections is also straightforward. Chapter 1 lists the steps involved in a sensory evaluation project, and Chapter 2 briefly reviews the workings of our senses. In Chapter 3, we list what is required of the equipment, the tasters, and the samples, while in Chapter 4, we have collected a list of those psychological pitfalls that invalidate many otherwise good studies. Chapter 5 discusses how sensory responses can be measured in quantitative terms. In Chapter 6, we describe all the common sensory tests for difference, the Triangle, Duo-trio, etc., and, in Chapter 7, the various attribute tests, such as ranking and numerical intensity scaling. Thresholds and just-noticeable differences are briefly discussed in Chapter 8, followed by what we consider the main chapters: Chapter 9 on selection and training of tasters, Chapters 10 and 11 on descriptive testing, and Chapter 12 on affective tests (consumer tests).

The body of text on statistical procedures is found in Chapters 13 and 14, but, in addition, each method (Triangle, Duo-trio, etc.) in Chapters 6 and 7 is followed by a number of examples showing how statistics are used in the interpretation of each. Basic concepts for tabular and graphical summaries, hypothesis testing, and the design of sensory panels are presented in Chapter 13. We refrain from detailed discussion of statistical theory, preferring instead to give examples. Chapter 14 discusses multifactor experiments that can be used, for example, to screen for variables that have large effects on a product, to identify variables that interact with each other in how they affect product characteristics, or to identify the combination of variables that maximize some desirable product characteristic, such as consumer acceptability. Chapter 14 also contains a discussion of multivariate techniques that can be used to summarize large numbers of responses with fewer, meaningful ones, to identify relationships among responses that might otherwise go unnoticed, and to group respondents of samples that exhibit similar patterns of behavior. New in the third edition is a detailed discussion of data-relationship techniques used to link data from diverse sources collected on the same set of samples. The techniques are used to identify relationships, for example, between instrumental and sensory data or between sensory and consumer data.

At the end of the book, the reader will find guidelines for the choice of techniques and for reporting results, plus the usual glossaries, indexes, and statistical tables.

With regard to terminology, the terms "assessor," "judge," "panelist," "respondent," "subject," and "taster" are used interchangeably, as are "he," "she," and "(s)he" for the sensory analyst (the sensory professional, the panel leader) and for individual panel members.

Morten Meilgaard
Gail Vance Civille
B. Thomas Carr

The Authors

Morten C. Meilgaard, M.Sc., D.Sc., F.I. Brew, is Visiting Professor (emeritus) of Sensory Science at the Agricultural University of Denmark and Vice President of Research (also emeritus) at the Stroh Brewery Co., Detroit, MI. He studied biochemistry and engineering at the Technical University of Denmark, to which he returned in 1982 to receive a doctorate for a dissertation on beer flavor compounds and their interactions. After 6 years as a chemist at the Carlsberg Breweries, he worked from 1957 to 1967 and again from 1989 as a worldwide consultant on brewing and sensory testing. He served for 6 years as Director of Research for Cervecería Cuauhtémoc in Monterrey, Mexico, and for 25 years with Stroh. At the Agricultural University of Denmark his task was to establish Sensory Science as an academic discipline for research and teaching.

Dr. Meilgaard's professional interest is the biochemical and physiological basis of flavor, and more specifically the flavor compounds of hops and beer and the methods by which they can be identified, namely, chemical analysis coupled with sensory evaluation techniques. He has published over 70 papers. He is the recipient of the Schwarz Award and the Master Brewers Association Award of Merit for studies of compounds that affect beer flavor. He is founder and past president of the Hop Research Council of the U.S., and is past chairman of the Scientific Advisory Committee of the U.S. Brewers Association. For 14 years he was chairman of the Subcommittee on Sensory Analysis of the American Society of Brewing Chemists. He has chaired the U.S. delegation to the ISO TC34/SC12 Subcommittee on Sensory Evaluation.

Gail Vance Civille is President of Sensory Spectrum, Inc., a management consulting firm involved in the field of sensory evaluation of foods, beverages, pharmaceuticals, paper, fabrics, personal care, and other consumer products. Sensory Spectrum provides guidance in the selection, implementation, and analysis of test methods for solving problems in quality control, research, development, production, and marketing. She has trained several flavor and texture descriptive profile panels in her work with industry, universities, and government.

As a Course Director for the Center for Professional Advancement and Sensory Spectrum, Ms. Civille has conducted several workshops and courses in basic sensory evaluation methods as well as advanced methods and theory. In addition, she has been invited to speak to several professional organizations on different facets of sensory evaluation.

Ms. Civille has published several articles on general sensory methods, as well as sophisticated descriptive flavor and texture techniques. A graduate of the College of Mount Saint Vincent, New York with a B.S. degree in Chemistry, Ms. Civille began her career as a product evaluation analyst with the General Foods Corporation.

B. Thomas Carr is Principal of Carr Consulting, a research consulting firm that provides project management, product evaluation, and statistical support services to the food, beverage, personal care, and home care industries. He has over 18 years of experience in applying statistical techniques to all phases of research on consumer products. Prior to founding Carr Consulting, Mr. Carr held a variety of business and technical positions in the food and food ingredient industries. As Director of Contract Research for NSC Technologies/NutraSweet, he identified and coordinated outside research projects that leveraged the technical capabilities of all the groups within NutraSweet R&D, particularly in the areas of product development, analytical services and sensory evaluation. Prior to that, as Manager of Statistical Services at both NutraSweet and Best Foods, Inc., he worked closely with the sensory, analytical, and product development groups on the design and analysis of a full range of research studies in support of product development, QA/QC, and research guidance consumer tests.

Mr. Carr is a member of the U.S. delegation to the ISO TC34/SC12. He is actively involved in the statistical training of scientists and has been an invited speaker to several professional organizations on the topics of statistical methods and statistical consulting in industry. Since 1979, Mr. Carr has supported the development of new food ingredients, consumer food products, and OTC drugs by integrating the statistical and sensory evaluation functions into the mainstream of the product development effort. This has been accomplished through the application of a wide variety of statistical techniques including design of experiments, response surface methodology, mixture designs, sensory/instrumental correlation, and multivariate analysis.

Mr. Carr received his B.A. degree in Mathematics from the University of Dayton, and his Master's degree in Statistics from Colorado State University.

Acknowledgments

The authors wish to thank our associates at work and our families at home for thoughts and ideas, for material assistance with typing and editing, and for emotional support. Many people have helped with suggestions and discussion over the years. Contributors at the concept stage were Andrew Dravnieks, Jean Eggert, Roland Harper, Derek Land, Elizabeth Larmond, Ann Noble, Rosemarie Pangborn, John J. Powers, Patricia Prell, and Elaine Skinner. Improvements in later editions were often suggested by readers and were given form with help from our colleagues from two Subcommittees on Sensory Evaluation, ASTM E-18 and ISO TC34/SC12, of whom we would like to single out Louise Aust, Donna Carlton, Sylvie Issanchou, Sandy MacRae, Magni Martens, Suzanne Pecore, Rick Schifferstein, and Pascal Schlich. We also thank our colleagues Clare Dus, Kathy Foley, Kernon Gibes, Stephen Goodfellow, Dan Grabowski, Marie Rudolph, and Barbara Pirmann for help with illustrations and ideas, and The Stroh Brewery Company, Sensory Spectrum, Inc., and The NutraSweet Co. for permission to publish and for the use of their facilities and equipment.

Acknowledgments

Dedication

to
Manon, Frank, and Cathy

Table of Contents

CHAPTER 13

STATISTICAL TABLES

1 Introduction to Sensory Techniques

CONTENTS

I. INTRODUCTION

This introduction is in three parts. The first part lists some reasons why sensory tests are done and traces briefly the history of their development. The second part introduces the basic approach of modern sensory analysis, which is to treat the panelists as measuring instruments. As such, they are highly variable and very prone to bias, but they are the only instruments that will measure what we want to measure, so we must minimize the variability and control the bias by making full use of the best existing techniques in psychology and psychophysics. In the third part we show how these techniques are applied with the aid of seven practical steps.

II. DEVELOPMENT OF SENSORY TESTING

Sensory tests of course have been conducted for as long as there have been human beings evaluating the goodness and badness of food, water, weapons, shelters, and everything else that can be used and consumed.

The rise of trading inspired slightly more formal sensory testing. A buyer, hoping that a part would represent the whole, would test a small sample of a shipload. Sellers began to set their prices on the basis of an assessment of the quality of goods. With time, ritualistic schemes of *grading* wine, tea, coffee, butter, fish, and meat developed, some of which survive to this day.

Grading gave rise to the professional taster and consultant to the budding industries of foods, beverages, and cosmetics in the early 1900s. A literature grew up which used the term "organoleptic testing" (Pfenninger, 1979) to denote supposedly objective measurement of sensory attributes. In reality, tests were often subjective, tasters too few, and interpretations open to prejudice.

Pangborn (1964) traces the history of systematic "sensory" analysis which is based on wartime efforts of providing acceptable food to American forces (Dove, 1946, 1947) and on the development of the Triangle test in Scandinavia (Bengtsson and Helm, 1946; Helm and Trolle, 1946). A major role in the development of sensory testing was played by the Food Science Department at the University of California at Davis, resulting in the book by Amerine, Pangborn, and Roessler (1965).

Scientists have developed sensory testing, then, very recently as a formalized, structured, and codified methodology, and they continue to develop new methods and refine existing ones. The current state of sensory techniques is recorded in the dedicated journals *Chemical Senses, Journal of Sensory Studies,* and *Journal of Texture Studies*; in the proceedings of the Pangborn Symposia (triennial) and the international Sensometrics Group (biannual), both usually published as individual papers in the journal *Food Quality & Preference*; and the proceedings of the Weurman Symposia (triennial, but published in book form, e.g., Martens et al., 1987; Bessière and Thomas, 1990). Sensory papers presented to the Institute of Food Technologists are usually published in the IFT's *Journal of Food Science* or *Food Technology*.

The methods that have been developed serve economic interests. Sensory testing can establish the worth of a commodity or even its very acceptability. Sensory testing evaluates alternative courses in order to select the one that optimizes value for money. The *principal uses* of sensory techniques are in quality control, product development, and research. They find application not only in characterization and evaluation of foods and beverages, but also in other fields such as environmental odors, personal hygiene products, diagnosis of illnesses, testing of pure chemicals, etc. The primary function of sensory testing is to conduct valid and reliable tests, which provide data on which sound decisions can be made.

III. HUMAN SUBJECTS AS INSTRUMENTS

Dependable sensory analysis is based on the skill of the sensory analyst in optimizing the four factors, which we all recognize because they are the ones which govern any measurement (Pfenninger, loc. cit.).

1. Definition of the problem: We must define precisely what it is we wish to measure; important as this is in "hard" science, it is much more so with senses and feelings.
2. Test design: Not only must the design leave no room for subjectivity and take into account the known sources of bias, but it also must minimize the amount of testing required to produce the desired accuracy of results.
3. Instrumentation: The test subjects must be selected and trained to give a reproducible verdict; the analyst must work with them until he/she knows their sensitivity and bias in the given situation.
4. Interpretation of results: Using statistics, the analyst chooses the correct null hypothesis and the correct alternative hypothesis, and draws only those conclusions which are warranted by the results.

Tasters, as measuring instruments, are (1) quite variable over time; (2) very variable among themselves; and (3) very prone to bias. To account adequately for these requires (1) that measurements be repeated; (2) that enough subjects (often 20 to 50) are made available so that verdicts are representative; and (3) that the sensory analyst respects the many rules and pitfalls which govern panel attitudes (see Chapter 4). Subjects vary innately in sensitivity by a factor of 2 to 10 or more (Meilgaard and Reid, 1979; Pangborn, 1981) and should not be interchanged halfway through a project. Subjects must be selected for sensitivity and must be trained and retrained (see Chapter 9) until they fully understand the task at hand. The annals of sensory testing are replete with results that are unreliable because many of the panelists did not understand the questions and/or the terminology used in the test, did not recognize the flavor or texture parameters in the products, or did not feel comfortable with the mechanics of the test or the numerical expressions used.

For these reasons and others, it is very important for the sensory analyst to be actively involved in the development of the scales used to measure the panelists' responses. A good scale requires much study, must be based on a thorough understanding of the physical and chemical factors that govern the sensory variable in question, and requires several reference points and thorough training of the panel. It is unreasonable to expect that even an experienced panelist would possess the necessary knowledge and skill to develop a scale that is consistently accurate and precise. Only through the direct involvement of a knowledgeable sensory professional in the development of scales can one obtain descriptive analyses, for instance, that will mean the same in 6 months time as they do today.

The chain of sensory perception — When sensory analysts study the relationship between a given physical stimulus and the subject's response, the outcome is often regarded as a one-step process. In fact there are at least three steps in the process, as shown below. The stimulus hits the sense organ and is converted to a nerve signal which travels to the brain. With previous experiences in memory, the brain then interprets, organizes, and integrates the incoming sensations into perceptions. Lastly, a response is formulated based on the subject's perceptions (Schiffman, 1990):

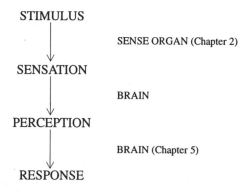

In dealing with the fact that humans often yield varied responses to the same stimulus, sensory professionals need to understand that differences between two people's verdicts can be caused either by a difference in the sensation they receive because their sense organs differ in sensitivity or by a difference in their mental treatment of the sensation, e.g., because of a lack of knowledge of the particular odor, taste, etc. or because of lack of training in expressing what they sense in words and numbers. Through training and the use of references we can attempt to shape the mental process so that subjects move toward showing the same response to a given stimulus.

IV. CONDUCTING A SENSORY STUDY

The best products are developed in those organizations where the sensory professional is more than the provider of a specialized testing service. Only through a process of total involvement can he or she be in the position of knowing what tests are necessary and appropriate at every point during the life of a research project. The sensory professional (like the statistician) must take an active role in developing the research program, collaborating with the other involved parties on the development of the experimental designs that ultimately will be used to answer the questions posed. Erhardt (1978) divides the role of the sensory analyst into the following seven practical tasks:

Determine the project objective — Defining the needs of the project leader is the most important requirement for conducting the right test. Were the samples submitted as a product improvement, to permit cost reduction or ingredient substitution, or as a match of a competitor's product? Is one sample expected to be similar or different from others, preferred or at parity, variable in one or more attributes? If this critical step is not carried out, the sensory analyst is unlikely to use the appropriate test or to interpret the data correctly.

Determine the test objective — Once the objective of the project can be clearly stated, the sensory analyst and the project leader can determine the test objective: overall difference, attribute difference, relative preference, acceptability, etc. Avoid attempting to answer too many questions in one single test. A good idea is for the sensory analyst and project leader to record in writing before the test is initiated the project objective, the test objective, and a brief statement of how the test results will be used.

Screen the samples — During the discussion of project and test objectives the sensory analyst should examine all of the sensory properties of the samples to be tested. This enables the sensory analyst to use test methods which take into account any sensory biases introduced by the samples. For example, visual cues (color, thickness, sheen) may influence overall difference responses, such as those provided in a Triangle test, e.g., to measure differences due to sweetness of sucrose vs. aspartame. In such a case, an attribute test would be more appropriate. In addition, product screening provides information on possible terms to be included in the scoresheet.

Design the test — After defining the project and test objectives and screening the samples, the sensory analyst can proceed to design the test. This involves selection of the test technique (see Chapters 6, 7, 8, 10, 11, 12, and 15); selecting and training subjects (see Chapter 9); designing the accompanying scoresheet (ballot, questionnaire); specifying the criteria for sample preparation and presentation (see Chapter 3); and determining the way in which the data will be analyzed (see Chapters 13 and 14). Care must be taken, in each step, to adhere to the principles of statistical design of experiments to ensure that the most sensitive evaluation of the test objective is obtained.

Conduct the test — Even when technicians are used to carry out the test, the sensory analyst is responsible for ensuring that all the requirements of the test design are met.

Analyze the data — As the procedure for analysis of the data was determined at the test design stage, the necessary expertise and statistical programs, if used, will be ready to begin data analysis as soon as the study is completed. The data should be analyzed for the main treatment effect (test objective) as well as other test variables, such as order of presentation, time of day, different days, and/or subject variables such as age, sex, geographic area, etc.*

Interpret and report results — The initial clear statement of the project and test objectives will enable the sensory analyst to review the results, express them in terms of the stated objectives, and make any recommendations for action that may be warranted. The latter should be stated clearly and concisely in a written report that also summarizes the data, identifies the samples, and states the number and qualification of subjects (see Chapter 16).

The main purpose of this book is to help the sensory analyst develop the methodology, subject pool, facilities, and test controls required to conduct analytical sensory tests with trained and/or experienced tasters. In addition, Chapter 12 discusses the organization of consumer tests, i.e., the use of naive consumers (nonanalytical) for large-scale evaluation, structured to represent the consumption and responses of the large population of the product market.

The role of sensory evaluation is to provide valid and reliable information to R&D, production, and marketing in order for management to make sound business decisions about the perceived sensory properties of products. The ultimate goal of any sensory program should be to find the most cost-effective and efficient method with which to obtain the most sensory information. When possible, internal laboratory difference or descriptive techniques are used in place of more expensive and time-consuming consumer tests to develop cost-effective sensory analysis. Further cost savings may be realized by correlating as many sensory properties as possible with instrumental, physical, or chemical analyses. In some cases it may be found possible to replace a part of routine sensory testing with cheaper and quicker instrumental techniques.

---

* It is assumed that computers will be used to analyze the data and possibly also in the booth to record the subject's verdict. Many paperless systems are available, but this field changes from month to month, and the reader is referred to the sensory literature, e.g., *Journal of Sensory Studies* and *Sensory Forum* (the newsletter of the Sensory Evaluation Division, Institute of Food Technologists [IFT]). Exhibitions at meetings of sensory professionals are another good source of information about available systems.

REFERENCES

Amerine, M.A., Pangborn, R.M., and Roessler, E.B., 1965. *Principles of Sensory Evaluation of Food*. Academic Press, New York, 602 pp.

Bengtsson, K. and Helm, E., 1946. Principles of taste testing. *Wallerstein Lab. Commun.* 9, 171.

Bessière, Y. and Thomas, A.F., Eds., 1990. *Flavour Science and Technology*. John Wiley & Sons, Chichester, 369 pp.

Dove, W.E., 1946. Developing food acceptance research. *Science* 103, 187.

Dove, W.E., 1947. Food acceptability: its determination and evaluation. *Food Technol.* 1, 39.

Erhardt, J.P., 1978. The role of the sensory analyst in product development. *Food Technol.* 32(11), 57.

Helm, E. and Trolle, B., 1946. Selection of a taste panel. *Wallerstein Lab. Commun.* 9, 181.

Martens, M., Dalen, G.A., and Russwurm, H., Jr., 1987. *Flavour Science and Technology*. John Wiley & Sons, Chichester, 566 pp.

Meilgaard, M.C. and Reid, D.S., 1979. Determination of personal and group thresholds and the use of magnitude estimation in beer flavour chemistry. In: *Progress in Flavour Research*, Land, D.G. and Nursten, H.E., Eds. Applied Science Publishers, London, 67–73.

Pangborn, R.M., 1964. Sensory evaluation of food: a look backward and forward. *Food Technol.* 18, 1309.

Pangborn, R.M., 1981. Individuality in response to sensory stimuli. In: *Criteria of Food Acceptance. How Man Chooses What He Eats,* Solms, J. and Hall, R.L., Eds. Forster-Verlag, Zürich, 177.

Pfenninger, H.B., 1979. Methods of quality control in brewing. *Schweizer Brauerei-Rundschau* 90, 121.

Schiffman, H.R., 1996. *Sensation and Perception. An Integrated Approach,* 4th ed. John Wiley & Sons, New York.

2 Sensory Attributes and the Way We Perceive Them

CONTENTS

I. INTRODUCTION

This chapter reviews: (1) the sensory attributes with which the book is concerned, e.g., the appearance, odor, flavor, and feel of different products and (2) the mechanisms by which we perceive those attributes, e.g., the visual, olfactory, gustatory, and tactile/kinesthetic senses. The briefness of the chapter is dictated by the scope of the book and is not an indication of the importance of the subject. We urge the sensory professional to study our references (pp. 21–22) and to build a good library of books and journals on sensory perception. Sensory testing is an inexact science. Experimental designs need to be based on a thorough knowledge of the physical and chemical factors behind the attributes of interest. Results of sensory tests as a rule have many possible explanations, and the chances of misinterpretation can be much reduced by every bit of new knowledge about the workings of our senses and the true nature of product attributes.

II. SENSORY ATTRIBUTES

We tend to perceive the attributes of a food item in the following order:

- Appearance
- Odor/aroma/fragrance
- Consistency and texture
- Flavor (aromatics, chemical feelings, taste)

However, in the process of perception, most or all of the attributes overlap, i.e., the subject receives a jumble of near-simultaneous sensory impressions, and without training he or she will not be able to provide an independent evaluation of each. This section gives examples of the types of sensory attributes that exist in terms of the way in which they are perceived and the terms which may be associated with them.

Flavor, in this book, is the combined impression perceived via the chemical senses from a product in the mouth, i.e., it does not include appearance and texture. The term "aromatics" is used to indicate those volatile constituents that originate from food in the mouth and are perceived by the olfactory system via the posterior nares.

A. APPEARANCE

As every shopper knows, the appearance is often the only attribute on which we can base a decision to purchase or consume. Hence, we become adept at making wide and risky inferences from small clues, and test subjects will do the same in the booth. It follows that the sensory analyst must pay meticulous attention to every aspect of the appearance of test samples (Amerine et al., 1965, p. 399; McDougall, 1983) and must often attempt to obliterate or mask much of it with colored lights, opaque containers, etc.

General appearance characteristics are listed below, and an example of the description of appearance with the aid of scales is given in Chapter 11, Appendix 11.1A, pp. 177–178.

Color	A phenomenon that involves both physical and psychological components: the perception by the visual system of light of wavelengths 400 to 500 nm (blue), 500 to 600 nm (green and yellow), and 600 to 800 nm (red), commonly expressed in terms of the hue, value, and chroma of the Munsell color system. The evenness of color as opposed to uneven or blotchy appearance is important. Deterioration of food is often accompanied by a color change. Good descriptions of procedures for sensory evaluation of appearance and color are given by Clydesdale (1984), McDougall (1988) and Lawless and Heymann (1998).
Size and shape	Length, thickness, width, particle size, geometric shape (square, circular, etc.), distribution of pieces, e.g., of vegetables, pasta, prepared foods, etc.; size and shape as indications of defects (Kramer and Twigg, 1973; Gatchalian, 1981).
Surface texture	The dullness or shininess of a surface, the roughness vs. evenness; does the surface appear wet or dry, soft or hard, crisp or tough?
Clarity	The haze (Siebert et al., 1981) or opacity (McDougall, 1988) of transparent liquids or solids, the presence or absence of particles of visible size.
Carbonation	For carbonated beverages, the degree of effervescence observed on pouring. This is commonly measured with Zahm-Nagel instruments* and may be judged as follows:

Carbonation (vols)	Carbonation (% weight)	Degree of effervescence	Examples
1.5 or less	0.27 or less	None	Still drinks
1.5 to 2.0	0.27 to 0.36	Light	Fruit drinks
2.0 to 3.0	0.36 to 0.54	Medium	Beer, cider
3.0 to 4.0	0.54 to 0.72	High	Soft drinks, champagne

---

* 74 Jewett Ave., Buffalo, NY, tel. 716-833-1532, or via Mangel, Scheuermann & Oeters, 107 Witmer Rd., Horsham, PA 19044.

B. Odor/Aroma/Fragrance

The odor of a product is detected when its volatiles enter the nasal passage and are perceived by the olfactory system. We talk of odor when the volatiles are sniffed through the nose (voluntarily or otherwise). Aroma is the odor of a food product, and fragrance is the odor of a perfume or cosmetic. Aromatics, as mentioned earlier, are the volatiles perceived by the olfactory system from a substance in the mouth. (The term smell is not used in this book because it has a negative connotation [= malodor] to some people while to others it is the same as odor.)

The amount of volatiles that escape from a product is affected by the temperature and by the nature of the compounds. The vapor pressure of a substance increases exponentially with temperature according to the following formula:

$$\log p = -0.05223a/T + b \tag{2.1}$$

where p is the vapor pressure in mmHg, T is the absolute temperature ($T = t°C + 273.1$), and a and b are substance constants that can be found in handbooks (Howard, 1996; Lyman et al., 1982). Volatility is also influenced by the condition of a surface: at a given temperature, more volatiles escape from a soft, porous, and humid surface than from a hard, smooth, and dry one.

Many odors are released only when an enzymic reaction takes place at a freshly cut surface (e.g., the smell of an onion). Odorous molecules must be transmitted by a gas, which can be the atmosphere, water vapor, or an industrial gas, and the intensity of the perceived odor is determined by the proportion of such gas which comes into contact with the observer's olfactory receptors (Laing, 1983).

The sorting of fragrance/aroma sensations into identifiable terms continues to challenge sensory professionals (see Chapter 10 on descriptive analysis and Civille and Lyon [1996] for a database of descriptors for many products). There is not at this point any internationally standardized odor terminology. The field is very wide; according to Harper (1972) some 17,000 odorous compounds are known, and a good perfumer can differentiate 150 to 200 odorous qualities. Many terms may be ascribed to a single compound (thymol = herb-like, green, rubber-like), and a single term may be associated with many compounds (lemon = α-pinene, β-pinene, α-limonene, β-ocimene, citral, citronellal, linalool, α-terpineol, etc.).

C. Consistency and Texture

The third set of attributes to be considered are those perceived by sensors in the mouth, other than taste and chemical feelings. By convention we refer to:

- Viscosity (for homogeneous Newtonian liquids)
- Consistency (for non-Newtonian or heterogeneous liquids and semisolids)
- Texture (for solids or semisolids)

"Viscosity" refers to the rate of flow of liquids under some force, such as gravity. It can be accurately measured and varies from a low of approximately 1 cP (centipoise) for water or beer to 1000s of cP for jelly-like products. "Consistency" (of fluids like purees, sauces, juices, syrups, jellies, and cosmetics) in principle must be measured by sensory evaluation (Kramer and Twigg, 1973a); in practice, some standardization is possible by the aid of consistometers (Kramer and Twigg, 1973b; Mitchell, 1984). "Texture" is much more complex, as shown by the existence of the *Journal of Texture Studies*. Texture can be defined as the sensory manifestation of the structure or inner makeup of products in terms of their:

- Reaction to stress, measured as mechanical properties (such as hardness/firmness, adhesiveness, cohesiveness, gumminess, springiness/resilience, viscosity) by the kinesthetic sense in the muscles of the hand, fingers, tongue, jaw, or lips
- Tactile feel properties, measured as geometrical particles (grainy, gritty, crystalline, flaky) or moisture properties (wetness, oiliness, moistness, dryness) by the tactile nerves in the surface of the skin of the hand, lips, or tongue

Table 2.1 lists general mechanical, geometrical, and moisture properties of foods, skincare products, and fabrics. Note that across such a wide variety of products the textural properties are all derived from the same general classes of texture terms measured kinesthetically or tactile-wise. Additional food texture terms are listed in Chapter 11, Appendices 11.2C, 11.2D, and 11.3. Recommended reviews of texture perception and measurement are those by De Man et al. (1976), Bourne (1982), and Brennan (1988).

D. FLAVOR

Flavor, as an attribute of foods, beverages, and seasonings, has been defined (Amerine et al., 1965, p. 549) as the sum of perceptions resulting from stimulation of the sense ends that are grouped together at the entrance of the alimentary and respiratory tracts, but for purposes of practical sensory analysis, the authors prefer to follow Caul (1957) and restrict the term to *the impressions perceived via the chemical senses from a product in the mouth*. Defined in this manner, flavor includes:

- The aromatics, i.e., olfactory perceptions caused by volatile substances released from a product in the mouth via the posterior nares
- The tastes, i.e., gustatory perceptions (salty, sweet, sour, bitter) caused by soluble substances in the mouth
- The chemical feeling factors, which stimulate nerve ends in the soft membranes of the buccal and nasal cavities (astringency, spice heat, cooling, bite, metallic flavor, umami taste)

A large number of individual flavor words are listed in Chapter 11, and in Civille and Lyon (loc. cit.).

E. NOISE

The noise produced during mastication of foods or handling of fabrics is a minor but not negligible sensory attribute (see the review by Brennan, 1966). It is common to measure the pitch, loudness, and persistence of sounds produced by foods or fabrics. The pitch and loudness of the sound contribute to the overall sensory impression. Differences in pitch of some rupturing foods (crispy, crunchy, brittle) provide sensory input, which we use in the assessment of freshness/staleness. Oscilloscopic measurements by Vickers and Bourne (1976a, 1976b) permitted sharp differentiation between products described as crispy and those described as crunchy. Kinesthetically these differences correspond to measurable differences in hardness, denseness, and the force of rupture (fracturability) of a product. A crackly or crisp sound on handling can bias a subject to expect stiffness in a fabric. The duration or persistence of sound from a product often suggests other properties, e.g., strength (crisp fabric), freshness (crisp apples, potato chips), toughness (squeaky clams), or thickness (plopping liquid). Table 2.2 lists common noise characteristics of foods, skincare products, and fabrics.

TABLE 2.1
The Components of Texture

MECHANICAL PROPERTIES: reaction to stress, measured kinesthetically
  Hardness: force to attain a given deformation

Foods	Skincare	Fabrics
Firmness (compression)	Force to compress	Force to compress
Hardness (bite)	Force to spread	Force to stretch

  Cohesiveness: degree to which sample deforms (rather than ruptures)

Foods	Skincare	Fabrics
Cohesive	Cohesive	Stiffness
Chewy	Short	
Fracturable (crispy/crunchy)		
Viscosity	Viscosity	

  Adhesiveness: force required to remove sample from a given surface

Foods	Skincare	Fabrics
Sticky (tooth/palate)	Tacky	Fabric/fabric friction
Tooth pack	Drag	Hand friction (drag)

  Denseness: compactness of cross-section

Foods	Skincare	Fabrics
Dense/heavy	Dense/heavy	Fullness/flimsy
Airy/puffy/light	Airy/light	

  Springiness: rate of return to original shape after some deformation

Foods	Skincare	Fabrics
Springy/rubbery	Springy	Resilient (tensile and compression)
		Cushy (compression)

GEOMETRICAL PROPERTIES: perception of particles (size, shape, orientation) measured
  by tactile means
  Smoothness: absence of all particles
  Gritty: small, hard particles
  Grainy: small particles
  Chalky/powdery: fine particles (film)
  Fibrous: long, stringy particles (fuzzy fabric)
  Lumpy/bumpy: large, even pieces or protrusions

MOISTURE PROPERTIES: perception of water, oil, fat, measured by tactile means
  Moistness: amount of wetness/oiliness present, when not certain whether oil and/or water
  Moisture release: amount of wetness/oiliness exuded

Foods	Skincare	Fabrics
Juicy	Wets down	Moisture release

  Oily: amount of liquid fat
  Greasy: amount of solid fat

TABLE 2.2
Common Noise Characteristics
of Foods, Skincare Products,
and Fabrics

Noise Properties[a]

Pitch: frequency of sound

Foods	Skincare	Fabric
Crispy	Squeak	Crisp
Crunchy		Crackle
Squeak		Squeak

Loudness: intensity of sound
Persistence: endurance of sound over time

[a] Perceived sounds (pitch, loudness, persistence) and auditory measurement.

III. THE HUMAN SENSES

The five senses are so well covered in textbooks (Piggott, 1988; Kling and Riggs, 1971; Sekuler and Blake, 1990; Geldard, 1972) that a description here is superfluous. We shall limit ourselves to pointing out some characteristics which are of particular importance in designing and evaluating sensory tests. A clear and brief account of the sensors and neural mechanisms by which we perceive odor, taste, vision, and hearing, followed by a chapter on intercorrelation of the senses, is found in *Basic Principles of Sensory Evaluation* (ASTM, 1968). Touch and kinesthesis are well described by Brennan (1988). Lawless and Heymann (1998, p. 67) review what is known about sensory interaction within and between the sensory modalities.

A. VISION

Light entering the lens of the eye (see Figure 2.1) is focused on the retina, where the rods and cones convert it to neural impulses which travel to the brain via the optic nerve. Some aspects of *color* perception which must be considered in sensory testing are:

- Subjects often give consistent responses about an object color even when filters are used to mask differences (perhaps because the filters mask hues but not always brightness and chroma).
- Subjects are influenced by adjoining or background color and the relative sizes of areas of contrasting color; blotchy appearance, as distinct from an even distribution of color, affects perception.
- The gloss and texture of a surface also affect perception of color.
- Color vision differs among subjects; degrees of color blindness exist, e.g., inability to distinguish red and orange, or blue and green; exceptional color sensitivity also exists, allowing certain subjects to discern visual differences which the panel leader cannot see.

The chief lesson to be learned from this is that attempts to mask differences in color or appearance are often unsuccessful and if undetected can cause the experimenter to erroneously conclude that a difference in flavor or texture exists.

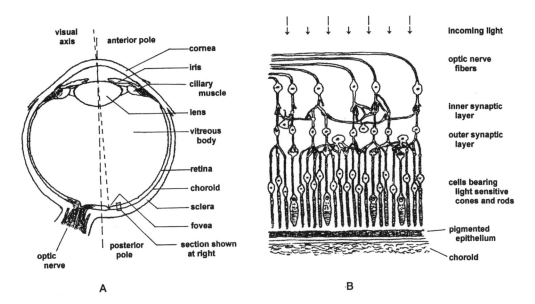

FIGURE 2.1   The eye, showing the lens, retina, and optic nerve. The entrance of the optic nerve is the blind spot. The fovea is a small region, central to the retina, which is highly sensitive to detail and consists entirely of cones. (Modified from Hochberger, J.E., *Perception*, Prentice-Hall, Englewood Cliffs, NJ, 1964.)

B. Touch

The group of perceptions generally described as the sense of touch can be divided into "somesthesis" (tactile sense, skinfeel) and "kinesthesis" (deep pressure sense or proprioception), both of which sense variations in physical pressure. Figure 2.2 shows the several types of nerve endings in the skin surface, epidermis, dermis, and subcutaneous tissue. These surface nerve ends are responsible for the somesthetic sensations we call touch, pressure, heat, cold, itching, and tickling. Deep pressure, kinesthesis, is felt through nerve fibers in muscles, tendons, and joints whose main purpose is to sense the tension and relaxation of muscles. Figure 2.3 shows how the nerve fibers are buried within a tendon. Kinesthetic perceptions corresponding to the mechanical movement of muscles (heaviness, hardness, stickiness, etc.) result from stress exerted by muscles of the hand, jaw, or tongue and the sensation of the resulting strain (compression, shear, rupture) within the sample being handled, masticated, etc. The surface sensitivity of the lips, tongue, face, and hands is much greater than that of other areas of the body, resulting in ease of detection of small force differences, particle size differences, and thermal and chemical differences from hand and oral manipulation of products.

C. Olfaction

Airborne odorants are sensed by the olfactory epithelium which is located in the roof of the nasal cavity (see Figure 2.4). Odorant molecules are sensed by the millions of tiny, hair-like cilia which cover the epithelium, by a mechanism which is one of the unsolved mysteries of science (see below). The anatomy of the nose is such that only a small fraction of inspired air reaches the olfactory epithelium via the nasal turbinates, or via the back of the mouth on swallowing (Maruniak, 1988). Optimal contact is obtained by moderate inspiration (sniffing) for 1 to 2 sec (Laing, 1983). At the end of 2 sec, the receptors have adapted to the new stimulus and one must allow 5 to 20 sec or longer for them to de-adapt before a new sniff can produce a full-strength sensation. A complication is that the odorant(s) can fill the location in which a stimulus is to be tested, thus reducing the ability of the subject to detect a particular odorant or differences among similar

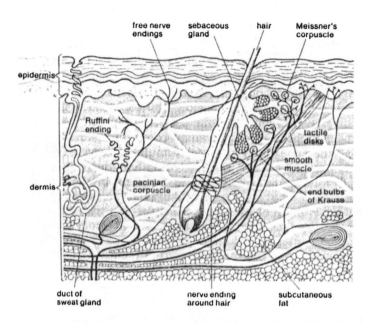

FIGURE 2.2 Composite diagram of the skin in cross section. Tactile sensations are transmitted from a variety of sites, e.g., the free nerve endings and the tactile discs in the epidermis, and the Meissner corpuscles, end bulbs of Krause, Ruffini endings, and Pacinian corpuscles in the dermis. (From Gardner, E., *Fundamentals of Neurology*, 5th ed., W.B. Saunders Company, Philadelphia, 1968. With permission.)

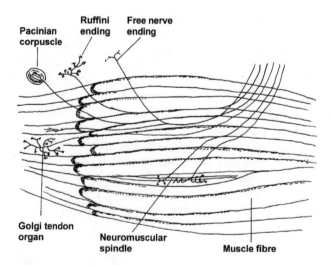

FIGURE 2.3 Kinesthetic sensors in a tendon and muscle joint. (Modified from Geldard, F.A., *The Human Senses*, John Wiley & Sons, New York, 1972.)

odorants. Cases of total odor blindness, anosmia, are rare, but specific anosmia, inability to detect specific odors, is not uncommon (Harper, 1972). For this reason, potential panelists should be screened for sensory acuity using odors similar to those to be tested eventually.

Whereas the senses of hearing and sight can accommodate and distinguish stimuli which are 10^4- to 10^5-fold apart, the olfactory sense has trouble accommodating a 10^2-fold difference between the threshold and the concentration which produces saturation of the receptors. On the other hand,

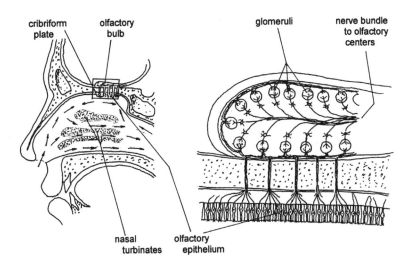

FIGURE 2.4   Anatomy of the olfactory system. Signals generated by the approx. 1000 types of sensory cells pass through the cribriform plate into the olfactory bulb where they are sorted through the glomeruli before passing on to the higher olfactory centers. (Modified from Axel, R., "The molecular logic of smell," in *Scientific American*, October 1995, 154–159.)

while the ear and the eye each can sense only one type of signal, namely, oscillations of air pressure and electromagnetic waves of 400 to 800 nm wavelength, the nose has enormous discriminating power: as mentioned previously, a trained perfumer can identify 150 to 200 different odor qualities (odor types) (Harper, loc. cit.).

The sensitivity of the receptors to different chemicals varies over a range of 10^{12} or more (Harper, loc. cit.; Meilgaard, 1975). Typical thresholds (see Table 2.3) vary from 1.3×10^{19} molecules per milliliter air for ethane to 6×10^7 molecules per milliliter for allyl mercaptan, and it is very likely that substances exist or will be discovered which are even more potent. Note that water and air are not in the list because these bathe the sensors and hence cannot be sensed.

The table illustrates how easily a chemical standard can be misflavored by impurities. For example, an average observer presented with a concentration of 1.5×10^{17} molecules per milliliter of methanol 99.99999% pure but containing 0.00001% ionone would perceive a $10 \times$ threshold of methanol but a $100 \times$ threshold odor of ionone. Purification by distillation and charcoal treatment might reduce the level of ionone impurity tenfold, but it would still be at $10 \times$ threshold, or as strong as the odor of methanol itself.

The most sensitive gas chromatographic method can detect approximately 10^9 molecules per milliliter. This means that there are numerous odor substances, probably thousands occurring in nature, for which the nose is 10- or 100-fold more sensitive than the gas chromatograph. We are a long way away from being able to predict an odor from gas chromatographic analysis.

We do not know how the receptors generate the signals which they send to the brain, but we have some ideas (see Maruniak, 1988). We know absolutely nothing definite about the way the brain handles the incoming information to produce in our minds the perception of a given odor quality and the strength of that quality, and even much less how the brain handles mixtures of different qualities whose signals arrive simultaneously via the olfactory nerve. Moncrieff (1951) lists 14 conditions which any theory of olfaction must fulfill. Beets (1978) envisaged the existence of patterns and subpatterns of molecules on the surface of the epithelium. Odorous molecular compounds on the incoming air, in their many orientations and conformations, are attracted and briefly interact with particular sites in the pattern. Buck and Axel (1991) found evidence in mammalian olfactory mucosa of a family of approximately 1000 genes, coding for as many

TABLE 2.3
Some Typical Threshold Values in Air

Chemical substance	Molecules/mL air
Allyl mercaptan	6×10^7
Ionone	1.6×10^8
Vanillin	2×10^9
sec-Butyl mercaptan	2×10^8
Butyric acid	1.4×10^{11}
	6.9×10^9
Acetaldehyde	9.6×10^{12}
Camphor	5×10^{12}
	6.4×10^{12}
	$\cdot 4 \times 10^{14}$
Trimethylamine	2.2×10^{13}
Phenol	7.7×10^{12}
	2.6×10^{13}
	1×10^{13}
	1.3×10^{15}
Methanol	1.1×10^{16}
	1.9×10^{16}
Ethanol	2.4×10^{15}
	2.3×10^{15}
	1.6×10^{17}
Phenyl ethanol	1.7×10^{17}
Ethane	1.3×10^{19}

From Harper, R., *Human Senses in Action*, Churchill Livingstone, London, 1972, 253. With permission. (The figures quoted should be treated as orders of magnitude only, since they may have been derived by different methods.)

different olfactory receptor proteins. This group then found (Axel, 1995) that each olfactory neuron expresses one and only one receptor protein, and the neurons that express a given protein all terminate in two and only two of the approximately 2000 glomeruli in the olfactory bulb. It seems to follow that the work of the brain is one of sorting and learning. For example, figuratively speaking, it may learn that if glomeruli nos. 205, 464, and 1723 are strongly stimulated, that equals the odor of geraniol.

Human sensitivity to various odors may be measured by dual flow olfactometry, using *n*-butanol as a standard (Moskowitz et al., 1974). Subjects show varying sensitivity to odors depending on hunger, satiety, mood, concentration, presence or absence of respiratory infections, and, in women, menstrual cycle and pregnancy (Maruniak, loc. cit.)

Given the complexity of the receptors and the enormous range shown by the thresholds for different compounds, it is not surprising that different people may receive very different perceptions from a given odorant. The largest study ever in this area was *The National Geographic Smell Survey*; see Gibbons and Boyd (1986); Gilbert and Wysocki (1987); Wysocki and Gilbert (1989); and Wysocki et al. (1991). The lesson to be learned from this is that if the job is to characterize or identify a new odor, one needs as large a panel as possible if the results are to have any validity for the general population. A panel of one can be very misleading.

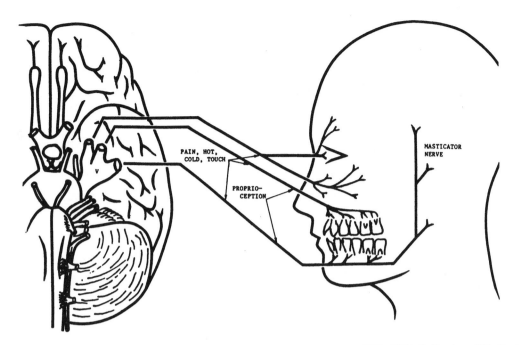

FIGURE 2.5　Pathway of the trigeminus (V) nerve. (Modified from Netter, F.H., *CIBA Collection of Medical Illustrations*, Vols. 1 and 3, Ciba-Geigy Corp., Summit, NJ, 1973.) Readers interested in greater detail are referred to Boudreau (1986).

D. CHEMICAL/TRIGEMINAL FACTORS

Chemical irritants such as ammonia, ginger, horseradish, onion, chili peppers, menthol, etc. stimulate the trigeminal nerve ends (see Figure 2.5), causing perceptions of burn, heat, cold, pungency, etc. in the mucosa of the eyes, nose, and mouth. Subjects often have difficulty separating trigeminal sensations from olfactory and/or gustatory ones. Experiments which seek to determine olfactory sensitivity among subjects can be confounded by responses to trigeminal rather than olfactory sensations.

For most compounds, the trigeminal response requires a concentration of the irritant which is orders of magnitude higher than that which stimulates the olfactory or gustatory receptors. Trigeminal effects assume practical significance: (1) when the olfactory or gustatory threshold is high, e.g., for short-chain compounds such as formic acid or for persons with partial anosmia or ageusia, and (2) when the trigeminal threshold is low, e.g., for capsaicin.

The trigeminal response to mild irritants (such as carbonation, mouthburn caused by high concentrations of sucrose and salt in confections and snacks, the heat of peppers and other spices) may contribute to, rather than distract from, acceptance of a product.

E. GUSTATION

Like olfaction, gustation is a chemical sense. It involves the detection of stimuli dissolved in water, oil, or saliva by the taste buds which are located primarily on the surface of the tongue as well as in the mucosa of the palate and areas of the throat. Figure 2.6 shows the taste system in three different perspectives. Compared with olfaction, the contact between a solution and the taste epithelium on the tongue and walls of the mouth is more regular in that every receptor is immersed for at least some seconds. There is no risk of the contact being too brief, but there is ample opportunity of oversaturation. Molecules causing strong bitterness probably bind to the receptor proteins, and some may remain for

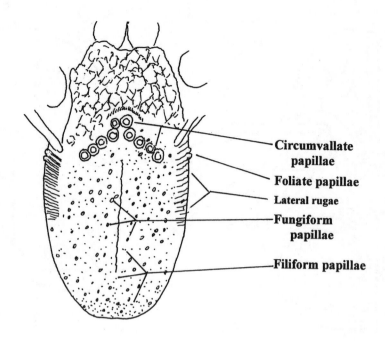

Circumvallate
papillae

Foliate papillae

Lateral rugae

Fungiform
papillae

Filiform papillae

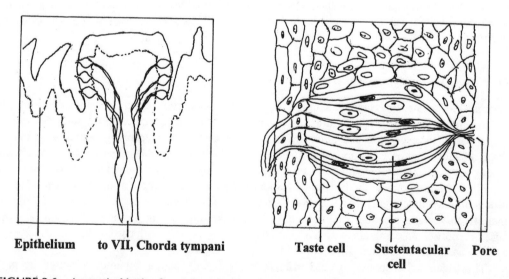

Epithelium   to VII, Chorda tympani   Taste cell   Sustentacular   Pore
cell

FIGURE 2.6 Anatomical basis of gustation, showing the tongue, a cross section of a fungiform papilla, and a section thereof showing a taste bud with receptor cells. The latter carry chemosensitive villi that protrude through the taste pore. At the opposite end their axons continue until they make synaptic contact with cranial nerve VII, the chorda tympani. The surrounding epithelial cells will eventually differentiate into taste receptor cells that renew the current ones as often as once a week.

hours or days (the cells of the olfactory and gustatory epithelium are renewed on average every 6 to 8 days [Beidler, 1960]). The prudent taster should take small sips and keep each sip in the mouth for only a couple of seconds, then wait (depending on the perceived strength) for 15 to 60 sec before tasting again. The first and second sip are the most sensitive, and one should train oneself to accomplish in those first sips all the mental comparisons and adjustments required by the task at hand. Where this is not possible, e.g., in a lengthy questionnaire with more than eight or ten questions and untrained subjects, the experimenter must be prepared to accept a lower level of discrimination.

The gustatory sensors are bathed in a complex solution, the saliva (which contains water, amino acids, proteins, sugars, organic acids, salts, etc.), and they are fed and maintained by a second solution, the blood (which contains an even more complex mixture of the same substances). Hence, we can only taste differences in the concentration of many substances, not absolute concentrations, and our sensitivity to levels (e.g., of salt) that are lower than those in saliva is low and ill defined. Typical thresholds for taste substances are shown in Figure 2.7.

The range between the weakest tastant, sucrose, and the strongest, Strophantin (a bitter alkaloid) is no more than 10^4, much smaller than the range of 10^{12} shown by odorants. The figure also shows the range of thresholds for 47 individuals, and it is seen that the most and least sensitive individuals generally differ by a factor of 10^2. In the case of phenylthiocarbamide (also phenylthiourea) a bimodal distribution is seen (Amerine et al., 1965, p. 109): the population consists of two groups, one with an average threshold of 0.16 g/100 mL and another with an average threshold of 0.0003 g/100 mL. Vanillin (Meilgaard et al., 1982) is another substance which appears to show two peaks, but the total number of compounds for which bimodal distributions have been reported (Amoore, 1977) is small, and their role in food preferences or in odor and taste sensitivity in general is a subject which has not been explored.

In addition to the concentration of a taste stimulus, other conditions in the mouth which affect taste perception are the temperature, viscosity, rate, duration, and area of application of the stimulus, the chemical state of the saliva, and the presence of other tastants in the solution being tasted. The incidence of ageusia, or the absence of the sense of taste, is rare. However, variability in taste sensitivity, especially for bitterness with various bitter agents, is quite common.

F. HEARING

Figure 2.8 shows a cross section of a human ear. Vibrations in the local medium, usually air, cause the eardrum to vibrate. The vibrations are transmitted via the small bones in the middle ear to create hydraulic motion in the fluid of the inner ear, the cochlea, a spiral canal covered in hair cells which when agitated send neural impulses to the brain. Students of crispness, etc. should familiarize themselves with the concepts of intensity, measured in decibels, and pitch, determined by the frequency of sound waves. A possible source of variation or error which must be controlled in such studies is the creation and/or propagation of sound inside the cranium but outside of the ear, e.g., by movement of the jaws or teeth and propagation via the bone structure.

Psychoacoustics is the science of building vibrational models on a sound oscilloscope to represent perceived sound stimuli such as pitch, loudness, sharpness, roughness, etc. These models work for simple sounds but not for more complex ones. They can be used to answer questions such as "What kind of sound?" and "How loud?", but they often fail to provide a sound that is appropriate to what the listener expects.

Recently academics and engineers who are responsible for sound characteristics of products have realized the need for a common vocabulary to describe sound attributes for complex sounds. This is because automobile, airframe, and industrial and consumer products manufacturers are concerned with sounds that their products produce, and how humans respond to those sounds. Author Civille is collaborating with an ANSI working group [ANSI S12/WG 36] to create a comprehensive list of words to describe different sounds and their component attributes along with a selected group of reference sounds comprised of real and synthetic auditory examples. Examples of sound attributes such as hiss, squeal, rumble, flutter, and buzz will be made available on a compact disc to be used as a tool to understand the complex sounds of products.

IV. PERCEPTION AT THRESHOLD AND ABOVE

Perhaps this is the place to warn the reader that a threshold is not a constant for a given substance, but rather a constantly changing point on the sensory continuum from nonperceptible to easily

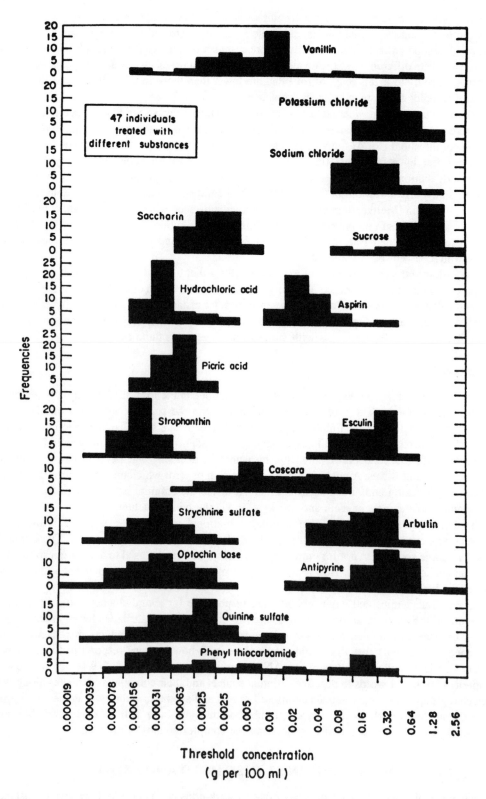

FIGURE 2.7   Distribution of taste thresholds for 47 individuals. (From Amerine et al., *Principles of Sensory Evaluation of Food*, Academic Press, New York, 1965, 109. With permission.)

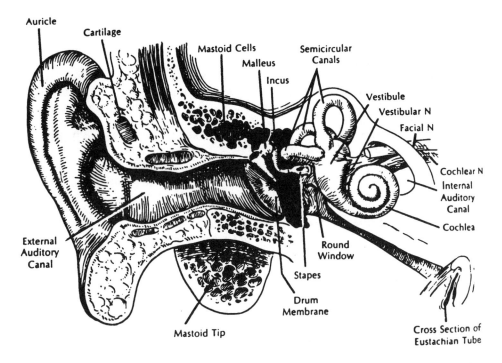

FIGURE 2.8 A semidiagrammatic drawing of the ear. (From Kling, J.W. and Riggs, L.A., Eds., *Woodworth & Schlosberg's Experimental Psychology*, 3rd ed., Holt, Rinehart & Winston, New York, 1971. With permission.)

perceptible (see Chapter 8). Our thresholds change with moods and the time of the biorhythm, and also with hunger and satiety. Compounds with identical thresholds can show very different rates of increase in intensity with concentration, hence the use of the threshold as a yardstick of intensity of perception must be approached with considerable caution (Bartoshuk, 1978; Pangborn, 1984). In practical studies involving products which emit mixtures of large numbers of flavor-active substances, in which the purpose is to detect those compounds which play a role in the flavor of the product, the threshold has some utility, provided the range covered does not extend too far from the threshold, e.g., from $0.5 \times$ threshold to $3 \times$ threshold. Above this range, intensity of odor or taste must be measured by scaling (see Chapter 5, p. 52).

REFERENCES

Amerine, M.A., Pangborn, R.M., and Roessler, E.B., 1965. *Principles of Sensory Evaluation of Food.* Academic Press, New York, 602 pp.
Amoore, J.E., 1977. Specific anosmia and the concept of primary odors. *Chem. Senses Flavor* 2, 267–281.
ASTM, 1968. *Basic Principles of Sensory Evaluation.* Standard Technical Publication 433, American Society for Testing and Materials, Philadelphia, 110 pp.
Axel, R., 1995. The molecular logic of smell. *Scientific American*, October 1995, 154–159.
Bartoshuk, L.M., 1978. The psychophysics of taste. *J. Am. Clin. Nutr.* 31, 1068.
Beets, M.G.J., 1978. *Structure-Activity Relationships in Human Chemoreception.* Applied Science, London, 408 pp.
Beidler, L.M., 1960. Physiology of olfaction and gustation. *Ann. Otol. Rhinol. Laryngol.* 69, 398.
Blakeslee, A.F. and Salmon, T.N., 1935. Genetics of sensory thresholds: individual taste reactions to different substances. *Proc. Natl. Acad. Sci. U.S.A.,* 21, 78.
Boudreau, J.C., 1986. Neurophysiology and human taste sensations. *J. Sensory Stud.* 1(3/4), 185.
Buck, L. and Axel, R., 1991. A novel multigene family may encode odorant receptors: A molecular basis for odor reception. *Cell* 65(1), 175–187.

Caul, J.F., 1957. The profile method of flavor analysis. *Adv. Food Res.* 7, 1.

Civille, G.V. and Lyon, B.G., Eds., 1996. *Aroma and Flavor Lexicon for Sensory Evaluation. Terms, Definitions, References, and Examples.* ASTM Data Series Publication DS 66, American Society for Testing and Materials, W. Conshohocken, PA, 158 pp. (plus diskette).

Clydesdale, F.M., 1984. Color measurement. In: *Food Analysis. Principles and Techniques,* Vol. 1, Gruenwedel, D.W. and Whitaker, J.R., Eds. Marcel Dekker, New York, p. 95ff.

De Man, J.M., Voisey, P.W., Rasper, V.F., and Stanley, D.W., 1976. *Rheology and Texture in Food Quality.* AVI Publishing, Westport, CT.

Gatchalian, M.M., 1981. *Sensory Evaluation Methods with Statistical Analysis.* College of Home Economics, University of the Philippines, Diliman, p. 34.

Geldard, F.A., 1972. *The Human Senses,* 2nd ed. John Wiley & Sons, New York.

Gibbons and Boyd, 1986. The intimate sense of smell. *National Geographic Magazine* 170, 324–361.

Gilbert, A.N. and Wysocki, C.J., 1987. The National Geographic Smell Survey Results. *National Geographic Magazine* 172, 514–525.

Harper, R., 1972. *Human Senses in Action.* Churchill Livingston, London, 358 pp.

Howard, P., 1996. *Handbook of Physical Properties of Organic Chemicals.* CRC Press, Boca Raton, FL, 2000 pp.

Kling, J.W. and Riggs, L.A., Eds., 1971. *Woodworth & Schlosberg's Experimental Psychology,* 3rd ed. Holt, Rinehart & Winston, New York, 1279 pp.

Kramer, A. and Twigg, B.A., 1966. *Quality Control for the Food Industry,* Vol. 1. AVI Publishing, Westport, CT.

Laing, D.G., 1983. Natural sniffing gives optimum odor perception for humans. *Perception* 12, 99.

Lawless, H.T., 1986. Sensory interaction in mixtures. *J. Sensory Stud.* 1(3/4), 259.

Lawless, H.T. and Heymann, H., 1998. *Sensory Evaluation of Food. Principles and Practices.* Chapman & Hall, New York, 819 pp.

Maruniak, J.A., 1988. The sense of smell. In: *Sensory Analysis of Foods,* 2nd ed., Piggott, J.R., Ed. Elsevier, London, p. 25 ff.

McDougall, D.B., 1983. Assessment of the appearance of food. In: *Sensory Quality in Foods and Beverages: Its Definition, Measurement and Control,* Williams, A.A. and Atkin, R.K., Eds. Ellis Horwood, Chichester/Verlag Chemie, Deerfield Beach, p. 121ff.

McDougall, D.B., 1988. Color vision and appearance measurement. In: *Sensory Analysis of Foods,* 2nd ed. Piggott, J.R., Ed. Elsevier, London, p. 103ff.

Meilgaard, M.C., 1975. Flavor chemistry of beer. II. Flavor and threshold of 239 aroma volatiles. *Tech. Q. Master Brew. Assoc. Am.* 12, 151–168.

Meilgaard, M.C., Reid, D.S., and Wyborski, K.A., 1982. Reference standards for beer flavor terminology system. *J. Am. Soc. Brewing Chem.* 40, 119–128.

Mitchell, J.R., 1984. Rheological techniques. In: *Food Analysis. Principles and Techniques,* Vol. 1, Gruenwedel, D.W. and Whitaker, J.R., Eds. Marcel Dekker, New York, p. 151ff.

Moncrieff, R.W., 1951. *The Chemical Senses.* Leonard Hill, London, pp. 32, 89, 220, 355.

Moskowitz, H.R., Dravnieks, A., Cain, W.S., and Turk, A., 1974. Standardized procedure for expressing odor intensity. *Chem. Senses Flavor* 1, 235–237.

Netter, F.H., 1973. *The CIBA Collection of Medical Illustrations,* Vols. 1 and 3. Ciba-Geigy Corp., Summit, NJ.

Pangborn, R.M., 1984. Sensory techniques of food analysis. In: *Food Analysis. Principles and Techniques,* Vol. 1, Gruenwedel, D.W. and Whitaker, J.R., Eds. Marcel Dekker, New York, pp. 37ff. (see p. 49).

Piggott, J.R., Ed., 1988. *Sensory Analysis of Foods,* 2nd ed. Elsevier, London, chapters 1–4.

Sekuler, R. and Blake, R., 1990. *Perception,* 2nd ed. McGraw-Hill, New York, 525 pp.

Siebert, K.J., Stenroos, L.E., and Reid, D.S., 1981. Characterization of amorphous-particle haze. *J. Am. Soc. Brewing Chem.* 39, 1–11.

Vickers, Z.M. and Bourne, M.C., 1976. Crispness in foods. A review. A psychoacoustical theory of crispness. *J. Food Sci.* 41, 1153 and 1158.

Wysocki, C.J. and Gilbert, A.N., 1989. National Geographic Smell Survey. Effects of age are heterogeneous. In: *Nutrition and the chemical senses in aging: recent advances and current needs. Ann. New York Acad. Sci.,* Vol. 561.

Wysocki, C.J., Pierce, J.D., and Gilbert, A.N., 1991. Geographic, cross-cultural, and individual variation in human olfaction. In: *Smell and Taste in Health and Disease,* Getchell, T.V., Ed. Raven Press, New York, pp. 287–314.

3 Controls for Test Room, Product, and Panel

CONTENTS

I. INTRODUCTION

Many variables must be controlled if the results of a sensory test are to measure the true product differences under investigation. It is convenient to group these variables under three major headings:

1. Test controls: The test room environment, the use of booths or a round table, the lighting, the room air, the preparation area, the entry and exit areas
2. Product controls: The equipment used, the way samples are screened, prepared, numbered, coded, and served
3. Panel controls: The procedure to be used by a panelist evaluating the sample in question

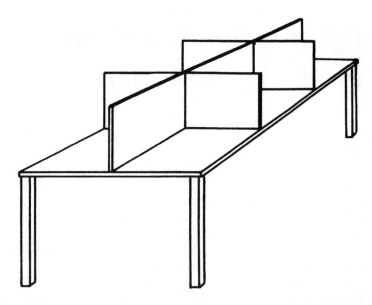

FIGURE 3.1   Simple booths consisting of a set of dividers placed on a table.

II. TEST CONTROLS

The physical setting must be designed so as to minimize the subjects' biases, maximize their sensitivity, and eliminate variables which do not come from the products themselves. Panel tests are costly because of the high cost of panelists' time. A high level of reduction of disturbing factors is easily justified. Dropoffs in panel attendance and panel motivation are universal problems, and management must clearly show the value it places on panel tests by the care and effort expended on the test area. The test area should be centrally located, easy to reach, and free of crowding and confusion, as well as comfortable, quiet, temperature controlled, and above all, free from odors and noise.

A. DEVELOPMENT OF TEST ROOM DESIGN

Since the first edition of this book (1987), test room design has matured, as reflected in national and international standards (ASTM, 1976; European Cooperation, 1995; ISO, 1988, 1998). A move toward requiring accreditation of sensory services under ISO 9000 has accelerated a trend toward uniformly high standards, e.g., with separate air exhausts from each booth.

Early test rooms made allowance for six to ten subjects and consisted of a laboratory bench or conference table on which samples were placed. The need to prevent subjects from interacting, thus introducing bias and distraction, led to the concept of the booth (see Figure 3.1).

In a parallel development, the Arthur D. Little organization (Caul, 1957) argued that panelists should interact and come to a consensus, which required a round table with a "lazy Susan" on which reference materials were used to standardize terminology and scale values.

Current thinking often combines these two elements into: (1) a booth area, which is the principal room used for difference tests as well as some descriptive tests, and (2) a round table area used for training and/or other descriptive tasks (see Figure 3.2). Convenience dictates that a sample preparation area be located nearby, but separate from, the test room. Installations above a certain size also require office area, sample storage area, and data processing area.

FIGURE 3.2 Top: Circular table with "lazy Susan" used for consensus-type descriptive analysis. (Courtesy of Ross Products Division, Columbus, OH.) Bottom: Round table discussion used for descriptive analysis ballot development. (Courtesy of NutraSweet/Kelco Inc., Mt. Pleasant, IL.)

B. LOCATION

The panel test area should be readily accessible to all. A good location is one which most panel members pass on their way to lunch or morning break. If panel members are drawn from the outside, the area should be near the building entrance. Test rooms should be separated by a suitable distance from congested areas because of noise and the opportunity this would provide for unwanted socializing. Test rooms should be away from other noise and from sources of odor such as machine shops, loading docks, production lines, and cafeteria kitchens.

C. TEST ROOM DESIGN

1. The Booth

It is customary for one sample preparation area to serve six to eight booths. The booths may be arranged all side-by-side, in an L-shape, or with two sets of three to four booths facing each other across the serving area. The L-shape represents the most efficient use of the "work triangle" concept in kitchen design, resulting in a minimum of time and distance covered by technicians in serving samples. One unit of six to eight booths will accommodate a moderate test volume of 300 to 400 sittings per year of panels up to 18 to 24 members. For higher volumes of testing and/or larger panels, multiple units served from one or several preparation areas are recommended. Consideration also should be given to placement of the technicians' monitor(s) and central processing unit(s) for any automated data handling system.

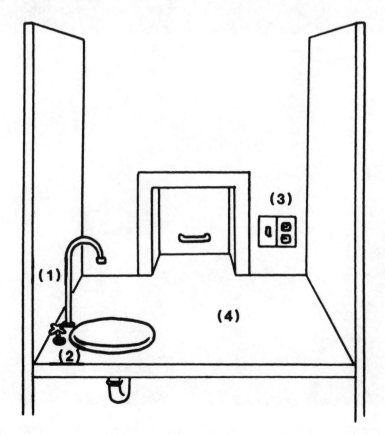

FIGURE 3.3   Sensory evaluation booth with hatch (in background) for receipt and return of sample tray. (1) Tap water; (2) small sink; (3) electrical outlet and signal switch to panel attendant; (4) table covered with odorless Formica or other easy-to-clean surface.

Figure 3.3 shows a typical booth which is 27 to 32 in. wide with an 18 to 22 in. deep counter installed at the same height as the sample preparation table (normally 36 in.). Space can be allowed for installation of a PC monitor and a keyboard, if required. The dividers should extend approximately 18 in. above the countertop in order to reduce visual and auditory distraction between booths. The dividers may extend from the floor to the ceiling/soffit for complete privacy (with the design allowing for adequate ventilation and/or cleaning), or it may be suspended from the wall enclosing only the torso and head of the assessor. The latter is preferred in most cases, as claustrophobia is a permanent problem whereas assessors soon learn to refrain from looking over shoulders or uttering loud comments on the quality of samples. A minimum free distance of 4 ft is recommended as a corridor to allow easy access to the booths.

Special booth features — A small stainless steel sink and a water faucet are usually included for rinsing. These are mandatory for evaluation of such products as mouthwashes, toothpastes, and household items, but are not recommended for solid foods which may plug the traps. Filtered water may be required if odor-free tap water is unavailable.

A signal system is sometimes included so that the panel supervisor knows when an assessor is ready for a sample or has a question. Usually this takes the form of a switch in each booth that will trigger a signal light for that booth in the sample preparation area. It may include an exterior light panel which indicates to incoming subjects those booths which are available.

A direct computer entry system located in each panel booth (Malek et al., 1982) requires a 32 in. width to accommodate the entry device (keypad, tablet digitizer, CRT terminal).

Sample trays may be carried to each booth if they consist of nonodorous items that will keep their condition for 10 to 20 min. If these conditions are not fulfilled, the sample preparation area must be located behind the booths and a hatch provided through which the tray can be passed once the subject is in place. Three types of pass-throughs are in use (see Figure 3.4). The sliding door (vertical or horizontal) requires the least space. The types known as the breadbox and the carousel are more effective in preventing passage of odors or visible cues from the preparation area to the subject.

The materials of construction in the booths and surrounding area should be odor-free and easy to clean. Formica and stainless steel are the most common surface materials.

2. Descriptive Evaluation and Training Area

As a minimum, this function may be filled by a table in the panel leader's office where standards may be served as a means of educating panel members. At the other extreme, if descriptive analysis is a common requirement or if needs for training and testing are large, the following equipment is recommended:

- A conference-style room with several tables which can be arranged as required by the size and objective of the group
- Audiovisual equipment which may include an "electronic white board" capable of making hard copies of results, etc. entered on it
- Separate preparation facilities for reference samples used to illustrate the descriptors; depending on type, these may include a storage space (frozen, refrigerated, or room temperature, perhaps sealed to prevent odors from escaping) and a holding area for preparing the references (perhaps hooded)

3. Preparation Area

The preparation area is a laboratory which must permit preparation of all of the possible and foreseeable combinations of test samples at the maximum rate at which they are required. Each booth area and descriptive analysis area should have a separate preparation laboratory so as to

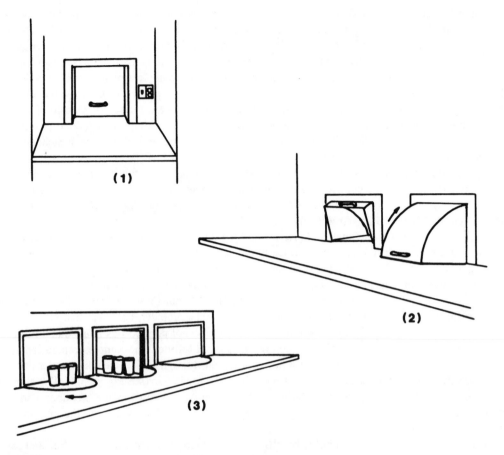

FIGURE 3.4 Three types of hatch for passing samples to and from the panelists. (1) Sliding door; (2) breadbox; (3) carousel.

maximize the technician's ability to prepare, present, and clean up each study. Typically, the preparation area includes immediate access to the following in addition to any specialized equipment dictated by the type of samples:

- A laboratory bench flush with the hatches so that sample trays will slide through
- Benches, kitchen range, ovens, etc. for preparation
- Refrigerator and freezer for storage of samples
- Storage for glassware, dishes, glasses, trays, etc.
- Dishwashers, disposers, trash compactors, wastebaskets, sinks, etc.
- Frozen storage for panel member treats, if used
- Large garbage containers for quick disposal of used product, etc.

Consideration should be given to company and local recycling policies so that appropriate receptacles are available in the preparation area.

4. Office Facilities

An office is usually situated within view of the panel booths as someone must be present while testing is in progress. It may be convenient to locate records, storage space, and any computer terminals and other hardware (printers, digitizers, plotters, etc.) in the same area so that the panel leader's time may be effectively utilized. Equipment such as paging phones and printers should be at a sufficient distance to avoid distracting the subjects.

5. Entrance and Exit Areas

In large facilities it is advisable to separate entrance and exit areas for assessors so as to prevent unwanted exchange of information. The exit area commonly contains a desk where assessors can study the identity of the day's samples and where they may receive a "treat" to encourage participation. If some of the panelists are nonemployees, the entrance/exit area should contain sufficient waiting room with comfortable seats, coat closet or coat rack, and separate restrooms.

6. Storage

Space must be allocated for storage of:

- Samples prior to preparation, after preparation, and at the time of serving
- Reference samples and controls or standards under the appropriate temperature and humidity conditions
- Large volumes of disposable containers and utensils
- Clean-up materials
- Paper scoresheets before and after use

Figures 3.5 and 3.6 show typical layouts of medium and large-scale installations showing various facilities which may be located around the booth area.

D. GENERAL DESIGN FACTORS

1. Color and Lighting

The color and lighting in the booths should be planned to permit adequate viewing of samples while minimizing distractions (Amerine et al., 1965; Malek et al., 1982; ASTM, 1986; Poste et al., 1991; European Cooperation, 1995; Chambers and Wolf, 1996; ISO, 1988, 1998). Walls should be off-white; the absence of hues of any color will prevent unwanted difference in appearance. Booths should have even, shadow-free illumination at 70 to 80 footcandles (fc) (typical of an office area). If appearance is critical, rheostat control may be used to vary the light intensity up to 100 fc. Incandescent lighting allows wider variation and permits the use of colored lights (see below), but more heat is generated requiring adequate cooling. Fluorescent lighting generates less heat and allows a choice of whiteness (i.e., cool white, warm white, simulated North daylight, see Figure 3.7).

Colored lights — A common feature of many panel booths is a choice of red, green, and/or blue lighting at low intensity obtained through the use of colored bulbs or special filters. The lights are used to mask visual differences between samples in difference tests calling for the subject to determine by taste (or by feel, if appropriate) which samples are identical.

Many colored bulbs emit sufficient white light to be ineffective in reducing color differences. Theater gel filters are quite effective and may be placed in frames over recessed spotlights. Another alternative is a low pressure sodium lamp which emits light at a single wavelength. Suitable color masking lamps are available custom made from Trimble House Corp., 4658 Old Peach Tree Road, Norcross, GA 30071, tel.: 770 448-1972. Both the theater gels and color masking lamps remove colors, but do not eliminate differences in color intensity. The effect is that of black and white television with degrees of gray still detectable.

Pangborn (1967) notes that an abnormal level of illumination may in itself influence the assessor's impressions. An alternative is to choose methods other than simultaneous presentation

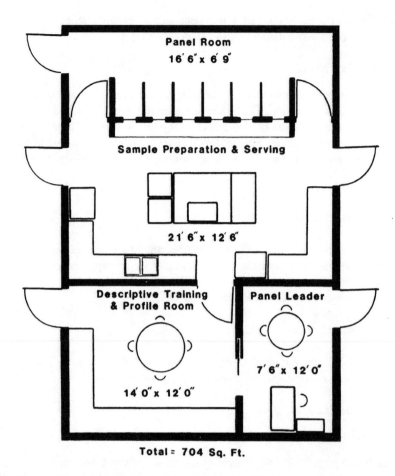

FIGURE 3.5 Layout for medium-size sensory evaluation area suitable for 300 to 400 tests per year. (Drawn by D. Grabowski.)

to accommodate the presence of visual differences between samples. For example, samples may be served sequentially and scored with reference to a common standard.

2. Air Circulation, Temperature, and Humidity

The sensory evaluation area should be air conditioned at 72 to 75°F and 45 to 55% relative humidity (RH). (For tactile evaluation of fabrics, paper nonwovens, and skin care products, tighter humidity control may be required, e.g., 50 ± 2% or 65 ± 2% RH.) Recirculated and makeup air should pass through a bank of activated carbon cannisters capable of removing all detectable odor. The cannisters may be placed outside the testing area in a location that allows easy replacement, e.g., every 2 or 3 months. Frequent monitoring is required to prevent the filters from becoming ineffective and/or becoming an odor source. A slight positive pressure should be maintained in the booth areas so as to prevent odor contamination from the sample preparation area or from outside. If testing of odorous materials such as sausages or cheese is a possibility, separate air exhausts must be provided from each booth.

3. Construction Materials

The materials used in the construction and furnishing of a sensory evaluation laboratory must be in accordance with the specific environment required for the products to be evaluated in the laboratory.

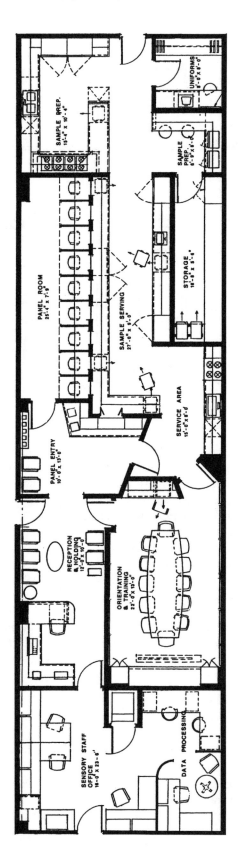

FIGURE 3.6 Layout for large sensory evaluation area suitable for preparation and evaluation of 600 to 1000 samples per year. (From ASTM, Physical Requirements for Sensory Evaluation Laboratories, American Society for Testing and Materials, Philadelphia, 1986. With permission.)

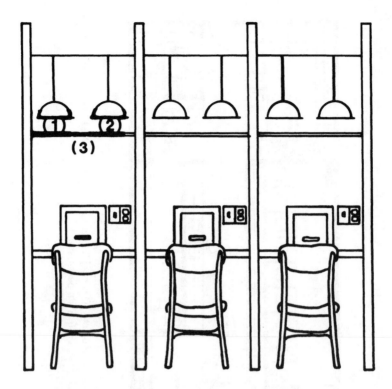

FIGURE 3.7 Panel booths showing arrangements for lighting. (1) Incandescent; (2) fluorescent; (3) holder for sheet filters. (Courtesy University of California, Davis and M.M. Gatchalian.)

Nonodorous — Paper, fabric, carpeting, porous tile, etc. must be avoided as they are either odorous in themselves or may harbor dirt, molds, etc. which will emit odor. Construction materials must be smooth, easy to clean, and nonabsorbing so that they do not retain odors from previous sessions. The materials which best meet these requirements are stainless steel, Teflon, and Formica. Nonodorous vinyl laminate is suitable for ceilings, walls, and floors.

Color — A neutral, unobtrusive color scheme using off-white colors and few patterns provides a background which is nondistracting to panelists. Especially for countertops it is important to choose a color that does not confound or bias evaluations. A white paper or fabric on a black benchtop will show visual flaws more dramatically, thus biasing both visual and tactile evaluations.

Plumbing — Product trapped in pipes causes distracting and confounding odors in a sensory laboratory. It is essential that all pipes and drains open to the room can be cleaned and flushed. If spit sinks are necessary for some tests (toothpaste, mouthwash), thought should be given to having them detachable, i.e., connected by flexible hose to water inlet and drain. When the sinks are not in use, they can be stored separately and the pipes can be closed off with caps.

III. PRODUCT CONTROLS

A. General Equipment

When a sensory evaluation test is conducted, the product researcher and the sensory analyst are looking for some treatment effect: effect of an ingredient change, a processing variable, a packaging change, a storage variable, etc. One of the primary responsibilities of the sensory analyst is to control the early handling, the preparation, and the presentation for each product. These controls ensure that extraneous variables are not introduced, and that no real treatment variables are obscured.

The preparation area should be situated adjacent to the test area. However, the air handling system should be structured so that the test area has positive pressure which feeds into the preparation area, which in turn contains the air return system as well as a supplementary exhaust.

B. SAMPLE PREPARATION

1. Supplies and Equipment

In addition to the necessary major appliances, the controlled preparation of products requires adequate supplies and equipment such as:

- Scales, for weighing products and ingredients
- Glassware, for measurement and storage of products
- Timers, for monitoring of preparation procedures
- Stainless steel equipment, for mixing and storing products, etc.

2. Materials

Equipment used for preparation and presentation of samples must be carefully selected to reduce the introduction of biases and new variables. Most plastic cutlery, storage containers, and wraps or bags are unsuitable for preparation of foods, beverages, or personal care products. The transfer of volatiles to and from the plastic can change the aroma and/or flavor characteristics of a product.

Wooden materials should not be used for cutting boards, bowls, mixing utensils, or pastry boards. They are porous and absorb aqueous and oil-based materials, which are then easily transferred from the wood to the next product which the wood contacts.

Containers used for storage, preparation, or serving should therefore be glass, glazed china, or stainless steel because of the reduced transfer of volatiles with these materials. Plastic, which has been pretested for low odor transfer, should be used only when the test product(s) will be held for less than 10 min in the container during and prior to the test.

3. Preparation Procedures

The controlled preparation of products requires careful regulation and monitoring of procedures used, with attention given to:

- Amount of product to be used, measured by weight or volume using precise equipment (volumetric cylinders, gram scales, etc.)
- Amount of each added ingredient (as above)
- The process of preparation, regulation of time (stopwatch), temperature (thermometers, thermocouples), rates of agitation (rpm), size, and type of preparation equipment
- Holding time defined as the minimum and maximum time after preparation that a product can be used for a sensory test

C. SAMPLE PRESENTATION

1. Container, Sample Size, and Other Particulars

The equipment and procedures used for product presentation during the test must be carefully selected to reduce introduction of biases and new variables. Attention should be given to control of the following.

Serving containers — Again, these are preferably glass or glazed china, not plastic unless tested.

Serving size — Extreme care must be given to regulating the precise amount of product to be given to each subject. Technicians should be carefully trained to deliver the correct amount of product with the least amount of handling. Special equipment may be advantageous for measuring precise amounts of a product for sensory testing.

Serving matrix — For most difference tests, the product under test is presented on its own, without additives. Products such as coffee, tea, peanut butter, vegetables, meats, etc. are served without condiments or other adjuncts that may normally be used by consumers, such as milk, bread, butter, spices, etc. In contrast, for consumer tests (preference/acceptance tests), products should be presented as normally consumed: coffee or tea with milk, sugar, or lemon, as required; peanut butter with bread or crackers; vegetables and meat with spices, according to the consumer's preference. Products which are normally tasted *in* or *on* other products (condiments, dressings, sauces, etc.) should be evaluated in or on a uniform carrier which does not mask the product characteristics. These include a flour roux (a cooked flour-and-water base used for sauces), a fondant (sugared candy base), and sweetened milk (for vanilla and similar spices and flavorings).

Serving temperature — After the sample is distributed into each serving container, and just before serving, the product should be checked to determine if it is at the appropriate temperature. Most sensory laboratories develop standard preparation procedures which determine the needed temperature in the preparation container, which is necessary to ensure the required temperature after delivery to the tasting/smelling container. The use of standard procedures greatly reduces the need for monitoring of each individual portion.

2. Order, Coding, and Number of Samples

As part of any test, the order, coding, and number of samples presented to each subject must be monitored.

The *order of presentation* should be *balanced* so that each sample appears in a given position an equal number of times. For example, these are the possible positions for three products, A, B, and C, to be compared in a ranking test:

$$ABC — ACB — BCA — BAC — CBA — CAB$$

Such a test should be set up with a number of subjects which is a multiple of six, so as to permit presentation of the six possible combinations an equal number of times (see Chapter 4, p. 40).

The presentation also should be *random*, which may be achieved by drawing sample cards from a bag or by using a compilation of random numbers (see Table Tl, p. 353).

The codes assigned to each product can be biasing: for example, subjects may subconsciously choose samples marked A over those marked with other letters. Therefore, single and double letters and digits are best avoided. In addition, letters or numbers that represent companies, area codes, and test numbers or samples should not be used. Most sensory analysts rely on the table of three-digit random numbers for product coding. Codes should not be very prominent, either on the product or on the scoresheet. They can be clearly yet discreetly placed on the samples and scoresheets to reduce confusion as to sample identification, and to reduce potential biases at the same time.

The number of samples that can be presented in a given session is a function of both sensory and mental fatigue in the subject. With cookies or bisquits, eight or ten may be the upper limit; with beer, six or eight. Products with a high carryover of flavor, such as smoked or spicy meats, bitter substances, or greasy textures may allow only one or two per test. On the other hand, visual evaluations can be done on series of 20 to 30 samples with mental fatigue as the limiting factor.

D. PRODUCT SAMPLING

The sensory analyst should determine how much of a product is required and should know the history of the products to be tested. Information about prior handling of experimental and control samples is important in the design of the test and interpretation of the results. A log book should be kept in the sensory laboratory to record pertinent sample data:

- The source of the product: when and where it was made. Sample identification is necessary for laboratory samples (lab notebook number) as well as production samples (date and machine codes).
- The testing needs: how much product will be required for all of the tests to be run, and possibly rerun, for this evaluation. All of the product representing a sample should come from one source (same place, same line, same date, etc.). If the product is not uniform, attempts should be made to blend and repackage the different batches.
- The storage: where the sample has been and under what conditions. If two products are to be compared for a processing or ingredient variable, it is not possible to measure the treatment effect if there are differences in age, storage temperature and humidity, shipping storage and humidity, packaging differences, etc. which can cloud the measurement.

IV. PANELIST CONTROLS

The way in which a panelist interacts with the environment, the product, and the test procedure are all potential sources of variation in the test design. Control or regulation of these interactions is essential to minimizing the extraneous variables, which may potentially bias the results.

A. PANEL TRAINING OR ORIENTATION

It goes without saying that panelists need careful instruction with respect to the handling of samples, the use of the scoresheet, and the information sought in the test. The training of panelists is discussed in detail in Chapter 9. As a minimum, panelists must be prepared to participate in a laboratory sensory test with no instruction from the sensory analysts once the test has started. They should be thoroughly familiar with:

- The test procedures, such as the amount of sample to be tasted at one time, delivery system (spoon, cup, sip, slurp), the length of time of contact with the product (sip/spit, short sniff, one bite/chew), and the disposition of the product (swallow, expectorate, leave in contact with skin or remove from skin) must be predetermined and adhered to by all panelists.
- The scoresheet design, which includes instructions for evaluation, and questions, terminology, and scales for expressing judgment, must be understood and familiar to all panelists.
- The type of judgment/evaluation required (difference, description, preference, acceptance) should be understood by the panelists as part of their test orientation.

Kelly and Heymann (1989) found no significant difference between ingestion and expectoration in tests, e.g., with added salt in kidney beans; ingestion did produce narrower confidence limits.

B. PRODUCT/TIME OF DAY

With panelists who are not highly trained it is wise to schedule the evaluation of certain product types at the time of day when that product is normally used or consumed. The tasting of highly flavored or alcoholic products in the early morning is not recommended. Product testing just after meals or coffee breaks also may introduce bias and should be avoided. Some preconditioning of the panelists' skin or mouth may be necessary in order to improve the consistency of verdicts.

C. PANELISTS/ENVIRONMENT

As discussed in Section II of this chapter, the test environment, as seen by the panelist, must be controlled if biases are to be avoided. Note, however, that certain controls, such as colored lights, high humidity, or enclosed testing area, may cause anxiety or distraction, unless panelists are given ample opportunity to become used to such "different" surroundings.

Again, it is necessary to prepare panelists for what they are to expect in the actual test situation, to give them the orientation and time to feel comfortable with the test protocols, and to provide them with enough information to respond properly to the variables under study.

REFERENCES

Amerine, M.A., Pangborn, R.M., and Roessler, E.B., 1965. *Principles of Sensory Evaluation of Food.* Academic Press, New York, pp. 299–300.

ASTM, 1986. *Physical Requirement Guidelines for Sensory Evaluation Laboratories*, Eggert, J. and Zook, K., Eds. ASTM Special Technical Publication 913, American Society for Testing and Materials, Philadelphia.

Chambers, E., IV and Baker Wolf, M., Eds., 1996. *Sensory Testing Methods*, 2nd ed. ASTM Manual 26, American Society for Testing and Materials, West Conshohocken, PA.

Caul, J.F., 1957. The profile method of flavor analysis. *Adv. Food Res.* 7,1–40.

European Cooperation for Accreditation of Laboratories, 1995. *EAL-G16, Accreditation for Sensory Testing Laboratories.* Available from national members of EAL, e.g., in the U.K., NAMAS. tel. 44 181 943-7068; fax 44 181 943-7134.

Gatchalian, M.M., 1981. *Sensory Evaluation Methods with Statistical Analysis.* College of Home Economics, University of the Philippines, Diliman, Quezon City, pp. 99–112.

ISO, 1988. *Sensory Analysis — General Guidance for the Design of Test Rooms.* International Standard ISO 8589. International Organization for Standardization, 1 rue Varembé, CH-1211 Génève 20, Switzerland (under revision 1999).

Kelly, F.B. and Heymann, H., 1989. Contrasting the effects of ingestion and expectoration in sensory difference tests. *J. Sensory Stud.* 3(4), 249.

Malek, D.M., Schmitt, D.J., and Munroe, J.H., 1982. A rapid system for scoring and analyzing sensory data. *J. Am. Soc. Brewing Chem.* 40, 133.

Pangborn, R.M., 1967. Use and misuse of sensory methodology. *Food Quality Control* 15,7–12.

Poste, L.M., Mackie, D.A., Butler, G., and Larmond, E., 1991. *Laboratory Methods for Sensory Analysis of Food.* Agriculture Canada, Ottawa, Publication 1864/E, pp. 4–13.

4 Factors Influencing Sensory Verdicts

CONTENTS

I. INTRODUCTION

Good sensory measurements require that we look at the tasters as measuring instruments, somewhat variable over time and among themselves, and very prone to bias. In order to minimize variability and bias, the experimenter must understand the basic physiological and psychological factors which may influence sensory perception. Gregson (1963) notes that perception of the real world is not a passive process, but an active and selective one. An observer records only those elements of a complex situation that he can readily see and associate as meaningful. The rest he eliminates even if it is staring him in the face. We must put the observer in a frame of mind to understand the characteristics we want him to measure. This is done through training (see Chapter 9), and by avoiding a number of pitfalls (Amerine et al., 1965; Poste et al., 1991; Lawless and Heymann, 1998) inherent in the presentation of samples, the text of the questionnaire, and the handling of the participants.

II. PHYSIOLOGICAL FACTORS

A. ADAPTATION

Adaptation is a decrease in or change in sensitivity to a given stimulus as a result of continued exposure to that stimulus or a similar one. In sensory testing this effect is an important unwanted source of variability of thresholds and intensity ratings.

In the following example of "cross-adaptation" (O'Mahony, 1986), the observer in condition B

	Adapting stimulus	Test stimulus
Condition A	H_2O	Aspartame
Condition B	Sucrose	Aspartame

is likely to perceive less sweetness in the test sample because the tasting of sucrose reduces his sensitivity to sweetness. The water used in condition A contains no sweetness and does not fatigue (or cause adaptation in the perception of sweet taste).

Condition A	H_2O	Quinine
Condition B	Sucrose	Quinine

Here, "cross-potentiation" or facilitation is likely to occur. In condition B, the observer perceives more bitterness in the test sample because the tasting of sucrose has heightened his sensitivity to quinine. A detailed discussion of adaptation phenomena in sensory testing is given by O'Mahony (loc. cit.).

B. ENHANCEMENT OR SUPPRESSION

Enhancement or suppression involves the interaction of stimuli presented simultaneously as mixtures.

Enhancement — The effect of the presence of one substance increasing the perceived intensity of a second substance.

Synergy — The effect of the presence of one substance increasing the perceived combined intensity of two substances, such that the perceived intensity of the mixture is greater than the sum of the intensities of the components.

Suppression — The effect of the presence of one substance decreasing the perceived intensity of a mixture of two or more substances.

Examples (see Key below):

1. Total perceived intensity of mixture

Situation	Name of effect
MIX < A + B (each alone)	Mixture suppression
MIX > A + B (each alone)	Synergy

2. Components of analyzable mixture:

Situation	Name of effect
A′ < A	Mixture suppression
A′ > A	Enhancement

Key: MIX = perceived intensity of mixture
  A = perceived intensity of unmixed component A
  A′ = perceived intensity of component A in mixture

III. PSYCHOLOGICAL FACTORS

A. EXPECTATION ERROR

Information given with the sample may trigger preconceived ideas. You usually find what you expect to find. In testing, such as the classic tests for threshold which consist of a series of ascending concentrations, the subject (through autosuggestion) anticipates the sensation and reports his response before it is applicable. A panelist who hears that an overage product has been returned to the plant will have a tendency to detect aged flavors in the samples of the day. A beer taster's verdict of bitterness will be biased if he knows the hop rate employed. Expectation errors can destroy the validity of a test and must be avoided by keeping the source of samples a secret and by not giving panelists any detailed information in advance of the test. Samples should be coded and the order of presentation should be random among the participants. It is sometimes argued that well-trained and well-motivated panelists should not let themselves be influenced by accidental knowledge about a sample, but in practice, the subject does not know how much to adjust his verdict for the expected autosuggestion, and it is much better for him/her to be ignorant of the history of the sample.

B. ERROR OF HABITUATION

Human beings have been described as creatures of habit. This description holds true in the sensory world and leads to an error, the error of habituation. This error results from a tendency to continue to give the same response when a series of slowly increasing or decreasing stimuli are presented, for example, in quality control from day to day. The panelist tends to repeat the same scores and hence to miss any developing trend or even accept an occasional defective sample. Habituation is common and must be counteracted by varying the types of product or presenting doctored samples.

C. STIMULUS ERROR

This error is caused when irrelevant criteria, such as the style or color of the container, influence the observer. If the criteria suggest differences, the panelist will find them even when they do not exist. For example (Amerine et al., 1965), knowing that wines in screw-capped bottles are usually less expensive, tasters served from such bottles may produce lower ratings than if served from cork-closure bottles. Panel sessions called urgently may trigger reports of known production defects. Samples served late in a test may be rated more flavorful because panelists know that the panel leader will present light-flavored samples first in order to minimize fatigue. The remedies in these cases are obvious: avoid leaving irrelevant (as well as relevant) cues, schedule panel sessions regularly, and make frequent and irregular departures from any usual order or manner of presentation.

D. LOGICAL ERROR

Logical errors occur when two or more characteristics of the samples are associated in the minds of the assessors. Knowledge that a darker beer tends to be more flavorful, or that darker mayonnaise tends to be stale, causes the observer to modify his verdict, thus disregarding his own perceptions. Logical errors must be minimized by keeping the samples uniform and by masking differences with the aid of colored glasses, colored lights, etc. Certain logical errors cannot be masked but may be avoided in other ways; for example, a more bitter beer will always tend to receive a higher score for hop aroma. With trained panelists the leader may attempt to break the logical association by occasionally doctoring a sample with quinine in order to produce high bitterness combined with low hop aroma.

E. Halo Effect

When more than one attribute of a sample is evaluated, the ratings will tend to influence each other (halo effect). Simultaneous scoring of various flavor aspects along with overall acceptability can produce different results rather than if each characteristic is evaluated separately. For example, in a consumer test of orange juice, subjects are asked not only to rate their overall liking, but also to rate specific attributes. When the product is generally well liked, all of its various aspects — sweetness, acidity, fresh orange character, flavor strength, mouthfeel — tend to be rated favorably as well. Conversely, if the product is not well liked, most of the attributes will be rated unfavorably. The remedy, when any particular variable is important, is to present separate sets of samples for evaluation of that characteristic.

F. Order of Presentation of Samples

At least five types of bias may be caused by the order of presentation.

Contrast effect — Presentation of a sample of good quality just before one of poor quality may cause the second sample to receive a lower rating than if it had been rated monadically (i.e., as a single sample). As an example, if one lives in Minneapolis in the winter and the thermometer hits 40°F, the city is having a heat wave. If one lives in Miami and the thermometer registers 40°F, the news media will report a severe cold spell. The converse is also true: a sample that follows a particularly poor one will tend to be rated higher.

Group effect — One good sample presented in a group of poor samples will tend to be rated lower than if presented on its own. This effect is the opposite of the contrast effect.

Error of central tendency — Samples placed near the center of a set tend to be preferred over those placed at the ends. In triangle tests, the odd sample is detected more often if it is in the middle position. (An error of central tendency is also found with scales and categories; see Chapter 5.)

Pattern effect — Panelists will use all available clues (this, of course, is legitimate on their part) and are quick to detect any pattern in the order of presentation.

Time error/positional bias — One's attitude undergoes subtle changes over a series of tests, from anticipation or even hunger for the first sample, to fatigue or indifference with the last. Often, the first sample is abnormally preferred (or rejected). A short-term test (sip and evaluate) will yield a bias for the sample presented first, whereas a long-term test (one-week home placement) will produce a bias for the sample presented last. Discrimination is greater with the first pair in a set than with subsequent pairs.

All of these effects must be minimized by the use of a balanced, randomized order of presentation. "Balanced" means that each of the possible combinations is presented an equal number of times. Each sample in a panel session should appear an equal number of times in 1st, 2nd... and nth position. If there are large numbers of samples to be presented, a balanced incomplete block design can be used (see Chapter 7, p. 116 and Chapter 13, p. 293).

"Randomized" means that the order in which the selected combinations appear was chosen according to the laws of chance. In practice, randomization is obtained by drawing sample cards from a bag, or it may be planned with the aid of a compilation of random numbers (see Table T1, p. 353, also Product Controls in Chapter 3, p. 34).

G. Mutual Suggestion

The response of a panelist can be influenced by other panelists. Because of this, panelists are separated in booths, thus preventing a judge from reacting to the facial expression registered by another judge. Vocalizing an opinion in reaction to samples is not permitted. The testing area also should be free from noise and distraction and separate from the preparation area.

H. LACK OF MOTIVATION

The degree of effort a panelist will make to discern a subtle difference, to search for the proper term for a given impression, or to be consistent in assigning scores is of decisive importance for the results. It is the responsibility of the panel leader to create an atmosphere in which assessors feel comfortable and do a good job. An interested panelist is always more efficient. Motivation is best in a well-understood, well-defined test situation. The interest of panelists can be maintained by giving them reports of their results. Panelists should be made to feel that the panels are an important activity. This can be subtly accomplished by running the tests in a controlled, efficient manner.

I. CAPRICIOUSNESS VS. TIMIDITY

Some people tend to use the extremes of any scale, thereby exerting more than their share of influence over the panel's results. Others tend to stick to the central part of the scale and to minimize differences between samples. In order to obtain reproducible, meaningful results, the panel leader should monitor new panelists' scores on a daily basis, giving guidance in the form of typical samples already evaluated by the panel and, if necessary, using doctored samples as illustrations.

IV. POOR PHYSICAL CONDITION

Panelists should be excused from sessions: (1) if they suffer from fever or the common cold, in the case of tasters, and if they suffer from skin or nervous system disorders in the case of a tactile panel; (2) if they suffer from poor dental hygiene or gingivitis; and (3) in the case of emotional upset or heavy pressure of work which prevents them from concentrating (conversely, panel work can be an oasis in a frantic day). Smokers can be good tasters but should refrain from smoking for 30 to 60 min before a panel. Strong coffee paralyzes the palate for up to an hour. Tasting should not take place the first 2 h after a major meal. The optimal time for panel work (for persons on the day shift) is between 10 a.m. and lunch. Generally the best time for an individual panelist depends on his biorhythm: it is that time of the day when one is most awake and one's mental powers are at their peak. Matthes (1986) reviews the many ways in which health or nutrition disorders affect sensory function and, conversely, how sensory defects can be used in the diagnosis of health or nutrition disorders.

REFERENCES

Amerine, M.A., Pangborn, R.M., and Roessler, E.B., 1965. *Principles of Sensory Evaluation of Food.* Academic Press, New York, Chapter 5.

Gregson, R.A.M., 1963. The effect of psychological conditions on preference for taste mixtures. *Food Technol.* 17(3), 44.

Lawless, H.T. and Heymann, H., 1998. *Sensory Evaluation of Food. Principles and Practices.* Chapman & Hall, New York, p. 49.

Matthes, R.D., 1986. Effects of health disorders and poor nutritional status on gustatory function. *J. Sensory Stud.* 1(3/4), 225.

O'Mahony, M., 1986. Sensory adaptation. *J. Sensory Stud.* 1(3/4), 237.

Pangborn, R.M., 1979. Physiological and psychological misadventures in sensory measurement or the crocodiles are coming. In: *Sensory Evaluation Methods for the Practicing Food Technologist,* Johnston, M.R., Ed. Institute of Food Technologists, Chicago, pp. 2-1 and 2-22.

Poste, L.M., Mackie, D.A., Butler, G., and Larmond, E., 1991. *Laboratory Methods for Sensory Analysis of Food.* Agriculture Canada, Ottawa, Publication 1864/E, pp. 4–13.

5   Measuring Responses

CONTENTS

I. INTRODUCTION

This chapter describes the various ways in which sensory responses can be measured. The purpose is to present the principle of each method of measuring responses and to discuss its advantages and disadvantages.

In the simplest of worlds, if tasters were really measuring instruments, we could set them up with a range of 0 to 100 and supply a couple of calibration points (doctored samples) for each attribute to be rated. Unfortunately, the real world of testing is not simple, and a much more varied approach is needed. The degree of complexity is such that the psychology departments of major universities maintain laboratories of psychophysics. (See Lawless and Heymann, 1998; Laming, 1994; Sekuler and Blake, 1990; Cardello and Maller, 1987; Baird and Noma, 1978; Anderson, 1974; Kling and Riggs, 1971.) Some of the factors to watch are outlined in this chapter.

When we ask panelists to assign numbers or labels to sensory impressions, they may do this in at least four ways (see Figure 5.1):

- Nominal data: (Latin: *nomen* = name): the items examined are placed in two or more groups which differ in name but do not obey any particular order or any quantitative relationship; example: the numbers carried by football players.
- Ordinal data: (Latin: *ordinalis* = order): the panelist places the items examined into two or more groups which belong to an ordered series; example: slight, moderate, strong.
- Interval data: (Latin: *inter vallum* = space between ramparts): panelists place the items into numbered groups separated by a constant interval; example: three, four, five, six.
- Ratio data: Panelists use numbers which indicate how many times the stimulus in question is stronger (or saltier, or more irritating) than a reference stimulus presented earlier.

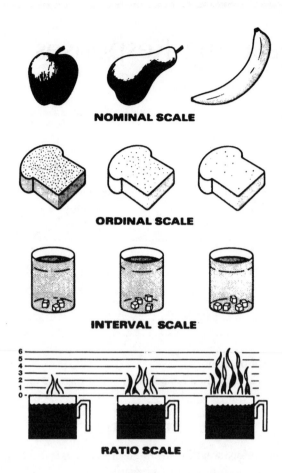

FIGURE 5.1 Pictorial illustration of scales. The names of the three food items (apple, pear, banana) provide nominal data. In the example of ordinal data, three rye breads are ranked from greatest to least number of caraway seeds. The three beverages form an interval scale in that they are separated by constant intervals of one unit of sucrose. In the last example, two volatiles from three cups of coffee are measured on a GC, and it is established that the first cup contains 3/4 of the volatiles of the second cup and only 1/2 of the third. Note that the illustration shows physical/chemical scales. A panelist's sensory scales may be different; for example, the sweetness of sugar increases less from 4 to 5 lumps than it does from 3 to 4 lumps.(From Cardello, A.V. and Maller, O., Psychophysical bases for the assessment of food quality, in *Objective Methods in Food Quality Assessment*, Kapsalis, J.G., Ed., CRC Press, Boca Raton, FL, 1987. With permission.)

Nominal data contain the least information. Ordinal data carry more information and can be analyzed by most nonparametric statistical tests. Interval and ratio data are even better because they can be analyzed by all nonparametric and often by parametric methods. Ratio data are preferred by some because they are free from end-of-scale distortions; however, in practice, interval data, which are easier to collect, appear to give equal results (see Section VI.C, p. 54).

The most frequently used methods of measuring sensory response to a sample are, in order of increasing complexity:

- Classification: The items evaluated are sorted into groups which differ in a *nominal* manner; example: marbles sorted by color.
- Grading: Time-honored methods used in commerce which depend on expert graders who learn their craft from other graders; example: "USDA Choice" grade of meat.

- Ranking: The samples (usually three to seven) are arranged in order of intensity or degree of some specified attribute; the scale used is *ordinal.*
- Scaling: The subjects judge the sample by reference to a scale of numbers (often from 0 to 10) which they have been trained to use; *category scaling* yields ordinal data or sometimes interval data, *line scales* usually yield interval data, and *magnitude estimation*, although designed to yield ratio data, in practice seems to produce mixed interval/ratio data.

A further method, the use of Odor Units based on thresholds, will be discussed in Chapter 8.

In choosing among these methods and training the panel to use them, the practicing panel leader needs to understand and then address the two major sources of variation in panel data: (1) the difference in the way test subjects perceive the stimulus and (2) the different ways the subjects express those perceptions (see Chapter 1, p. 3).

Actual differences in perception are part of the considerable variability in sensory data which sensory analysts learn to live with and psychophysicists learn to measure. Sensory thresholds vary from one person to another (Pangborn, 1981). Meilgaard (1993), in a study of difference thresholds for substances added to beer, found that panels of 20 trained tasters tend to contain 2 who show a threshold 4 × below the median for the panel, and 2 whose threshold is 5 × above. For panels of 200 + healthy but untrained individuals, Amoore (1977), who studied solutions of pure compounds in water, found differences of 1000-fold between the most and the least sensitive, excluding anosmics. It follows that the verdict of a small panel of four or seven people can be seriously at variance with the population in general, hence the tendency in this book to recommend panel sizes of 20 to 30 or preferably many more. *A small panel is representative only of itself.*

The second source of variation, the way in which the subjects express a given sensory impression, can be many times greater again, but luckily it can be minimized by thorough training and by careful selection of the terminology and scaling techniques provided to panelists. The literature is replete with examples of sensory verdicts which can only be explained by assuming that many panel members were quite "at sea" during the test: they probably did perceive the attribute under study, but they did not have a clear picture in their mind of what aspect they were asked to measure and/or they were unfamiliar with the mechanics of the test.

In choosing a way of measuring responses, generally the sensory analyst should select the simplest sensory method which will measure the expected differences between the samples, thus minimizing panel training time. Occasionally a more complex method will be used, one that uses more terminology and more sophisticated scales, thus requiring more training and evaluation time. For example, there may be sample differences which were not taken into account at the planning stage and which would have been missed with the simpler method. Overall training time may end up being less, because once the panel has reached the higher level of training, it can tackle many types of samples without the need for separate training sessions for each.

II. PSYCHOPHYSICAL THEORY

Psychophysics is that branch of experimental psychology devoted to studying the relationships between sensory stimuli and human responses, that is, to improving our understanding of how the human sensory system works. University psychophysicists are constantly refining the methods by which a response can be measured and, as sensory analysts, we need to study their techniques and indeed cooperate in their experiments. This chapter will provide an overview of psychophysics as applied to sensory testing. Those interested in more detail should read the references listed at the beginning of Section I of this chapter (p. 43); see also Lawless (1990).

A major focus of psychophysics is to discover the form of the psychophysical function, the relationship between a stimulus, C, and the resulting sensation, R, preferably expressed as a mathematical function, $R = f(C)$ (see Figure 5.2).

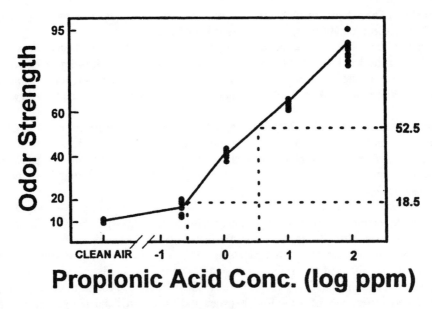

FIGURE 5.2 Example of a psychophysical function. Odor strength was rated 0 to 99 with zero = no odor or nasal irritation sensation. (From Kendal-Reed et al., Human responses to propionic acid, in *Chemical Senses* 23, Oxford University Press, 1998, 71–82. With permission.)

While the stimulus is either known (an added concentration) or easy to measure (a peak height, an Instron reading), it is the sensation that causes difficulty. We have to ask the subject questions such as: "Judge this odor on a scale of 0 to 99." "Is this sensation 2× as strong or 3× as strong?" "Which of these solutions has the strongest taste of quinine?", but no one can answer such questions reproducibly and precisely. A variety of experimental techniques are being used, e.g., comparison with a second, better known sensation such as the loudness of a tone (this is called cross-modality matching, see p. 56), or direct electrical measurement of the nerve impulse generated in the chorda tympani (taste nerve) in persons undergoing inner ear operations (Borg et al., 1967).

Over the past century, two forms of the psychophysical function have been in use: Fechner's law and Stevens' law. Although neither is perfect, each (when used within its limits of validity) provides a much better guide for experiment design than simple intuition. For a thorough discussion of the two, see Lawless and Heymann (1998). Two other reviews of psychophysical theory, Laming (1994) and Norwich and Wong (1997), include worthwhile attempts at reconciling Fechner's and Stevens' laws. More recently, the Michaelis-Menten equation known from enzyme chemistry, or the Beidler equation derived from it, have been used to model the dose–response relationship (see Section C below and Chastrette et al., 1998).

A. FECHNER'S LAW

Fechner selected as his measure of the strength of sensation the Just Noticeable Difference (JND); see Figure 5.3. For example, he would regard a perceived sensation of 8 JNDs as twice as strong as one of 4 JNDs. JNDs had just become accessible to measurement through difference testing, which Fechner learned from Ernst Weber at the University of Leipzig in the mid-1800s. Weber found that difference thresholds increase in proportion to the initial perceived absolute stimulus intensity at which they are measured:

$$\frac{\Delta C}{C} = k \quad \text{(Weber's law)} \qquad (5.1)$$

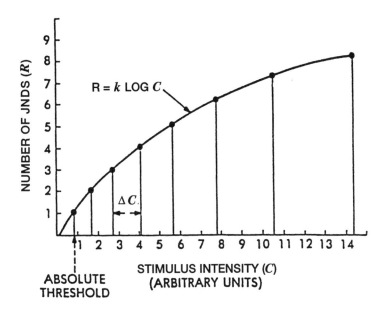

FIGURE 5.3 Derivation of Fechner's law by the method of summing JNDs. (Adapted from Cardello, A.V. and Maller, O., Psychophysical bases for the assessment of food quality, in *Objective Methods in Food Quality Assessment*, Kapsalis, J.G., Ed., CRC Press, Boca Raton, FL, 1987. With permission.)

where C is the absolute intensity of the stimulus, e.g., concentration, ΔC is the change in intensity of the stimulus that is necessary for 1 JND, and k is a constant, usually between 0 and 1. Weber's law states, e.g., that the amount of an added flavor which is just detectable depends on the amount of that added flavor that is already present. If the k has been determined, we can calculate how much extra flavorant is needed.

The actual derivation of Fechner's law,

$$R = k \, log \, C \quad \text{(Fechner's law)} \tag{5.2}$$

is complex and depends on a number of assumptions, some of which may not hold (Norwich and Wong, 1997). Support for Fechner's law is provided by common category scaling. When panelists score a number of samples that vary along one dimension (say, sweetness) using a scale such as 0 to 9, the results plot out as a logarithmic curve like that of Figure 5.3. One tangible outcome of Fechner's theories was a logarithmic scale of sound intensity, the Decibel scale.

B. Stevens' Law

S.S. Stevens, working at Harvard a century after Fechner, pointed out that if Equation (5.2) were correct, a tone of 100 dB should only sound twice as loud as one of 50 dB. He then showed, with the aid of *Magnitude Estimation Scaling* (see p. 54), that subjects found the 100-dB tone to be 40× as loud as the one of 50 dB (Stevens, 1970). Stevens' main contention (1957), that perceived sensation magnitude grows as a power function of stimulus intensity, can be expressed mathematically as:

$$R = k \, C^n \quad \text{(Stevens' power law)} \tag{5.3}$$

where k is a constant that depends on the units in which we choose to measure R and C, and n is the exponent of the power function, that is, n is a measure of the rate of growth of perceived intensity as a function of stimulus intensity. Figure 5.4 shows power functions with $n = 0.5$, 1.0,

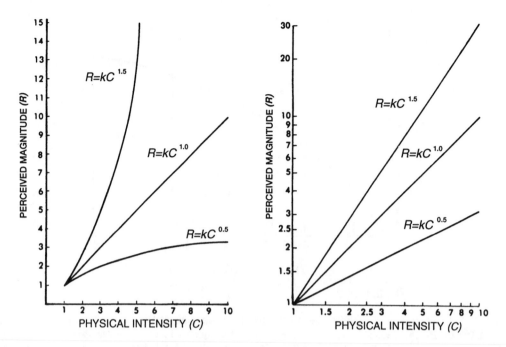

FIGURE 5.4 Plots of power functions with k = 1 and n = 0.5, 1.0, and 1.5 in linear (left) and logarithmic (right) coordinates. (Adapted from Cardello, A.V. and Maller, O., Psychophysical bases for the assessment of food quality, in *Objective Methods in Food Quality Assessment*, Kapsalis, J.G., Ed., CRC Press, Boca Raton, FL, 1987. With permission.)

and 1.5, and Table 5.1 lists typical exponents for a variety of sensory attributes. The finding that the exponent for visual length is 1.0, i.e., simple proportionality, has led to the common use of line scales for rating sensory intensity (Einstein, 1976).

When *n* is larger than 1.0, the perceived sensation grows faster than the stimulus; an extreme example is electric shock (Table 5.1). Conversely, when *n* is smaller than 1.0, as for many odors, the sensation grows more slowly than the stimulus, and a curve results that is superficially similar to Figure 5.3.

Stevens proposed that only ratio scales are valid for the measurement of perceived sensation, and his magnitude estimation scales are probably widely used in psychophysical laboratories. However, many authors have pointed out (Cardello and Maller, 1987) that for sensory evaluation of foods and fragrances, there are serious shortcomings. The exponents vary with the range of stimuli in the test and with the modulus used, and worse, the exponents differ greatly among investigators and among individuals, because of the subjects' idiosyncratic use of numbers.

C. THE BEIDLER MODEL

The log function and the power function are merely mathematical equations that happen to fit observed sensory data. There is nothing physiological about them. McBride (1987) has suggested that the equation below, which Beidler (1954, 1974) derived from animal experiments and the Michaelis-Menton equation for the kinetics of enzyme-substrate relationships in biological systems, can be used to describe human taste response. McBride proposes that we get away from the dependence on subjects' use of numbers or scales and simply assume that human psychophysical response is proportional to the underlying neurophysiological response:

$$\frac{R}{R_{max}} = \frac{C}{k+C} \quad \text{(The Beidler equation)} \quad (5.4)$$

TABLE 5.1
Representative Exponents of Power
Functions for a Variety of Sensory Attributes

Attribute	Exponent	Stimulus
Bitter taste	0.65	Quinine, sipped
	0.32	Quinine, flowed
Brightness	0.33	5° field
Cold	1.0	Metal on arm
Duration	1.1	White noise
Electric shock	3.5	Current through fingers
Hardness	0.8	Squeezed rubber
Heaviness	1.45	Lifted weights
Lightness (visual)	1.20	Gray papers
Loudness	0.67	1000-Hz tone
Salt taste	1.4	NaCl, sipped
	0.78	NaCl, flowed
Smell	0.55	Coffee
	0.60	Heptane
Sour taste	1.00	HCl, sipped
Sweet taste	1.33	Sucrose, sipped
Tactual roughness	1.5	Emery cloths
Thermal pain	1.0	Radiant heat on skin
Vibration	0.95	60 Hz on finger
	0.6	250 Hz on finger
Viscosity	0.42	Stirring fluids
Visual area	0.7	Projected squares
Visual length	1.00	Projected line
Warmth	1.6	Metal on arm

(From Cardello, A.V. and Maller, O., Psychophysical bases for the assessment of food quality, in *Objective Methods in Food Quality Assessment*, Kapsalis, J.G., Ed., CRC Press, Boca Raton, FL, 1987. With permission.)

The equation states that the response, R, divided by the maximal response, R_{max}, shows a sigmoidal relationship to the stimulus, C (the molar concentration), when C is plotted on a logarithmic scale (see Figure 5.5). The constant, k, is the concentration at which the response is half-maximal. Beidler (1974) calls it the association constant, or binding constant, and notes that it can be seen as a measure of the affinity with which the stimulus molecule binds to the receptor. The Beidler model works best for the middle and high range of sensory impressions, e.g., for the sweetness of sweet foods or beverages. Unlike Fechner's and Stevens' models, it assumes that the response has an upper limit, R_{max}, which is not exceeded no matter how great the concentration of the stimulus. It is seen as that concentration when all the receptors are saturated.

McBride shows, with a number of examples for sugars, salt, citric acid, and caffeine, that the Beidler equation provides a good description of human taste response, as obtained by two psychophysical methods, JND cumulation and category rating. Application of the Beidler equation allows estimation of the hitherto unobtainable parameters for human taste response, R_{max} and k. Therefore, unlike the empirical Fechner and Stevens laws, the Beidler equation offers the potential for quantitative estimation of human taste response, that is, of the psychophysical function. Details of how this may be done for studies of the biophysics of the sensory mechanism are given by Beidler (1974), Maes (1985), and Chartrette et al. (1998).

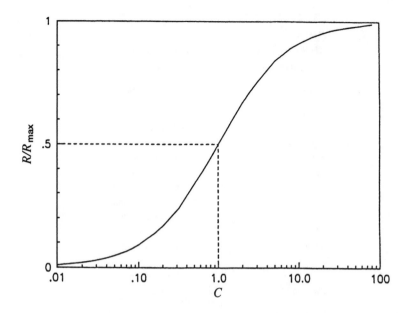

FIGURE 5.5   The sigmoidal relationship between taste response, R/R_{max}, and stimulus concentration, C, as specified by the Beidler equation; k is set equal to 1 for convenience. The inflexion point (maximum slope) of the curve occurs at $R = 0.5R_{max}$, when $C = k$.

Other techniques that attempt to model the assessor's decision process, which are frequently used by psychophysicists, are finding application in sensory evaluation, especially in threshold and discrimination testing. These other psychophysical models include the Thurston-Ura model and the Signal Detection model (see Lawless and Heymann, loc. cit., Chapter 5; and Chapter 8.III, p. 126).

III. CLASSIFICATION

In classification tests, the subjects are asked to select an attribute or attributes which describe the stimulus. In a beverage test, for example, subjects place a mark next to the term(s) which best describe(s) the sample:

_____ sweet	_____ sour	_____ lemony
_____ blended	_____ thick	_____ refreshing
_____ pulpy	_____ natural	_____ aftertaste

No attempt is made to standardize the terms, and the results are reported as the number of check marks for each term. Such data are nominal: no numbers are used, and there is no increasing or decreasing series expressed in the data. For example, the apples in a lot may be characterized by predominant color (red, green, yellow).

The proper selection of the right terms is essential for the correct interpretation of the description of the stimulus. If panelists are not trained, as is the case with consumers, common nontechnical terms must be used. A source of confusion is that subjects often erroneously associate individual common terms with degrees of goodness or badness. The *caveats* below describe situations using classification in which selection of the proper words/terms/classes is the critical first step. Selection of the best possible terminology is important not only in classification tests, it is needed in *all* measuring techniques which use a term or descriptor to define the perceived property being investigated.

The selection of sensory attributes and the corresponding definition of these attributes should be related closely to the real chemical and physical properties of a product which can be perceived. Adherence to an understanding of the actual rheology or chemistry of a product makes the data easier to interpret and more useful for decision making.

Caveats:

1. If a product has noticeable defects, such as staleness or rancidity, and terms to describe such defects have not been included in the list, panelists (especially if untrained) will use another term in the list to express the off-note.
2. If a list of terms provided to panelists fails to mention some attribute which describes real differences between products, or which describes important characteristics in one product, panelists again will use another term from the list provided to express what they perceive.
3. It follows that if results are to be useful, selection of the terms for classification (and for the various forms of scaling discussed in Section VI, below) must be based on actual product characteristics. This in turn requires preexamination of the samples by a well-trained panel to make sure that all appropriate attributes are listed. The use of a list or lists of terms taken from previous studies may neglect to include attributes which are important in the current study, or the "old" list may include terms which are irrelevant to the current samples and thus confusing to the panelists.

Following are some examples of word lists that have been used for classification tasks or for subsequent rating tasks:

1. Afterfeel of skincare products, e.g., soaps, lotions, creams: tacky, smooth, greasy, supple, grainy, waxy, oily, astringent, taut, dry, moist, creamy. Note that no relationship is introduced between the attributes which may, in fact, be facets of the same parameter (moist/dry, smooth/grainy, etc.)
2. Spice notes (subjects may be asked to define which spices or herbs contribute to one overall spice complex): oregano, basil, thyme, sage, rosemary, marjoram, and/or clove, cinnamon, nutmeg, mace, cardamom.
3. Hair color/hair condition: panelists/hairdressers are asked to classify the hair color and hair condition of men and women who are to serve as subjects for half-head shampoos; such sorting may be necessary to balance all treatments.

For each subject, check the most appropriate descriptor(s) from each column:

Color of hair	Condition of hair
Blond	Healthy
Brown	Damaged
Red	Dull
Black	Split
Tinted/frosted	Oily scalp
	Dandruff

IV. GRADING

Grading is a method of evaluation much used in commerce which depends on expert "graders" who learn the scale used from other graders. Scales usually have four or five steps such as "Choice," "Extra," "Regular," and "Reject." Examples of items subjected to sensory grading are coffee, tea, spices, butter, fish, and meat.

Sensory grading most often involves a process of integration of perceptions by the grader. The grader is asked to give one overall rating of the combined effect of the presence of the positive attributes, the blend or balance of those attributes, the absence of negative characteristics, and/or the comparison of the products being graded with some written or physical standard.

Grading systems can be quite elaborate and useful in commerce where they protect the consumer against being offered low-quality products at a high price, while permitting the producer to recover the extra costs associated with provision of a high-quality product. However, grading suffers from the considerable drawback that statistical correlation with measurable physical or chemical properties is difficult or impossible.

Consequently, many of the time-honored grading scales are being replaced by the methods described in this book. Examples of good grading methods still in use are the Torry scale for freshness of fish (Sanders and Smith, 1976), and the USDA scales for butter (USDA, 1977) and meat (USDA, undated).

V. RANKING

In ranking, subjects receive three* or more samples which are to be arranged in order of intensity or degree of some specified attribute. For example, four samples of yogurt are to be ranked for degree of sensory acidity, or five samples of breakfast cereal may be ranked for preference. A full description of ranking tests and their statistical treatment will be found in Chapter 7.

For each subject, the sample ranked first is accorded a "1," that ranked second a "2," and so on. The rank numbers received by each sample are summed, and the resulting rank sums indicate the overall rank order of the samples. Rank orders cannot meaningfully be used as a measure of intensity, but they are amenable to significance tests such as the χ^2-test (see Chapter 13, pp. 284–285) and Friedman's test (see Chapter 7, pp. 104–105).

Ranking tests are rapid and demand relatively little training, although it should not be forgotten that the subjects must be thoroughly familiarized with the attribute under test. Ranking tests have wide application, but with sample sets above three they do not discriminate as well as tests based on the use of scales.

VI. SCALING

Scaling techniques involve the use of numbers or words to express the intensity of a perceived attribute (sweetness, hardness, smoothness) or a reaction to such attribute (e.g., too soft, just right, too hard). If words are used, the analyst may assign numerical values to the words (e.g., like extremely = 9, dislike extremely = 1) so that the data can be treated statistically. As this is written, methods of scaling are under intensive study around the world (ISO, 1999; Muñoz and Civille, 1998) and the recommendations which follow should be seen as preliminary.

The validity and reliability of a scaling technique are highly dependent upon:

- the selection of a scaling technique that is broad enough to encompass the full range of parameter intensities and also has enough discrete points to pick up all the small differences in intensity between samples;
- the degree to which the panel has or has not been taught to associate a particular sensation (and none other) with the attribute being scaled; and
- the degree to which the panel has or has not been trained to use the scale in the same way across all samples and across time (see Chapter 9 on panelist training).

---

* Ranking of two samples is covered by the Paired Comparison test (see Chapter 7. II, p. 100).

Compared with difference testing, scaling is a more informative and therefore a more useful form of recording the intensity of perception. As with ranking, the results are critically dependent on how well the panelists have been familiarized with the attribute under test and with the scale being used. In this respect, three different philosophies have been applied (Muñoz and Civille, loc. cit.):

- Universal scaling, in which panelists consider all products and intensities they have experienced as their highest intensity reference point (example: the Spectrum aromatics scale uses the cinnamon impact of Big Red chewing gum as a 13 in intensity on a 15-point scale);
- Product specific scaling, in which panelists consider only their experience within the selected product category in setting their highest reference point (example: the vanilla impact of typical vanilla cookies was set at 10 on a 15-point product specific scale); and
- Attribute specific scaling, in which panelists consider their experience of the selected attribute across all products in setting their highest reference point (example: a specific toothpaste is assigned the top value of 13 for the peppermint aromatic in any product).

A common problem with scales is that panelists tend to use only the middle section. For example, if ciders are judged for intensity of "appley" flavor on a scale of 0 to 9, subjects will avoid the numbers 0, 1, and 2 because they tend to keep these in reserve for hypothetical samples of very low intensity, which may never come. Likewise, the numbers 7, 8, and 9 are avoided in anticipation of future samples of very high intensity, which may never come. The result is that the scale is distorted: for example, a cider of outstanding apple intensity may be rated 6.8 by the panel while a cider which is only just above the average may receive a 6.2.

Although the properties of data obtained from any response scale may vary with the circumstances of the test (e.g., experience of judges in the test, familiarity of the attribute), it is typically assumed that:

- Category scaling (ISO term: rating) yields ordinal (or interval) data;
- Line scaling (ISO term: scoring) yields interval data;
- Magnitude estimation scaling (often called ratio scaling) sometimes, but not always, yields ratio data.

A. CATEGORY SCALING

A category (or partition) scale is a method of measurement in which the subject is asked to "rate" the intensity of a particular stimulus by assigning it a value (category) on a limited, usually numerical, scale. Category scale data are generally considered to be at least ordinal level data. They do not generally provide values which measure the degree (how much) one sample is more than another. On a 7-point category scale for hardness, a product rated a 6 is not necessarily twice as hard as a product with a 3 hardness rating. The hardness difference between 3 and 6 may not be the same as that between 6 and 9. Although attempts are made to encourage panelists to use all intervals as equal, panelists may also tend to use the categories with equal frequency, except that they usually avoid the use of the two scale end points in order to save them for "real extremes." Here are four examples of category scales of proven usefulness in descriptive analysis:

Number	Category scales	Word category scale I	Word category scale II
0	0	None	None at all
1	1	Threshold	Just detectable
2	2.5	Very slight	Very mild
3	5	Slight	Mild
4	7.5	Slight-moderate	Mild-distinct
5	10	Moderate	Distinct
6	12.5	Moderate-strong	Distinct-strong
7	15	Strong	Strong

Generally, even word category scales are converted to numbers. The numbers used in the above list are typical of the conversions which are made.

The Flavor Profile® and Texture Profile® descriptive analysis methods use a numerical type category scale anchored with words:

Numerical value	Word anchor
0	None
)(	Threshold, just detectable
½	Very slight
1	Slight
1½	Slight-moderate
2	Moderate
2½	Moderate-strong
3	Strong

Unless the scale represents a very small range of sensory perception or the number of samples to be tested is small (less than five), panel leaders should consider using at least a 10- to 15-point category scale. Data for category scales can be analyzed using χ^2-tests to compare the proportion of responses occurring in each category among a group of samples. Alternatively, if it is reasonable to assume that the categories are equally spaced, parametric techniques such as t-tests, analysis of variance, and regression can be applied to the data. Riskey (1986) discusses the use and abuse of category scales in considerable detail.

The practical steps involved in the construction of a scale are discussed in Chapters 9 and 10. Appendix 11.1 contains a wide selection of terms of proven usefulness as scale end points, and Appendix 11.2 gives reference points on a scale of 0 to 15 for the four basic tastes and for the intensity of selected aroma, taste, and texture characteristics of items readily available in supermarkets, such as Hellmann's Mayonnaise.

B. Line Scales

With a linear or line scale, the panelist "rates" the intensity of a given stimulus by making a mark on a horizontal line which corresponds to the amount of the perceived stimulus. The lengths most used are 15 cm and 6 in. with marks ("anchors") either at the ends or ½ in. or 1.25 cm from the two ends (see Figure 5.6). The use of more than two anchors tends to reduce the line scale to a category scale, which may or may not be desired. Normally the left end of the scale corresponds to "none" or zero amount of the stimulus while the right end of the scale represents a large amount or a very strong level of the stimulus (Anderson, 1970; Stone and Sidel, 1992). In some cases the scale is bipolar, i.e., opposite types of stimuli are used to anchor the end points.

Panelists use the line scale by placing a mark on the scale to represent the perceived intensity of the attribute in question. The marks from line scales are converted to numbers by manually measuring the position of each mark on each scale using a ruler, a transparent overlay, or a digitizer which is hooked up to a computer. The digitizer converts the position of the mark to a number, based on a preset program, and feeds the data to the computer for analysis.

C. Magnitude Estimation Scaling

Magnitude estimation (Moskowitz, 1977; ISO, 1994) or free number matching is a scaling technique based on Stevens' law (see Section II.B, p. 47). The first sample a panelist receives is assigned a freely chosen number. (The number can be assigned by the experimenter, in which case it is referred to as a modulus; or the number can be chosen by the panelist.) Panelists are then asked to assign all subsequent ratings of subsequent samples in proportion to the first sample rating. If the second sample appears three times as strong as the first, the assigned rating should be three times the rating

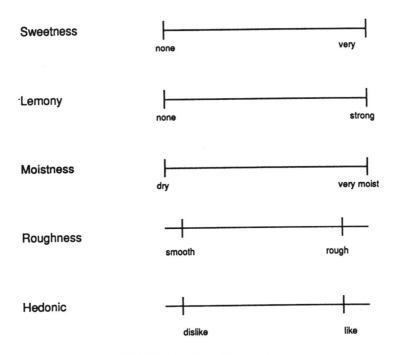

FIGURE 5.6 Typical line scales.

assigned to the first stimulus. Panelists are instructed to keep the number ratings in proportion to the ratios between sensations. Examples:

1. With a modulus: The first cookie which you taste has an assigned "crispness" rating of 25. Rate all other samples for crispness in proportion to that 25. If the crispness of any sample is half that of the first sample, assign it a crispness value of 12.5.

   First sample <u>25</u>
   Sample 549 _____
   Sample 306 _____

2. Without a modulus: Taste the first cookie; assign any number to the "crispness" of that cookie. Rate the crispness of all other samples in proportion to the rating given the first sample.

   Sample 928 _____ (first sample)
   Sample 549 _____
   Sample 306 _____

The results are evaluated as described by Moskowitz (loc. cit.) and ISO (loc. cit.). Alternative methods of evaluation are reviewed by Butler et al. (1987) and Lawless (1989).

Magnitude estimation vs. category scaling — A good discussion of the advantages and disadvantages of the two methods is given by Pangborn (1984). The data produced by magnitude estimation (ME) have ratio properties, like the standard forms of technical measurement (length, weight, volume, etc.). ME gets around the problem that panelists avoid the ends of scales so as to leave room for another stimulus. Adherents of ME also cite the fact that users of category scaling (CS) must spend time and effort on preparation of standards and on teaching the panel to use them. Those favoring CS note that ME is incapable of providing stable and reproducible values for flavor intensity. In practice, ME panelists require a good deal of training if they are to use the method with any facility; many judges rate in "nickles and dimes" using whole and half numbers or preferring the 10s or 5s in a series such as 15, 20, 25, etc., and they have trouble thinking in pure ratio terms such as "six times stronger" or "1.3 times weaker." In a number of applications (Powers et al., 1981; Giovanni and Pangborn, 1983; Pearce et al., 1986; Lawless and Malone, 1986), ME has provided no greater discrimination than CS. Further, ME is less suitable for scaling degree of liking (Pangborn et al., 1989). Where ME does offer more points of discrimination and separation is in academic applications with few judges (20 or less) studying a unidimensional system such as sucrose in water, one aromatic chemical in a diluent, or one increasing tone.

Magnitude matching (cross-modality matching) — In this technique, subjects match the intensity of an attribute (1), such as the sourness of acid solutions, to the intensity of another attribute (2), such as the loudness of 1000-Hz tones. If the two intensities are governed by the functions

$$R_1 = k_1 C_1^{n1} \quad \text{or} \quad \log R_1 = \log k_1 + n_1 \log C_1$$

and

$$R_2 = k_2 C_2^{n2} \quad \text{or} \quad \log R_2 = \log k_2 + n_2 \log C_2$$

then by matching them we obtain:

$$\log k_1 + n_1 \log C_1 = \log k_2 + n_2 \log C_2$$

or

$$\log C_1 = n_2/n_1 \log C_2 \text{ plus a constant}$$

In other words, a power function has been obtained which describes the intensity of sourness, and the exponent of the function is equal to the ratio of the original exponents (Cardello and Maller, 1987; Lawless and Heymann, 1998; Marks et al., 1988). The advantage of this approach is that no numbers are assigned, so it gets around the tendency of subjects to use numbers differently, as mentioned on p. 48.

REFERENCES

Amoore, J.E., 1977. Specific anosmia and the concept of primary odors. *Chem. Senses and Flavor* 2, 267–281.
Anderson, N.H., 1970. Functional measurement and psychological judgment. *Psychol. Rev.* 77, 153.
Anderson, N.H., 1974. Algebraic models in perception. In: *Handbook of Perception,* Vol. 2: *Psychophysical Judgment and Measurement,* Carterette, E.C. and Friedman, M.P., Eds. Academic Press, New York, 215–298.
Baird, J.C. and Noma, E., 1978. *Fundamentals of Scaling and Psychophysics.* Wiley-Interscience, New York.
Beidler, L.M., 1954. A theory of taste stimulation. *J. Gen. Physiol.* 38, 133.
Beidler, L.M., 1974. Biophysics of sweetness. In: *Symposium: Sweeteners,* Inglett, G.E., Ed. Avi Publishing, Westport, CT, 10.
Borg, G., Diamant, H., Strom, L., and Zotterman, Y., 1967. The relation between neural and perceptual intensity: a comparative study on the neural and psychophysical responses to gustatory stimuli. *J. Physiol.* 13, 192.
Butler, G., Poste, L.M., Wolynetz, M.S., Ayar, V.E., and Larmond, E., 1987. Alternative analyses of magnitude estimation data. *J. Sensory Stud.* 2(4), 243–257.
Cardello, A.V. and Maller, O., 1987. Psychophysical bases for the assessment of food quality. In: *Objective Methods in Food Quality Assessment,* Kapsalis, J.D., Ed. CRC Press, Boca Raton, FL, 61–125.
Chastrette, M., Thomas-Danguin, T. and Rallet, E., 1998. Modelling the human olfactory stimulus — response function. *Chem. Senses* 23, 181–196.
Doty, R.L., Ed., 1995. *Handbook of Gustation and Olfaction. Section B: Human Psychophysics and Measurement of Odor-Induced Responses.* Marcel Dekker, New York, pp. 191–298.
Einstein, M.A., 1976. Use of linear rating scales for the evaluation of beef flavor by consumers. *J. Food Sci.* 41, 383.

Giovanni, M.E. and Pangborn, R.M., 1983. Measurement of taste intensity and degree of liking of beverages by graphic scales and magnitude estimation. *J. Food Sci.* 48, 1175.

ISO, 1994. *Sensory Analysis — Methodology — Magnitude Estimation.* International Standard ISO 11056, available from International Organization for Standardization, Case Postale 56, 1 rue Varembé, CH1211 Génève 20, or from American National Standards Institute, 11 West 42nd St., New York, NY 10036.

ISO, 1999. *Sensory Analysis — Guidelines for the use of quantitative response scales.* Draft International Standard ISO CD 4121, available from ISO, c/o AFNOR, Tour Europe, Cedex 7, 92049 Paris La Défense.

Kling, J.W. and Riggs, L.A., Eds., 1971. *Woodworth & Schlosberg's Experimental Psychology*, 3rd ed. Holt, Rinehart & Winston, New York.

Laming, D., 1994. Psychophysics. In: *Companion Encyclopedia of Psychology*, A.M. Colman, Ed. Routledge, London and New York, Chapter 3.5, pp. 251–277.

Lawless, H.T., 1990. Applications of Experimental Psychology in Sensory Evaluation, In: *Psychological Basis of Sensory Evaluation*, McBride, R.L. and MacFie, H.J.H., Eds. Elsevier Applied Science, London, pp. 69–91.

Lawless, H.T. and Heymann, H., 1998, *Sensory Evaluation of Food: Principles and Practices.* Chapman & Hall, New York.

Lawless, H.T. and Malone, G.J., 1986. The discriminative efficiency of common scaling methods. *J. Sensory Stud.* 1(1), 85.

Maes, F.W., 1985. Improved best-stimulus classification of taste neurons. *Chem. Senses* 10, 35–44.

Marks, L.E., Stevens, J.C., Bartoshuk, L.M., Gent, J.F., Rifkin, B., and Stone, V.K., 1988. Magnitude-matching: the measurement of taste and smell. *Chem. Senses* 13(1), 63–87.

McBride, R.L., 1987. Taste psychophysics and the Beidler equation. *Chem. Senses* 12, 323–332.

Meilgaard, M.C., 1993. Individual differences in sensory threshold for aroma chemicals added to beer. *Food Quality and Preference* 4, 153–167.

Meilgaard, M.C. and Reid, D.S., 1979. Determination of personal and group thresholds and the use of magnitude estimation in beer flavour chemistry. In: *Progress in Flavour Research*, Land, D.G. and Nursten, H.E., Eds. Applied Science, London, p. 67–73.

Moskowitz, H.R., 1977. Magnitude estimation: notes on what, how and why to use it. *J. Food Qual.* 1, 195–228.

Muñoz, A.A. and Civille, G.V., 1998. Universal, product and attribute specific scaling and the development of common lexicons in descriptive analysis. *J. Sensory Studies* 13, 57–76.

Norwich, K.H. and Wong, W., 1997. Unification of psychophysical phenomena: The complete form of Fechner's law. *Perception & Psychophysics* 59, 929–940.

Pangborn, R.M., 1981. Individuality in responses to sensory stimuli. In: *Criteria of Food Acceptance. How Man Chooses What He Eats,* Solms, J. and Hall, R.L., Eds. Forster-Verlag, Zürich, p. 177–219.

Pangborn, R.M., 1984. Sensory techniques of food analysis. In: *Food Analysis. Principles and Techniques,* Vol. I: *Physical Characterization,* Gruenwedel, D.W. and Whitaker, J.R., Eds. Marcel Dekker, New York, p. 37ff; see pp. 61–68.

Pangborn, R.M., Guinard, J.-X., and Meiselman, H.L., 1989. Evaluation of bitterness of caffeine in hot chocolate drink by category, graphic, and ratio scaling. *J. Sensory Stud.* 4(1), 31–53.

Pearce, J.J., Warren, C.B., and Korth, B., 1986. Evaluation of three scaling methods for hedonics. *J. Sensory Stud.* 1(1), 27–46.

Powers, J.J., Warren, C.B., and Masurat, T., 1981. Collaborative trials involving the methods of normalizing magnitude estimations. *Lebensm.-Wiss.& Technol.* 14, 86–93.

Riskey, D.R., 1986. Use and abuse of category scales in sensory measurement. *J. Sensory Stud.* 1(3/4), 217.

Sanders. H.R. and Smith, G.L., 1976. The construction of grading schemes based on freshness assessment of fish. *J. Food Technol.* 11, 365.

Sekuler, R. and Blake, R., 1990. *Perception,* 2nd ed. McGraw-Hill, New York.

Stevens, S.S., 1957. On the psychophysical law. *Psychol. Rev.* 64, 153–181.

Stevens, S.S., 1970. Neural events and the psychophysical law. *Science* 170, 1043.

Stone, H. and Sidel, J.L., 1992. *Sensory Evaluation Practices,* 2nd ed. Academic Press, Orlando, FL.

USDA, undated. United States Grading and Certification Standards. Meats, Prepared Meat and Meat Products. *Regulations*, Title 7 CFR, Part 54.

USDA, 1977. United States Standards for Grades of Butter. *Regulations*, Title 7 CFR, Part 58 (published in *Federal Register*, February 1, 1977).

6 Overall Difference Tests: Does a Sensory Difference Exist Between Samples?

CONTENTS

I. INTRODUCTION

Chapters 6 and 7 contain "cookbook-style" descriptions of individual difference tests, with examples. The underlying theory will be found in Chapter 5, Measuring Responses, and in Chapter 13, Basic Statistical Methods. Guidelines for the choice of a particular test will be found under "Scope and Application" for each test, and also in summary form in Chapter 15, Guidelines for Choice of Technique.

Difference tests can be set up legitimately in hundreds of different ways, but in practice the procedures described here have acquired individual names and a history of use. There are two groups of difference tests with the following characteristics:

Overall difference tests (Chapter 6): Does a sensory difference exist between samples? These are tests, such as the Triangle and the Duo-trio, which are designed to show whether subjects can detect any difference at all between samples.

Attribute difference tests (Chapter 7): How does attribute X differ between samples? Subjects are asked to concentrate on a single attribute (or a few attributes), e.g., "Please rank these samples according to sweetness." All other attributes are ignored. Examples are the paired comparison tests, the n-AFC tests (Alternative Forced Choice), and various types of multiple comparison tests. The intensity with which the selected attribute is perceived may be measured by any of the methods described in Chapter 5, e.g., ranking, line scaling, or magnitude estimation (ME).

The 2- and 3-AFC tests are used often in threshold determinations (see Chapter 8). Affective tests (preference tests, e.g., consumer tests) are also attribute difference tests (see Chapter 12).

In the second edition of this book, similarity tests were treated as a separate subject, for which separate tables were provided. This third edition adopts a contemporary, unified approach in which a single set of tables covers all situations from difference to similarity.

II. THE UNIFIED APPROACH TO DIFFERENCE AND SIMILARITY TESTING

Discrimination tests can be used to address a variety of practical objectives. In some cases researchers are interested in demonstrating that two samples are perceptibly different. In other cases researchers want to determine if two samples are sufficiently similar to be used interchangeably. In yet another set of cases some researchers want to demonstrate a difference while other researchers involved in the same study want to demonstrate similarity. All of these situations can be handled in a unified approach through the selection of appropriate values for the test-sensitivity parameters, α, β, and p_d. What values are appropriate depend on the specific objectives of the test.

A spreadsheet application has been developed in Microsoft Excel to aid researchers in selecting values for α, β, and p_d that provide the best compromise between the desired test sensitivity and available resources (see Chapter 13.III.E, pp. 285–286). The "Test Sensitivity Analyzer" allows researchers to quickly run a variety of scenarios with different combinations of the number of assessors, n, the number of correct responses, x, and the maximum allowable proportion of distinguishers, p_d, and in each case observe the resulting impacts on α-risk and β-risk.

The Unified Approach also applies to paired-comparison tests, such as the 2-AFC (see Chapter 7.II, p. 100).

In the basic Triangle test for difference the objective is merely to discover whether a perceptible difference exists between two samples. The statistical analysis is made under the tacit assumption that only the α-risk matters (the probability of concluding that a perceptible difference exists when one does not). The number of assessors is determined by looking at the α-risk table and taking into account material concerns, such as availability of assessors, available quantity of test samples, etc. The β-risk (the probability of concluding that no perceptible difference exists when one does) and the proportion of distinguishers, p_d, on the panel are ignored or, rather, are assumed to be unimportant. As a result, in testing for difference, the researcher selects a small value for the α-risk and, by ignoring them, accepts arbitrarily large values for the β-risk and p_d in order to keep the required number of assessors within reasonable limits.

In testing for similarity the sensory analyst wants to determine that two samples are sufficiently similar to be used interchangeably. Reformulating for reduced costs and validating alternate suppliers are just two examples of this common situation. In designing a test for similarity, the analyst determines what constitutes a meaningful difference by selecting a value for p_d and then specifies a small value for β-risk to ensure that there is only a small chance of missing that difference if it really exists. The α-risk is allowed to become large in order to keep the number of assessors within reasonable limits.

In some cases, it may be important to balance the risk of missing a difference that exists (β-risk) with the risk of concluding that a difference exists when it does not (α-risk). In this case, the analyst chooses values for all three parameters, α, β, and p_d to arrive at the number of assessors required to deliver the desired sensitivity for the test (see Example 6.4).

As a rule of thumb, a statistically significant result at

- an α-risk of 10–5% (0.10–0.05) indicates moderate evidence that a difference is apparent;
- an α-risk of 5–1% (0.05–0.01) indicates strong evidence that a difference is apparent;
- an α-risk of 1–0.1% (0.01–0.001) indicates very strong evidence that a difference is apparent; and
- an α-risk below 0.1%(<0.001) indicates extremely strong evidence that a difference is apparent.

For β-risks, the strength of the evidence that a difference is not apparent is assessed using the same criteria as above (substituting "is not apparent" for "is apparent").

The maximum allowable proportion of distinguishers, p_d, falls into three ranges:

- $p_d < 25\%$ represent small values;
- $25\% < p_d < 35\%$ represent medium-sized values; and
- $p_d > 35\%$ represent large values.

III. TRIANGLE TEST

The section on the Triangle test, being the first in this book, is rather complex and includes many details which (1) all sensory analysts should know; (2) are common to many methods; and (3) are therefore omitted in subsequent methods. The application of the unified approach is described in Examples 6.3 and 6.4.

A. SCOPE AND APPLICATION

Use this method when the test objective is to determine whether a sensory difference exists between two products. This method is particularly useful in situations where treatment effects may have produced product changes, which cannot be characterized simply by one or two attributes. Although it is statistically more efficient than the paired comparison and duo-trio methods, the Triangle test has limited use with products that involve sensory fatigue, carryover, or adaptation, and with subjects who find testing three samples too confusing. This method is effective in certain situations:

1. To determine whether product differences result from a change in ingredients, processing, packaging, or storage
2. To determine whether an overall difference exists, where no specific attribute(s) can be identified as having been affected
3. To select and monitor panelists for ability to discriminate given differences

B. PRINCIPLE OF THE TEST

Present to each subject three coded samples. Instruct subjects that two samples are identical and one is different (or odd). Ask the subjects to taste (feel, examine) each product from left to right and select the odd sample. Count the number of correct replies and refer to Table T8* for interpretation.

C. TEST SUBJECTS

Generally, 20 to 40 subjects are used for Triangle tests, although as few as 12 may be employed when differences are large and easy to spot. Similarity testing, on the other hand, requires 50 to 100 subjects. As a minimum, subjects should be familiar with the Triangle test (the format, the task, the procedure for evaluation), and with the product being tested, especially because flavor memory plays a part in triangle testing.

An orientation session is recommended prior to the actual taste test to familiarize subjects with the test procedures and product characteristics. Care must be taken to supply sufficient information to be instructive and motivating, while not biasing subjects with specific information about treatment effects and product identity.

---

* See the final section of this book, "Statistical Tables." Tables are numbered T1 to T14.

Triangle Test		
Name _____ Date _____		
Type of Sample _____		
_____		
Instructions		
Taste samples from left to right. Two are identical; determine which is the odd sample.		
If no difference is apparent, you must guess.		
Sets of three samples	Which is the odd sample?	Comments
___ ___ ___	_____	_____
___ ___ ___	_____	_____
___ ___ ___	_____	_____

FIGURE 6.1   Example of scoresheet for three Triangle tests.

D. TEST PROCEDURE

The test controls (explained in detail in Chapter 3) should include a partitioned test area in which each subject can work independently. Control of lighting may be necessary to reduce any color variables. Prepare and present samples under optimum conditions for the product type investigated, e.g., samples should be appetizing and well presented.

Offer samples simultaneously, if possible; however, samples which are bulky, leave an aftertaste, or show slight differences in appearance may be offered sequentially without invalidating the test.

Prepare equal numbers of the six possible combinations (ABB, BAA, AAB, BBA, ABA, and BAB) and present these at random to the subjects. Ask subjects to examine (taste, feel, smell, etc.) the samples in the order from left to right, with the option of going back to repeat the evaluation of each, while the test is in progress.

The scoresheet, shown in Figure 6.1, could provide for more than one set of samples. However, this can only be done if sensory fatigue is minimal. Do not ask questions about preference, acceptance, degree of difference, or type of difference after the initial selection of the odd sample. This is because the subject's choice of the odd sample may bias his/her responses to these additional questions. Responses to such questions may be obtained through additional tests. See Chapter 12 p. 231 for Preference and Acceptance tests and Chapter 7, p. 99 for difference tests related to size or type (attribute) of difference.

E. ANALYSIS AND INTERPRETATION OF RESULTS

Count the number of correct responses (correctly identified odd samples) and the number of total responses. Determine if the number correct for the number tested is equal to or larger than the number indicated in Table T8.

Do not count "no difference" replies as valid responses. Instruct subjects to guess if the odd sample is not detectable.

F. Example 6.1: Triangle Difference Test — New Malt Supply

A test beer "B" is brewed using a new lot of malt, and the sensory analyst wishes to know if it can be distinguished from control beer "A" taken from current production. A 5% risk of error is accepted and 12 trained assessors are available; 18 glasses of "A" and 18 glasses of "B" are prepared to make 12 sets which are distributed at random among the subjects, using two each of the combinations ABB, BAA, AAB, BBA, ABA, and BAB.

Eight subjects correctly identify the odd sample. In Table T8, the conclusion is that the two beers are different at the 5% level of significance.

G. Example 6.2: Detailed Example of Triangle Difference Test — Foil vs. Paper Wraps for Candy Bar

Problem/situation — The director of packaging of a confection company wishes to test the effectiveness of a new foil-lined packaging material against the paper wrap currently being used for candy bars. Preliminary observation shows that paper-wrapped bars begin to show harder texture after 3 months while foil-wrapped bars remain soft. The director feels that if he can show a significant difference at 3 months, he can justify a switch in wrap for the product.

Project objective — To determine if the change in packaging causes an overall difference in flavor and/or texture after 3 months of shelf storage.

Test objective — To measure if people can differentiate between the two 3-month-old products by tasting them.

Test design — A Triangle difference test with 30 to 36 subjects. The test will be conducted under normal white lighting to allow for differences in appearance to be taken into account. The subjects will be scheduled in groups of six to ensure full randomization within groups. Significance for a difference will be determined at an α risk of 5%, that is, this test will falsely conclude a difference only 5% of the time.

Screen samples — Inspect samples initially (before packaging) to ensure that no gross sensory differences are noticeable from sample to sample. Evaluate test samples at 3 months to ensure that no gross sensory characteristics have developed which would render the test invalid.

Conduct the test — Code two groups each of 54 plates with three-digit random numbers from Table Tl. Remove samples from package; cut off ends of each bar and discard; cut bar into bite-size pieces and place on coded plates. Keep plates containing samples that were paper wrapped (P) separate from those containing samples that were foil wrapped (F). For each subject, prepare a tray marked by his/her number and containing three plates which are P or F according to the worksheet in Figure 6.2. Record the three plate codes on the subject's ballot (see Figure 6.3).

Analyze results — Of the 30 subjects who showed up for the test, 17 correctly identified the odd sample.

Number of subjects	30
Number correct	17

Table T8 indicates that this difference is significant at an α-risk of 1% (probability $p \leq 0.01$).

Test report — The full report should contain the project objective, the test objective, and the test design as described previously. Examples of worksheet and scoresheet may be enclosed. Any information or recommendations given to the subjects (for example, about the origin of samples) must be reported. The tabulated results (17 correct out of 30) and the α-risk (meets the objective of 5%) follow. In the conclusion, the results are tied to the project objective: "A significant difference was found between the paper- and foil-wrapped candies. The foil does produce a perceived effect. There were 10 comments about softer texture in the foil-wrapped samples."

```
┌─────────────────────────────────────────────────────────────────────┐
│ Date      6-2-99          WORKSHEET           Test code    587 FF03   │
├─────────────────────────────────────────────────────────────────────┤
│ Post this sheet in the area where trays are prepared. Code            │
│ scoresheets ahead of time. Label serving containers ahead of          │
│ time.                                                                 │
├─────────────────────────────────────────────────────────────────────┤
│ Type of samples :        Candy bars                                   │
│ Type of test :           Triangle  test                               │
├─────────────────────────────────────────────────────────────────────┤
│    Sample  identification                              Code           │
│        Pkg 4736  (paper)                                P             │
│                                                                       │
│        Pkg 3987  (foil)                                 F             │
├─────────────────────────────────────────────────────────────────────┤
│ Code serving containers as follows:                                   │
│                                                                       │
│        Panelist #               Order of Presentation                 │
│    1,7,13, 19,25,31                    P - F - F                       │
│    2,8,14, 20,26,32                    F - P - F                       │
│    3,9,15, 21,27,33                    F - F - P                       │
│    4,10,16,22,28,34                    F - P - P                       │
│    5,11,17,23,29,35                    P - F - P                       │
│    6,12,18,24,30,36                    P - P - F                       │
├─────────────────────────────────────────────────────────────────────┤
│    1.   Place stickers with panelist's number on tray.                │
│    2.   Select plates "P" or "F" from those previously coded and      │
│         place on tray from left to right.                             │
│    3.   Write codes selected on panelist's scoresheet.                │
│    4.   Serve samples.                                                │
│    5.   Receive filled-in scoresheet and note on it the order of      │
│         presentation used, and whether  reply was correct (c) or      │
│         incorrect (i).                                                │
└─────────────────────────────────────────────────────────────────────┘
```

FIGURE 6.2   Worksheet for a Triangle test. Example 6.2: foil vs. paper wraps for candy bar.

H. EXAMPLE 6.3: TRIANGLE TEST FOR SIMILARITY. DETERMINING PANEL SIZE USING α, β, AND p_d — BLENDED TABLE SYRUP

Problem/situation — A manufacturer of blended table syrup has learned that his supplier of corn syrup is raising the price of this ingredient. The research team has identified an alternate supplier of high quality corn syrup whose price is more acceptable. The sensory analyst is asked to test the equivalency of two samples of blended table syrup, one formulated with the current supplier's product and the other with the less expensive corn syrup from the alternate supplier.

Project objective — Determine if the company's blended syrup can be formulated with the less expensive corn syrup from the alternate supplier without a perceptible change in flavor.

```
┌─────────────────────────────────────────────────────────────┐
│              Triangle  Test        Test Code:                 │
├─────────────────────────────────────────────────────────────┤
│   Taster No. ____     Name: _____     Date: _____       │
│   Type of Sample: _____            │
├─────────────────────────────────────────────────────────────┤
│   Instructions                                                │
│   Taste the samples on the tray from left to right.  Two      │
│   samples are identical; one is different.  Select the        │
│   odd/different  sample and indicate by placing an X next to  │
│   the code of the odd sample.                                 │
├─────────────────────────────────────────────────────────────┤
│   Samples          Indicate         Remarks                   │
│   on tray          odd sample                                 │
│   _____            □            _____           │
│   _____            □            _____           │
│   _____            □            _____           │
├─────────────────────────────────────────────────────────────┤
│   If you wish to comment on the reasons for your choice or if │
│   you wish to comment on the product characteristics, you     │
│   may do so under Remarks.                                    │
└─────────────────────────────────────────────────────────────┘
```

FIGURE 6.3   Scoresheet for Triangle test. Example 6.2: foil vs. paper wraps for candy bar. The subject places an X in one of the three boxes but may write remarks on more than one line.

Test objective — To test for similarity of the blended table syrup produced with corn syrups from the current and alternate suppliers.

Number of assessors. Choice of α, β, and p_d — The sensory analyst and the project director, looking at Table T7, note that to obtain maximum protection against falsely concluding similarity, for example by setting β at 0.1% (i.e., $\beta = 0.001$) relative to the alternative hypothesis, that the true proportion of the population able to detect a difference between the samples is at least 20% (i.e., $p_d = 0.20$), then to preserve a modest α-risk of 0.10 they need to have at least 260 assessors. They decide to compromise at $\alpha = 0.20$, $\beta = 0.01$, and $p_d = 30\%$ which requires 64 assessors.

Test design — The sensory analyst conducts a 66-response Triangle test according to the established test protocol for blended table syrups. The sensory booths are prepared with red-tinted filters to mask color differences. Twelve panelists are scheduled for each of five consecutive sessions and six panelists are scheduled for the sixth and final session. Figure 6.4 shows the analyst's worksheet for a typical session.

Analyze results — Out of 66 respondents, 21 correctly picked the odd sample. Referring to Table T8, in the row corresponding to $n = 66$ and the column corresponding to $\alpha = 0.20$, one finds that the minimum number of correct responses required for significance is 26. Therefore, with only 21 correct responses, it can be concluded that any sensory difference between the two syrups is sufficiently small to be ignored, that is, the two samples are sufficiently similar to be used interchangeably.

Interpret results — The analyst informs the project manager that the test resulted in 21 correct selections out of 66, indicating with 99% confidence that the proportion of the population who can perceive a difference is less than 30% and probably much lower. The alternate supplier's product can be accepted.

Confidence limits on p_d — If desired, analysts can calculate confidence limits on the proportion of the population that can distinguish the samples. The calculations are as follows, where $c =$ the number of correct responses and $n =$ the total number of assessors.

```
Date   11-5-98              WORKSHEET           No.  35-0032-01

Post this sheet in the area where trays are prepared. Code
scoresheets ahead of time. Label serving containers ahead of
time.

Type of samples :        Blended table syrups
Type of test :           Triangle similarity test

                                       Codes used for :
  Sample  identification:        Sets with      Sets with
                                    2 A's          2 B's

  A:    Lab code 47-3651        587    246         413
  B:    Lab code 026 (Control)         894      365    751

Code serving containers as follows:
Subject #          Codes in order            Underlying
                                             pattern*
     1           587  246  894                  AAB
     2           413  365  751                  ABB
     3           751  413  365                  BAB
     4           246  587  894                  AAB
     5           751  365  413                  BBA
     6           587  894  246                  ABA
     7           413  751  365                  ABB
     8           246  894  587                  ABA
     9           894  587  246                  BAA
    10           365  751  413                  BBA
    11           894  246  587                  BAA
    12           365  413  751                  BAB
*Each pattern is repeated twice to allow for each code in each
position.
```

FIGURE 6.4   Worksheet for Triangle test for similarity. Example 6.3: blended table syrup.

p_c (proportion correct) = c/n

p_d (proportion distinguishers) = $1.5p_c - 0.5$

s_d (standard deviation of p_d) = $1.5\sqrt{p_c(1-p_c)/n}$

one-sided upper confidence limit = $p_d + z_\beta s_d$

one-sided lower confidence limit = $p_d - z_\alpha s_d$

z_α and z_β are critical values of the standard normal distribution. Commonly used values of z for one-sided confidence limits include:

Confidence Level	z
75%	0.674
80%	0.842
85%	1.036
90%	1.282
95%	1.645
99%	2.326

For the data in the example, the upper 99% one-sided confidence limit on the proportion of distinguishers is calculated as:

$$p_{max} = p_d + z_\beta s_d = \left[1.5(21/66) - 0.5\right] + (2.326)(1.5)\sqrt{(21/66)(1-(21/66))/66}$$

$$= \left[-0.023\right] + 2.326(1.5)(0.05733)$$

$$= 0.177 \text{ or } 18\%$$

while the lower 80% one-sided confidence limit falls at

$$p_{min} = p_d - z_\alpha s_d = \left[-0.023\right] - 0.842(1.5)(0.05733) = -0.095 \text{ (i.e., 0.0, it cannot be negative)}$$

or, in words, the sensory analyst is 99% sure that the true proportion of the population that can distinguish the samples is no greater than 18% and may be as low as 0%.*

I. EXAMPLE 6.4: BALANCING α, β, AND p_d. SETTING EXPIRATION DATE FOR A SOFT DRINK COMPOSITION

Problem/situation — A producer of a soft drink composition wishes to choose a recommended expiration date to be stamped on bottled soft drinks made with it. It is known that in the cold (2°C), bottled samples can be stored for more than one year without any change in flavor, whereas at higher temperatures, the flavor shelf life is shorter. A test is carried out in which samples are stored at high ambient temperature (30°C) for 6, 8, and 12 months, then presented for difference testing.

Project objective — To choose a recommended expiration date for a bottle product made with the composition.

Test objective — To determine whether a sensory difference is apparent between the product stored cold and each of the three products stored warm.

Number of assessors. Choice of α, β, and p_d — The producer would like to see the latest possible expiration date and decides he is only willing to take a 5% chance of concluding that there is a difference when there is not (i.e., $\alpha = 0.05$). The QA Manager, on the other hand, wishes to be reasonably certain that customers cannot detect an "aged" flavor until after the expiration date, so he agrees to accept 90% certainty (i.e., $\beta = 0.10$) that no more than 30% of the population (i.e., $p_d = 30\%$) can detect a difference. Entering Table T7 in the column under $\beta = 0.10$ and the section for $p_d = 30\%$, the sensory analyst finds that a panel of 53 is needed for the tests. However, only 30 panelists can be made available for the duration of the tests. Therefore, the three of them renegotiate the test sensitivity parameters to provide the maximum possible risk protection with the number of available assessors. Consulting Table T7 again, they decide that a compromise of $p_d = 30\%$, $\beta = 0.20$, and $\alpha = 0.10$ provides acceptable sensitivity given the number of assessors available.

Test design — The analyst prepares and conducts Triangle tests using a panel of 30.

Analyze results — The number of correct selections turns out as follows: at 6 months, 11; at 8 months, 13; at 12 months, 15. Entering Table T8, the analyst concludes that, at 6 months, no proof of difference exists. At 8 months, the difference is larger. Table T8 shows that proof of

---

* Unified Approach vs. Similarity Tables — Notice that the unified approach used in this third edition does not include similarity tables such as those found in the second edition. As the present example illustrates, Table T8 merely shows that proof of similarity exists. In order to learn how strong the evidence of similarity is, i.e., that "p_d is no greater than 18% and may be as low as 0%," the analyst needs to calculate the confidence limits. See Chapter 13.II.C, p. 270, for the derivation of confidence intervals.

difference would have existed had a higher $\alpha = 0.20$ been used. Finally, at 12 months, the table shows that proof of a difference exists at $\alpha = .05$.

Interpretation — The group decides that an expiration date at 8 months provides adequate assurance against occurrences of "aged" flavor in product that has not passed this date. As an added check on their conclusion, the 80% one-sided confidence limits are calculated for each test. It is found that they can be 80% sure that no more than 16% of consumers can detect a difference at 6 months, no more than 26% at 8 months, but possibly as many as 37% at 12 months. The product is safely under the $p_d = 30\%$ limit at 8 months.*

IV. DUO-TRIO TEST

A. SCOPE AND APPLICATION

The Duo-trio test is statistically inefficient compared with the Triangle test because the chance of obtaining a correct result by guessing is 1 in 2. On the other hand, the test is simple and easily understood. Compared with the Paired Comparison test, it has the advantage that a reference sample is presented which avoids confusion with respect to what constitutes a difference, but a disadvantage is that three samples, rather than two, must be tasted.

Use this method when the test objective is to determine whether a sensory difference exists between two samples. This method is particularly useful in situations:

1. To determine whether product differences result from a change in ingredients, processing, packaging, or storage
2. To determine whether an overall difference exists, where no specific attributes can be identified as having been affected

The Duo-trio test has general application whenever more than 15, and preferably more than 30, test subjects are available. Two forms of the test exist: the *constant reference* mode, in which the same sample, usually drawn from regular production, is always the reference, and the *balanced reference* mode, in which both of the samples being compared are used at random as the reference. Use the constant reference mode with trained subjects whenever a product well known to them can be used as the reference. Use the balanced reference mode if both samples are unknown or if untrained subjects are used.

If there are pronounced aftertastes, the Duo-trio test is less suitable than the Paired Comparison test. (See Chapter 7.II, p. 100.)

B. PRINCIPLE OF THE TEST

Present to each subject an identified reference sample, followed by two coded samples, one of which matches the reference sample. Ask subjects to indicate which coded sample matches the reference. Count the number of correct replies and refer to Table T10 for interpretation.

C. TEST SUBJECTS

Select, train, and instruct the subjects as described under Section III.C, p. 61. As a general rule, the minimum is 16 subjects, but for less than 28, the β-error is high. Discrimination is much improved if 32, 40, or a larger number can be employed.

---

* An example of the confidence limit calculation using the 6 month results is:

$$p_d = \left(1.5(11/30) - 0.5\right) + 0.84(1.5)\sqrt{(11/30)(1-(11/30))/30} = 0.16$$

```
┌─────────────────────────────────────────────────────────┐
│  ┌───────────────────────────────────────────────────┐  │
│  │                                        Test No.     │  │
│  │          DUO-TRIO TEST                              │  │
│  │                                                     │  │
│  ├───────────────────────────────────────────────────┤  │
│  │  Taster No. _____   Name: _____   Date: __ │  │
│  │  Type of Sample: _____ │  │
│  ├───────────────────────────────────────────────────┤  │
│  │                                                     │  │
│  │  Instructions: Taste samples from left to right.   │  │
│  │  The left hand sample is a reference. Determine    │  │
│  │  which of the two samples matches the reference    │  │
│  │  and indicate by placing an X.                     │  │
│  │                                                     │  │
│  │  If no difference is apparent between the two      │  │
│  │  unknown samples, you must guess.                  │  │
│  │                                                     │  │
│  ├───────────────────────────────────────────────────┤  │
│  │   Reference      Code ____        Code ____        │  │
│  │     ▨              □                □              │  │
│  ├───────────────────────────────────────────────────┤  │
│  │  Comments: _____ │  │
│  │            _____ │  │
│  └───────────────────────────────────────────────────┘  │
└─────────────────────────────────────────────────────────┘
```

FIGURE 6.5   Scoresheet for Duo-trio test.

D. TEST PROCEDURE

For test controls and product controls, see p. 62. Offer samples simultaneously, if possible, or else sequentially. Prepare equal numbers of the possible combinations (see examples) and allocate the sets at random among the subjects. An example of a scoresheet (which is the same in the balanced reference and constant reference modes) is given in Figure 6.5. Space for several Duo-trio tests may be provided on the scoresheet, but do not ask supplementary questions (e.g., the degree or type of difference or the subject's preference) as the subject's choice of matching sample may bias his response to these additional questions. Count the number of correct responses and the total number of responses and refer to Table T10. Do not count "no difference" responses; subjects must guess if in doubt. Three examples follow, all using the unified approach.

E. EXAMPLE 6.5: BALANCED REFERENCE — FRAGRANCE FOR FACIAL TISSUE BOXES

Problem/situation — A product development fragrance chemist needs to know if two methods of fragrance delivery for boxed facial tissues, fragrance delivered directly to the tissues, or fragrance delivered to the inside of the box, will produce differences in perceived fragrance quality or quantity.

Project objective — To determine if the two methods of fragrance delivery produce any difference in the perceived fragrance of the two tissues after they have been stored for a period of time comparable to normal product age at time of use.

Test objective — To determine if a fragrance difference can be perceived between the two tissue samples after storage for 3 months.

Test design — A Duo-trio test requires less repeated sniffing of samples than Triangle tests or attribute difference testing, when the stimuli are complex. This reduces the potential confusion caused by odor adaptation and/or the difficulty in sorting out three sample inter-comparisons. The test is conducted with 40 subjects who have some experience in odor evaluation. The samples are prepared by the fragrance chemist, using the same fragrance and the same tissues on the same day. The boxed tissues are then stored under identical conditions for 3 months. Test tissues are taken from the center 50% of the box; each tissue is placed in a sealed glass jar 1 h prior to evaluation.

```
┌─────────────────────────────────────────────────────────────┐
│                                         Test No.              │
│              DUO-TRIO TEST               230S                 │
│                                                               │
├─────────────────────────────────────────────────────────────┤
│  Panelist No. _21_  Name:_____  Date:_____      │
│  Type of Sample:  Facial tissue in a glass jar                │
├─────────────────────────────────────────────────────────────┤
│  Instructions                                                 │
│    1. Please sniff each sample, starting at the left. Remove  │
│       the cap only briefly  and take short, shallow sniffs.   │
│                                                               │
│    2. The left hand sample is a reference. Determine which    │
│       of the two coded samples matches the fragrance of the   │
│       reference.                                              │
│                                                               │
│    3. Indicate the matching sample by placing an X in the     │
│       corresponding box.                                      │
│                                                               │
│    If no difference is apparent between the two unknown samples, you must │
│    guess.                                                     │
├─────────────────────────────────────────────────────────────┤
│  Reference       Code _____        Code _____            │
│                                                               │
│      ▨                  □                  □                  │
│                                                               │
│  Comments:_____          │
│  _____         │
└─────────────────────────────────────────────────────────────┘
```

FIGURE 6.6   Scoresheet for Duo-trio test. Example 6.5: balanced reference mode.

This allows for some fragrance to migrate to the headspace, and the use of the closed container reduces the amount of fragrance buildup in the testing booths. Each of the two samples is used as the reference in half (20) of the evaluations. Figure 6.6 shows the scoresheet used.

Analyze results — Only 21 out of the 40 subjects chose the correct match to the designated reference. According to Table T10, 26 correct responses are required at an α-risk of 5%. In addition, when the data are reviewed for possible effects from the position of each sample as reference, the results show that the distribution of correct responses is even (10 and 11). This indicates that the quality and/or quantity of the two fragrances have little, if any, additional biasing effect on the results.

Interpret results — The sensory analyst informs the fragrance chemist that the odor Duo-trio test failed to detect any significant odor differences between the two packing systems given the fragrance, the tissue, and the storage time used in the study.

Sensitivity of the test — For planning future studies of this type, note that choosing 40 subjects for a Duo-trio test yields the following values for the test-sensitivity parameters:

Proportion of Distinguishers (p_d)	Probability of Detecting	
	$(1 - \beta)$ @ $\alpha = 0.05$	$(1 - \beta)$ @ $\alpha = 0.10$
10%	0.13	0.21
15%	0.21	0.32
20%	0.32	0.44
25%	0.44	0.57
30%	0.57	0.69
35%	0.70	0.80
40%	0.81	0.88
45%	0.89	0.94
50%	0.95	0.97

For example, using 40 subjects and testing at the $\alpha = 0.05$ level yields a test that has a 44% chance $(1 - \beta = 0.44)$ of detecting the situation where 25% of the population can detect a difference $(p_d = 25\%)$. Increasing the number of subjects increases the likelihood of detecting any given value of p_d. Testing at larger values of α also increases the chances of detecting a difference at a given p_d.

F. EXAMPLE 6.6: CONSTANT REFERENCE — NEW CAN LINER

Problem/situation — A brewer is faced with two supplies of cans, "A" being the regular supply he has used for years and "B" a proposed new supply said to provide a slight advantage in shelf life. He wants to know whether any difference can be detected between the two cans. The brewer feels that it is important to balance the risk of introducing an unwanted change to his beer against the risk of passing up the extended shelf life offered by can "B."

Project objective — To determine if the package change causes any perceptible difference in the beer after shelf storage, as normally experienced in the trade.

Test objective — To determine if any sensory difference can be perceived between the two beers after 8 weeks of shelf storage at room temperature.

Number of assessors — The brewer knows from past experience that if no more than $p_d = 30\%$ of his panel can detect a difference he assumes no meaningful risk in the marketplace. He is slightly more concerned with introducing an unwanted difference than he is with passing up the slightly extended shelf life offered by can "B." Therefore, he decides to set the β-risk at 0.05 and his α-risk at 0.10. Referring to Table T9 in the section for $p_d = 30\%$, the column for $\beta = 0.05$ and the row for $\alpha = 0.10$, he finds that 96 respondents are required for the test.

Test design — A Duo-trio test in the constant reference mode is appropriate because the company's beer in can "A" is familiar to the tasters. A separate test is conducted at each of the brewer's three testing sites. Each test is set up with 32 subjects, with "A" as the reference; 64 glasses of beer "A" and 32 of beer "B" are prepared and served to the subjects in 16 combinations AAB and 16 combinations ABA, the left-hand sample being the reference.

Analyze results — 18, 20, and 19 subjects correctly identified the sample that matched the reference. According to Table T10, significance at the 10% level requires 21 correct.

Note: In many cases it is permissible to combine two or more tests so as to obtain improved discrimination. In the present case, the cans were samples of the same lot, and the subjects were from the same panel, so combination is permissible. $18 + 20 + 19 = 57$ correct out of $3 \times 32 = 96$ trials. From Table T10, the critical numbers of correct replies with 96 samples are 55 at the 10% level of significance, and 57 at the 5% level.

Interpret results — Conclude that a difference exists, significant at the 5% level on the basis of combining three tests. Next, examine any notes made by panelists, describing the difference. If none is found, submit the samples to a descriptive panel. Ultimately, if the difference is neither pleasant nor unpleasant, a consumer test may be required to determine if there is preference for one can or the other.

G. EXAMPLE 6.7: DUO-TRIO SIMILARITY TEST — REPLACING COFFEE BLEND

Problem/situation — A manufacturer of coffee has learned that one coffee bean variety, which has long been a major component of its blend, will be in short supply for the next 2 years. A team of researchers has formulated three "new" blends, which they feel are equivalent in flavor to the current blend. The research team has asked the sensory evaluation analyst to test the equivalency of these new blends to the current product.

Project objective — To determine which of the three blends can best be used to replace the current blend.

Test objective — To test for similarity between the current blend and each of the project blends.

```
Date  3-4-99  Cell no.  3    WORKSHEET        No.    2803-30

   Post this sheet in the area where trays are prepared.  Code
   score sheets ahead of time.  Label serving containers ahead.

   Type of samples:  cups of coffee
   Type of test:   Duo-trio similarity test (balanced reference)

Samples
served     A =  Control    B = Blend 62-A  C = Blend 223B  D = Blend 211

                         Codes Used:

           For B versus A      For C versus A      For D versus A
           Sets w/  Sets w/    Sets w/  Sets w/    Sets w/  Sets w/
           2 A's    2 B's      2 A's    2 C's      2 A's    2 D's

Sample A   317 543     986     866 581    541      121 225    965
Sample B     314    393 737
Sample C                       674      373 158
Sample D                                            221     499 134

              Code serving containers as follows:

Subject Pat-              Pat-               Pat-
No.     tern Codes in order tern Codes in order tern Codes in order

 37    ABA  R - 314-543   AAC  R - 581-674   DAD  R - 965-134
 38    BBA  R - 737-986   ACA  R - 674-866   AAD  R - 225-221
 39    BAB  R - 986-393   CCA  R - 158-541   ADA  R - 221-121
 40    AAB  R - 317-314   CAC  R - 541-373   DDA  R - 499-965
 41    ABA  R - 314-317   AAC  R - 866-674   DAD  R - 965-499
 42    BBA  R - 393-986   ACA  R - 674-581   AAD  R - 121-221
 43    BAB  R - 986-737   CCA  R - 373-541   ADA  R - 221-225
 44    AAB  R - 543-314   CAC  R - 541-158   DDA  R - 134-965
 45    ABA  R - 314-543   AAC  R - 581-674   DAD  R - 965-134
 46    BBA  R - 737-986   ACA  R - 674-866   AAD  R - 225-221
 47    BAB  R - 986-393   CCA  R - 158-541   ADA  R - 221-121
 48    AAB  R - 317-314   CAC  R - 541-373   DDA  R - 499-965
```

FIGURE 6.7 Worksheet for Duo-trio similarity test. Example 6.7: replacing coffee blend.

Test design — Preliminary tests have shown that differences are small and not particularly related to a specific attribute. Therefore, use of the Duo-trio test for similarity is appropriate. In order to reduce the risk of missing a perceptible difference, the sensory analyst proposes the tests be run using 60 panelists each (an increase from the customary 36 used in testing for difference). Using her spreadsheet test-sensitivity analyzer* (see Chapter 13.III.E, pp. 285–287), she has determined that a 60-respondent Duo-trio test has a 90% (i.e., $\beta = 0.10$) probability of detecting the situation where $p_d = 25\%$ of the panelists can detect a difference, with an accompanying α-risk of approximately 0.25. The analyst accepts the large α-risk because she is much more concerned with incorrectly approving a blend that is different from the control and she only has 60 panelists available for the tests. For each blend, the sensory analyst plans to conduct one 60-response coffee test spaced over 1 week. As the preparation and holding time of the product is a critical factor which influences flavor, subjects must be carefully scheduled to arrive within 10 min after preparation of the

* Available on request in Excel as an e-mail attachment from Tom_Carr@msn.com.

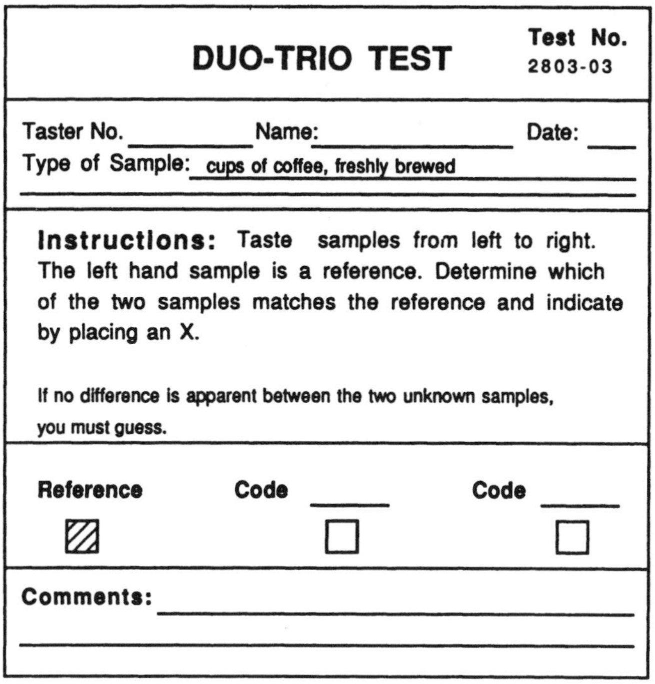

DUO-TRIO TEST Test No. 2803-03

Taster No. ________ Name: ______________ Date: ___

Type of Sample: cups of coffee, freshly brewed

Instructions: Taste samples from left to right. The left hand sample is a reference. Determine which of the two samples matches the reference and indicate by placing an X.

If no difference is apparent between the two unknown samples, you must guess.

Reference Code _____ Code _____

Comments: ______________

FIGURE 6.8 Scoresheet for Duo-trio similarity test. Example 6.7: replacing coffee blend.

products. Using the 12 booths in the sensory lab, prepared with brown-tinted filters on the lights, the analyst schedules 12 different subjects for each cell of each test. The use of 12 panelists per session permits balanced presentation of each sample as the reference sample, as well as a balanced order of presentation of the two test samples within the cell. Figure 6.7 shows the analyst's worksheet.

Samples are presented without cream and sugar. The pots are kept at 175°F and poured into heated (130°F) ceramic cups, which are coded as per the worksheet and placed in the order which it indicates. Scoresheets (see Figure 6.8) are prepared in advance to save time, and samples are poured when the subject is already sitting in the booth.

Analyze results — The number of correct responses for the three test blends were

Cell no. (of 12 subjects)	Blend B	Blend C	Blend D
1	3	6	8
2	4	5	8
3	5	7	5
4	7	7	7
5	5	5	7
Total	24	30	35

From her spreadsheet test-sensitivity analyzer, the analyst knows that 33 correct responses are necessary to conclude that a significant difference exists at the α-risk chosen for the test (approximately 0.25), so 32 or fewer correct responses from the 60-respondent test is evidence of adequate similarity.*

* In using the test-sensitivity analyzer, do not accept values of α, β, and p_d that result from a "Number of Correct Responses" that is less than what would be expected by chance alone (i.e., n/3 for Triangle tests, n/2 for Duo-trio tests, etc.). An observed number of correct responses less than what would be expected by chance alone, in fact, may suggest that some extraneous factor is biasing the selection of the odd sample.

Output from Test-Sensitivity Analyzer

INPUTS					OUTPUT		
Number of Respondents	Number of Correct Responses	Probability of a Correct Guess	Proportion Distinguishers	Probability of a Correct Response @ p_d	TYPE I Error	TYPE II Error	Power
n	x	p_0	p_d	p_{max}	α-risk	β-risk	$1-\beta$
60	33	0.50	0.25	0.625	0.2595	0.0923	0.9077

Interpretation:

33	or more correct responses is evidence of a difference at the $a = 0.26$ level of significance.
32	or fewer correct responses indicates that you can be 91% sure that no more than 25% of the panelists can detect a difference — that is, evidence of similarity relative to $p_d = 25\%$ at the $\beta = 0.09$ level of significance.

Therefore, it is concluded that test blends B and C are sufficiently similar to the control to warrant further consideration, but that test blend D, with 35 correct answers, is not. The 90% upper one-tailed confidence interval on the true proportion of distinguishers for test blend D (based on the Duo-trio test method) is

$$P_{max(90\%)} = \left[2(x/n) - 1\right] + z_\beta \sqrt{\left[4(x/n)(1-(x/n))\right]/n}$$

$$= \left[2(35/60) - 1\right] + 1.282 \sqrt{\left[4(35/60)(1-(35/60))\right]/60}$$

$$= \left[0.1667\right] + 1.282(0.1273)$$

$$= 0.33, \text{ or } 33\%$$

The sensory analyst concludes with 90% confidence that the true proportion of the population that can distinguish test blend D from the control may be as large as 33%, thus exceeding the prespecified critical limit (p_d) of 25% by as much as 8%.

The sensory analyst may have an additional concern. Only 24 of the 60 respondents correctly identified test blend B. In a Duo-trio test involving 60 respondents, the expected number of correct selections when all of the respondents are guessing ($p_d = 0$) is $n/2 = 30$. The less than expected number of correct responses may indicate that some extraneous factor was active during the testing of blend B that biased the respondents away from making the correct selection, for example, mislabeled samples or poor preparation or handling of the samples before serving. The sensory analyst tests the hypothesis that the true probability of a correct response is at least 50% (H_0: $p \geq 0.5$) against the alternative that it is less than 50% (H_a: $p < 0.5$) using the normal approximation to the binomial with the one-tailed confidence level set at 95% (i.e., $\alpha = 0.05$, lower tail). The test statistic is

$$z = \left[(x/n) - p_o\right] / \sqrt{p_o(1-p_o)/n}$$

$$= \left[(24/60 - 0.50)\right] / \sqrt{0.50(1-0.50)/60}$$

$$= \left[-0.10\right] / (0.06455)$$

$$= -1.55$$

Using Table T3 (noting that $\Pr[z < -1.55] = \Pr[z > 1.55]$), the sensory analyst finds that the probability of observing a value of the test statistic no larger than -1.55 is $(0.5 - 0.4394) = 0.0606$. This probability is greater than the value of $\alpha = 0.05$, and the analyst concludes that there is not sufficient evidence to reject the null hypothesis at the 95% level. The 24 correct responses were not sufficiently off the mark (of 30) for us to conclude that an extraneous factor was active.

V. TWO-OUT-OF-FIVE TEST

A. Scope and Application

This method is statistically very efficient because the chances of correctly guessing two out of five samples are 1 in 10, as compared with 1 in 3 for the Triangle test. By the same token, the test is so strongly affected by sensory fatigue and by memory effects that its principal use has been in visual, auditory, and tactile applications, and not in flavor testing.

Use this method when the test objective is to determine whether a sensory difference exists between two samples, and particularly when only a small number of subjects is available (e.g., ten).

As with the Triangle test, the Two-out-of-five test is effective in certain situations:

1. To determine whether product differences result from a change in ingredients, processing, packaging, or storage
2. To determine whether an overall difference exists, where no specific attribute(s) can be identified as having been affected
3. To select and monitor panelists for ability to discriminate given differences

in test situations where sensory fatigue effects are small.

B. Principle of the Test

Present to each subject five coded samples. Instruct subjects that two samples belong to one type and three to another. Ask the subjects to taste (feel, view, examine) each product from left to right and select the two samples that are different from the other three. Count the number of correct replies and refer to Table T14 for interpretation.

C. Test Subjects

Select, train, and instruct the subjects as described on p. 61. Generally 10 to 20 subjects are used. As few as five to six may be used when differences are large and easy to spot. Use only trained subjects.

D. Test Procedure

For test controls and product controls, see p. 62. Offer samples simultaneously if possible; however, samples which are bulky, or show slight differences in appearance, may be offered sequentially without invalidating the test. If the number of subjects is other than 20, select the combinations at random from the following, taking equal numbers of combinations with 3 A's and 3 B's:

AAABB	ABABA	BBBAA	BABAB
AABAB	BAABA	BBABA	ABBAB
ABAAB	ABBAA	BABBA	BAABB
BAAAB	BABAA	ABBBA	ABABB
AABBA	BBAAA	BBAAB	AABBB

```
┌─────────────────────────────────────────────────────────┐
│                  Two-Out-Of-Five Test                      │
├─────────────────────────────────────────────────────────┤
│  Name: _____  Date: _____      │
│  Type of Sample: _____     │
├─────────────────────────────────────────────────────────┤
│  Instructions                                              │
│                                                            │
│  1.  Examine the samples from left to right. Two are of   │
│  one type, and the other three of another.                 │
│                                                            │
│  2.  Identify the group of two samples by placing an X in │
│  the corresponding boxes.                                  │
├──────────────────┬──────────────────┬───────────────────┤
│       Test 1     │      Test 2      │      Test 3        │
│  left  _____  │    _____      │    _____        │
│        _____  │    _____      │    _____        │
│        _____  │    _____      │    _____        │
│        _____  │    _____      │    _____        │
│  right _____  │    _____      │    _____        │
├──────────────────┴──────────────────┴───────────────────┤
│                       Comments                             │
├──────────────────┬──────────────────┬───────────────────┤
│  left  _____  │    _____      │    _____        │
│        _____  │    _____      │    _____        │
│        _____  │    _____      │    _____        │
│        _____  │    _____      │    _____        │
│  right _____  │    _____      │    _____        │
└──────────────────┴──────────────────┴───────────────────┘
```

FIGURE 6.9   Scoresheet for three Two-out-of-five tests.

An example of a scoresheet is given in Figure 6.9. Count the number of correct responses and the number of total responses and refer to Table T14. Do not count "no difference" responses; subjects must guess if in doubt.

E. EXAMPLE 6.8: COMPARING TEXTILES FOR ROUGHNESS

Problem/situation — A textile manufacturer wishes to replace an existing polyester fabric with a polyester/nylon blend. He has received a complaint that the polyester/nylon blend has a rougher and scratchier surface.

Project objective — To determine whether the polyester/nylon blend needs to be modified because it is too rough.

Test objective — To obtain a measure of the relative difference in surface feel between the two fabrics.

Test design — As sensory fatigue is not a large factor, the Two-out-of-five test is the most efficient for assessing differences. A small panel of 12 will be able to detect quite small differences. Choose at random 12 combinations of the two fabrics from the table of 20 combinations previously presented. Ask the panelists: "Which two samples feel the same and different from the other three?"

Conduct the test — Place the anchored or loosely mounted fabric swatches each inside a cardboard tent in a straight line in front of each panelist (see Figure 6.10) who must be able to feel the fabrics but cannot see them. Assign sample codes from a list of random three-digit numbers (see Table Tl). Use the scoresheet in Figure 6.11.

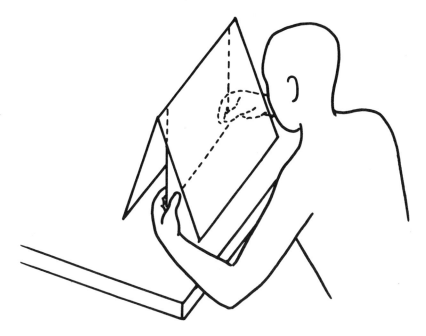

FIGURE 6.10   Two-out-of-five test. Example 6.8: arrangement of fabric samples in front of panelist.

Two-out-of-Five Test   Test Code

Name: _____   Date: _____

Type of Sample: _____

Type of Difference: _____

_____

Instructions

1. Examine the samples in the order listed below. Two are of one type and the other three of another. Feel the surface <u>gently</u> with fingers or palm of hand.

2. Identify the two samples that feel the same by placing an X in the corresponding boxes.

Codes	X	Comments
_____	☐	_____
_____	☐	_____
_____	☐	_____
_____	☐	_____
_____	☐	_____

FIGURE 6.11   Scoresheet for Two-out-of-five test. Example 6.8: comparing textiles for roughness.

Analyze results — Of the 12 subjects, 9 were able to correctly group the fabric samples. Reference to Table T14 shows that the difference in surface feel was detectable at a level of significance of $\alpha = 0.001$.

Interpret results — The fabric manufacturer is informed that a difference in surface feel between the two fabric types is easily detectable.

F. EXAMPLE 6.9: EMOLLIENT IN FACE CREAM

Problem/situation — The substitution of one emollient for another in the formula for a face cream is desirable because of a significant saving in cost of production. The substitution appears to reduce the surface gloss of the product.

Project objective — The marketing group wishes to determine whether a visually detectable difference exists between the two formulas before going to consumers to determine any effect on acceptance.

Test objective — To determine whether a statistically significant difference in appearance exists between the two formulas of face cream.

Test design/screen samples — Use ten subjects who have been screened for color blindness and impaired vision. Test 2 ml of product under white incandescent light on a watch glass against a white background. Pretest samples to be sure that surfaces do not change (crust, weep, discabor) within 30 min after exposure, the maximum length of one test cell.

Date	3-05-99	**WORKSHEET**	Test code	TO-AF88

Post this sheet in the area where trays are prepared. Code scoresheets ahead of time. Label serving containers ahead of time.

Type of samples :   Face cream for viewing

Type of test :   Two-out -of -five test

Sample identification	Code
Px-2316 (control)	A
Px-2602 (new emollient)	B

Arrange samples as follows in the front of each subject:

Judge No.	Order of Samples
1	A A B B B
2	A B B A B
3	B A A B B
4	B A B B A
5	B B A B A
6	B B A A A
7	B A A B A
8	A B B A A
9	A B A A B
10	A A B A B

FIGURE 6.12 Worksheet for Two-out-of-five test. Example 6.9: emollient in face cream. Arrangement of samples for viewing.

Conduct test — Arrange samples in a straight line from left to right according to the plan shown on the worksheet (see Figure 6.12); use a scoresheet similar to the one in Figure 6.11. Ask the subjects to "identify the two samples which are the same in appearance and different from the other three."

Analyze results — Five subjects group the samples correctly. According to Table T14, this corresponds to 1% significance for a difference.

VI. SAME/DIFFERENT TEST
(OR SIMPLE DIFFERENCE TEST)

A. SCOPE AND APPLICATION

Use this method when the test objective is to determine whether a sensory difference exists between two products, particularly when these are unsuitable for triple or multiple presentation, e.g., when the Triangle and Duo-trio tests cannot be used. Examples of such situations are comparisons between samples of strong or lingering flavor, samples which need to be applied to the skin in half-face tests, and samples which are very complex stimuli and are mentally confusing to the panelists.

As with other overall difference tests, the Same/Different test is effective in situations:

1. To determine whether product differences result from a change in ingredients, processing, packaging, or storage
2. To determine whether an overall difference exists, where no specific attribute(s) can be identified as having been affected

This test is somewhat time consuming because the information on possible product differences is obtained by comparing responses obtained from different pairs (A/B and B/A) with those obtained from matched pairs (A/A and B/B). The presentation of the matched pair enables the sensory analyst to evaluate the magnitude of the "placebo effect" of simply asking a difference question.

B. PRINCIPLE OF THE TEST

Present each subject with two samples, asking whether the samples are the same or different. In half the pairs present the two different samples; in half the pairs present a matched pair (the same sample, twice). Analyze results by comparing the number of "different" responses for the matched pairs to the number of "different" responses for the different pairs, using the χ^2-test.

C. TEST SUBJECTS

Generally, 20 to 50 presentations of each of the four sample combinations (A/A, B/B, A/B, B/A) are required to determine differences. Up to 200 different subjects can be used, or 100 subjects may receive two of the pairs. If the Same/Different has been chosen because of the complexity of the stimuli, then no more than one pair should be presented to any one subject at a time. Subjects may be trained or untrained but panels should not consist of mixtures of the two.

D. TEST PROCEDURE

For test controls and product controls, see p. 62. Offer samples simultaneously if possible, or else successively. Prepare equal numbers of the four pairs and present them at random to the subjects, if each is to evaluate one pair only. If the test is designed so that each subject is to evaluate more than one pair (one matched and one different or all four combinations), then records of each subject's test scores must be kept. Typical worksheets and scoresheets are given in Example 6.10.

E. ANALYSIS AND INTERPRETATION OF RESULTS

See Example 6.10.

F. EXAMPLE 6.10: REPLACING A PROCESSING COOKER FOR BARBECUE SAUCE

Problem/situation — In an attempt to modernize a condiment plant a manufacturer must replace an old cooker used to process barbecue sauce. The plant manager would like to know if the product produced in the new cooker tastes the same as that made in the old cooker.

Project objective — To determine if the new cooker can be put into service in the plant in place of the old cooker.

Test objective — To determine if the two barbecue sauce products, produced in different cookers, can be distinguished by taste.

Test design — The products are spicy and will cause carryover effects when tested. Therefore, the Same/Different test with a bland carrier, such as white bread, is an appropriate test to use. A total of 60 responses, 30 matched and 30 unmatched pairs, are collected from 60 subjects. Each subject evaluates either a matched pair (A/A or B/B) or an unmatched pair (A/B or B/A) in a single session. The worksheet and the scoresheet for the test are shown in Figures 6.13 and 6.14. The test is conducted in the booth area under red lights to mask any color differences.

Screen samples — Preliminary tests are made with five experienced tasters to determine if the samples are easier to taste plain or on a carrier, such as white bread. The carrier is used to make comparison easier without introducing extraneous sensory factors. The pretest is also helpful in determining the appropriate amount of product (by weight or volume) relative to bread (by size) for the test.

Conduct test — Just before each subject is to taste, add the premeasured sauce to the precut bread pieces, which had been stored cold in an airtight container. Place samples on labeled plates in the order indicated on the worksheet for each panelist.

Analyze results — In the table below, the columns indicate the samples which were tested; the rows indicate how they were identified by the subjects:

		Subjects received		
		Matched pair AA or BB	Unmatched pair AB or BA	Total
Subjects said:	Same	17	9	26
	Different	13	21	34
	Total	30	30	60

The χ^2-analysis (see Chapter 13.III.D.6, p. 284) is used to compare the placebo effect (17/13) with the treatment effect (9/21). The χ^2-statistic is calculated as:

$$\chi^2 = \sum \frac{(O - E)^2}{E}$$

where O is the observed number and E is the expected number, in each of the four boxes same/matched, same/unmatched, different/matched, and different/unmatched. For example, for the box same/matched:

$$E = (26 \times 30)/60 = 13, \text{ i.e.,}$$

$$\chi^2 = \frac{(17-13)^2}{13} + \frac{(9-13)^2}{13} + \frac{(13-17)^2}{17} + \frac{(21-17)^2}{17} = 4.34$$

Date	2-26-99	WORKSHEET	Test code	84-46F09

Post this sheet in the area where trays are prepared. Code scoresheets ahead of time. Label serving containers ahead of time.

Type of samples :   Barbecue sauce on white bread pieces
Type of test :   Same/Different test

Sample identification	Code
5-117-36 (old cooker)	36
5-117-39 (new cooker)	39

Code serving containers with 3-digit random numbers and divide into two lots, one lot to receive sample 36, the other sample 39.

When preparing panelists' trays, place samples from left to right in the following order :

Panelist Code	Sample Order
1 - 15	36 - 36
16 - 30	36 - 39
31 - 45	39 - 36
46 - 60	39 - 39

FIGURE 6.13   Worksheet for Same/Different test. Example 6.10: replacing a processing cooker for barbecue sauce.

SAME/DIFFERENT TEST   Test No. 84-4639

Taster No. _____   Name: _____   Date: ____
Type of Sample: Barbecue sauce on white bread pieces _____

Instructions

1.   Taste   the two samples from left to right.

2.   Determine if samples are the same/identical or different.

3.   Mark your response below.

Note that some of the sets consist of two identical samples.

_____   Products are the same

_____   Products are different

Comments: _____
_____

FIGURE 6.14   Scoresheet for Same/Different test. Example 6.10: replacing a processing cooker for barbecue sauce.

which is greater than the value in Table T5 (df = 1, probability = 0.05, $\chi^2 = 3.84$), i.e., a significant difference exists.

Interpret results — The results show a significant difference between the barbecue sauces prepared in the two different cookers. The sensory analyst informs the plant manager that the equipment supplier's claim is not true. A difference has been detected between the two products. The analyst suggests that if the substitution of the new cooker remains an important cost/efficiency item in the plant, the two barbecue sauces should be tested for preference among users. A consumer test resulting in parity for the two sauces or in preference for the sauce from the new cooker would permit the plant to implement the process.

Note: If Example 6.10 had been run with 30 subjects rather than 60, and with each of the 30 receiving both a matched and an unmatched pair in separate sessions, the results could have been the same as above, but the χ^2-test would have been inappropriate and a McNemar test would be indicated (Conover, 1980). In order to perform the McNemar procedure, the analyst must keep track of both responses from each panelist and tally them in the following format:

		Subject received A/B or B/A and responded:	
		Same	**Different**
Subject received A/A or B/B and responded:	Same	a = 2	b = 15
	Different	c = 7	d = 6

The test statistic is

$$\text{McNemar's } T = (b - c)^2/(b + c)$$

For $(b + c) \geq 20$, the assumption of no difference is rejected if T is greater than the critical value of a χ^2 with one degree of freedom from Table T5. For $(b + c) < 20$, a binomial procedure is applied (see Conover, loc. cit.). For the present example:

$$\text{McNemar's } T = (15 - 7)^2/(15 + 7) = 2.91$$

which is less than $\chi^2_{1,0.05} = 3.84$. Therefore, we cannot conclude that the samples are different.

If we had treated the paired data from the 30 panelists as if they were individual observations from 60 panelists, we would have obtained the data as presented under "Analyze results," p. 80. The standard χ^2-analysis would have led us to the incorrect conclusion that a statistically significant difference existed between the samples.

VII. "A" – "NOT A" TEST

A. SCOPE AND APPLICATION

Use this method (ISO, 1985) when the test objective is to determine whether a sensory difference exists between two products, particularly when these are unsuitable for dual or triple presentation, i.e., when the Duo-trio and Triangle tests cannot be used. Examples of such situations are comparisons of products with a strong and/or lingering flavor, samples which need to be applied to the skin in half-face tests, products which differ slightly in appearance, and samples which are very complex stimuli and are mentally confusing to the panelists. Use the "A" – "Not A" test in preference to the Same/Different test (Section VI) when one of the two products has importance as a standard or reference product, is familiar to the subjects, or is essential to the project as the current sample against which all others are measured.

As with other overall difference tests, the "A" – "Not A" test is effective in situations:

1. To determine whether product differences result from a change in ingredients, processing, packaging, or storage
2. To determine whether an overall difference exists, where no specific attribute(s) can be identified as having been affected

The test is also useful for screening of panelists, e.g., determining whether a test subject (or group of subjects) recognizes a particular sweetener relative to other sweeteners, and it can be used for determining sensory thresholds by a Signal Detection method (Macmillan and Creelman, 1991).

B. PRINCIPLE OF THE TEST

Familiarize the panelists with samples "A" and "not A." Present each panelist with samples, some of which are product "A" while others are product "not A"; for each sample the subject judges whether it is "A" or "not A." Determine the subjects' ability to discriminate by comparing the correct identifications with the incorrect ones using the χ^2-test.

C. TEST SUBJECTS

Train 10 to 50 subjects to recognize the "A" and the "not A" samples. Use 20 to 50 presentations of each sample in the study. Each subject may receive only one sample ("A" or "not A"), two samples (one "A" and one "not A"), or each subject may test up to ten samples in a series. The number of samples allowed is determined by the degree of physical and/or mental fatigue they produce in the subjects.

Note: A variant of this method, in which subjects are not familiarized with the "not A" sample, is not recommended. This is because subjects, lacking a frame of reference, may guess wildly and produce biased results.

D. TEST PROCEDURE

For test controls and product controls, see p. 62. Present samples with scoresheet one at a time. Code all samples with random numbers and present them in random order so that the subjects do not detect a pattern of "A" vs. "not A" samples in any series. Do not disclose the identity of samples until after the subject has completed the test series.

Note: In the standard version of the procedure, the following protocol is observed:

1. Products "A" and "not A" are available to subjects only until the start of the test.
2. Only one "not A" sample exists for each test.
3. Equal numbers of "A" and "not A" are presented in each test.

These protocols may be changed for any given test, but the subjects must be informed before the test is initiated. Under No. 2, if more than one "not A" samples exist, each must be shown to the subjects before the test.

E. ANALYSIS AND INTERPRETATION OF RESULTS

The analysis of the data with four different combinations of sample vs. response is somewhat complex and can best be understood by referring to Example 6.11.

```
Date    1-15-99      WORKSHEET        Test code   612A83
────────────────────────────────────────────────────────
Post this sheet in the area where trays are prepared. Code
scoresheets ahead of time. Label serving containers ahead of
time.
────────────────────────────────────────────────────────
Type of samples :        Sweetened beverage
Type of test :              "A" - "Not A" test
────────────────────────────────────────────────────────
Sample   identification                        Code

 Beverage with 0.1% sweetener      ("A")         A

 Beverage with 5%  sucrose        ("Not A")      B
────────────────────────────────────────────────────────
Code 200 6-oz cups with random 3-digit numbers and divide
into two lots of 100 each.  Use sample "A" for the first 100
cups and sample "Not A" for the second 100 cups.
When preparing panelists' trays, place samples from left to
right in the following order:
Panelist                    Sample Order
  1 - 5     A  A  B  B  A  B  A  B  B  A
  6 - 10    B  A  B  A  A  B  A  A  B  B
 11 - 15    A  B  A  B  B  A  B  B  A  A
 16 - 20    B  B  A  A  B  A  B  A  A  B
```

FIGURE 6.15   Worksheet for "A"–"Not A" test. Example 6.11: new sweetener compared with sucrose.

F. EXAMPLE 6.11: NEW SWEETENER COMPARED WITH SUCROSE

Problem/situation — A product development chemist is researching alternate sweeteners for a beverage which uses sucrose as 5% of the current formula. Preliminary taste tests have established 0.1% of the new sweetener as the level equivalent to 5% sucrose but have also shown that if more than one sample is presented at a time, discrimination suffers because of carryover of the sweetness and other taste and mouthfeel factors. The chemist wishes to know whether the two beverages are distinguishable by taste.

Project objective — Determine if the alternate sweetener at 0.1% can be used in place of 5% sucrose.

Test objective — To compare the two sweeteners directly while reducing carryover and fatigue effects.

Test design — The "A" – "Not A" test allows the samples to be indirectly compared, and it permits the subjects to develop a clear recognition of the flavors to be expected with the new sweetener. Solutions of the sweetener at 0.1% are shown repeatedly to the subjects as "A," and 5% sucrose solutions are shown as "not A"; 20 subjects each receive 10 samples to evaluate in one 20-min test session. Subjects are required to taste each sample once, record the response ("A" or "not A"), rinse with plain water, and wait 1 min before tasting the next sample. Figure 6.15 shows the test worksheet and Figure 6.16 shows the scoresheet.

```
┌─────────────────────────────────────────────────────────────┐
│                                          Test Code            │
│         " A" - "Not A" TEST                                   │
│                                                               │
│ ┌───────────────────────────────────────────────────────────┐│
│   Taster No. _____  Name: _____ Date: _____        │
│   Type of Sample: ___Sweetened beverage_____      │
│ ├───────────────────────────────────────────────────────────┤│
│                                                               │
│   Instructions                                                │
│                                                               │
│     1.   Before taking this test, familiarize yourself        │
│          with the flavor of the samples "A" and "Not A"       │
│          which are available from the attendant.              │
│                                                               │
│     2.   Taste the test samples from left to right. After     │
│          each sample, record your response below, rinse       │
│          your palate with water, and wait one full minute     │
│          between samples.                                     │
│                                                               │
│     Note:   You have received approximately equal numbers of "A" and │
│     "Not A" samples.                                          │
│ ├───────────────────────────────────────────────────────────┤│
│   Sample   The sample is:   Sample    The sample is:          │
│   No.  Code  "A"  "Not A"   No.  Code  "A"   "Not A"          │
│    1   ___   □    □          6   ___   □     □                │
│    2   ___   □    □          7   ___   □     □                │
│    3   ___   □    □          8   ___   □     □                │
│    4   ___   □    □          9   ___   □     □                │
│    5   ___   □    □         10   ___   □     □                │
│ ├───────────────────────────────────────────────────────────┤│
│   Comments: _____         │
│   _____        │
│                                                               │
└─────────────────────────────────────────────────────────────┘
```

FIGURE 6.16   Scoresheet for "A"–"Not A" test. Example 6.11: new sweetener compared with sucrose.

Analyze results — In the table below, the columns show how the samples were presented and the rows, how the subjects identified them:

		Subject received		
		A	Not A	Total
Subject said:	A	60	35	95
	Not A	40	65	105
	Total	100	100	200

The χ^2-statistic is calculated as in Section VI, p. 80.

$$\chi^2 = \frac{(60-47.5)^2}{47.5} + \frac{(35-47.5)^2}{47.5} + \frac{(40-52.5)^2}{52.5} + \frac{(65-52.5)^2}{52.5} = 12.53$$

which is greater than the value in Table T5 (df = 1, α-risk = 0.05, χ^2 = 3.84), i.e., a significant difference exists.

Note: The χ^2-analysis just presented is not entirely appropriate because of the multiple evaluations performed by each respondent. However, no computationally convenient alternative method is currently available. The levels of significance obtained from this test should be considered approximate values.

Interpret results — The results indicate that the 0.1% sweetener solution is significantly different from the 5% sucrose solution. The sensory analyst informs the development chemist that the particular sweetener is likely to cause a detectable change in flavor of the beverage. The next logical step may be a descriptive analysis in order to characterize the difference.

One might ask: What would it take for the difference to be nonsignificant? This would be the case if results had been:

60	50
40	50

for which χ^2 equals 2.02, a value less than 3.84. See ISO, 1985 for a number of similar examples.

VIII. DIFFERENCE-FROM-CONTROL TEST

A. SCOPE AND APPLICATION

Use this test when the project or test objective is twofold, both: (1) to determine whether a difference exists between one or more samples and a control and (2) to estimate the size of any such differences. Generally one sample is designated the "control," "reference," or "standard," and all other samples are evaluated with respect to *how different* each is from that control.

The Difference-from-control test is useful in situations in which a difference may be detectable, but the size of the difference affects the decision about the test objective. Quality assurance/quality control and storage studies are cases in which the relative size of a difference from a control are important for decision making. The Difference-from-control test is appropriate where the Duo-trio and Triangle tests cannot be used because of the normal heterogeneity of products such as meats, salads, and baked goods.

The Difference-from-control test can be used as a two-sample test in situations where multiple sample tests are inappropriate because of fatigue or carryover effects. The Difference-from-control test is essentially a simple difference test with an added assessment of the size of the difference.

B. PRINCIPLE OF THE TEST

Present to each subject a control sample plus one or more test samples. Ask subjects to rate the size of the difference between each sample and the control and provide a scale for this purpose. Indicate to the subject that some of the test samples may be the same as the control. Evaluate the resulting mean difference-from-control estimates by comparing them to the difference-from-control obtained with the blind controls.*

C. TEST SUBJECTS

Generally 20 to 50 presentations of each of the samples and the blind control with the labeled control are required to determine a degree of difference. If the Difference-from-control test is chosen because of a complex comparison or fatigue factor, then no more than one pair should be given to any one subject at a time. Subjects may be trained or untrained, but panels should not consist of a mixture of the two. All subjects should be familiar with the test format, the meaning of the scale, and the fact that a proportion of test samples will be blind controls.

---

* The use of the estimate obtained with the blind controls amounts to obtaining a measure of the placebo effect. This estimate represents the numerical effect of simply asking the difference question, when in fact no difference exists.

D. Test Procedure

For test controls and product controls, see p. 62. When possible, offer the samples simultaneously with the labeled control evaluated first. Prepare one labeled control sample for each subject plus additional controls to be labeled as test samples. If the test is designed to have all subjects eventually test all samples but this cannot be done in one test session, a record of subjects by sample must be kept to ensure that remaining samples are presented in subsequent sessions.

The scale used may be any of those discussed in Chapter 5, pp. 53–56. For example:

Verbal category scale	Numerical category scale
No difference	0 = No difference
Very slight difference	1
Slight/moderate difference	2
Moderate difference	3
Moderate/large difference	4
Large difference	5
Very large difference	6
	7
	8
	9 = Very large difference

(When calculating results with the verbal category scale, convert each verdict to the number placed opposite, e.g., large difference = 5.)

E. Analysis and Interpretation of Results

Calculate the mean difference-from-control for each sample and for the blind controls, and evaluate the results by analysis of variance (or paired *t*-test if only one sample is compared with the control), as shown in the examples.

F. Example 6.12: Analgesic Cream — Increase of Viscosity

Problem/situation — The home health care division of a pharmaceutical company plans to increase the viscosity of its analgesic cream base. The two proposed prototypes are instrumentally thicker in texture than the control. Sample F requires more force to initiate flow/movement while Sample N initially flows easily but has higher overall viscosity. The product researchers wish to know how different the samples are from the control. As this type of test is best done on the back of the hands, evaluation is limited to two samples at a time.

Project objective — To decide whether Sample F or Sample N is closest overall to the current product.

Test objective — To measure the perceived overall sensory difference between the two prototypes and the regular analgesic cream.

Test design — A preweighed amount of each product is placed on a coded watch glass. The same amount (the weight of product which is normally used on a 10-cm² area) is weighed out for each sample. A 10-cm² area is traced on the back of the subjects' hands. The test uses 42 subjects and requires 3 subsequent days for each. On each of the 3 days, a subject sees one pair, which may be

- Control vs. Product F
- Control vs. Product N
- Control vs. blind control

```
┌──────────────────────────────────────────────────────────────────┐
│  Date        10-2-98        WORKSHEET          No.    13-625       │
├──────────────────────────────────────────────────────────────────┤
│  Post this sheet in the area where trays are prepared. Code        │
│  scoresheets ahead of time. Label serving containers ahead of      │
│  time.                                                             │
├──────────────────────────────────────────────────────────────────┤
│  Type of samples :  _____ Analgesic cream _____         │
│  Type of test :  _____  Difference from Control test _____    │
├──────────────────────────────────────────────────────────────────┤
│  Sample   Description                        Sample Code           │
│    C      Control                               C                  │
│    F      Experimental 10A3  (thixotropic)     Random #s under 500 │
│    N      Experimental 2-6X  (high viscosity)  Random #s over 500  │
├──────────────────────────────────────────────────────────────────┤
│  Serve in the following order:                                     │
│  Subject #              Day 1        Day 2        Day 3            │
│    1 - 7                C - F        C - N        C - C            │
│    8 - 14               C - N        C - F        C - C            │
│   15 - 21               C - F        C - C        C - N            │
│   22 - 28               C - N        C - C        C - F            │
│   29 - 35               C - C        C - N        C - F            │
│   36 - 42               C - C        C - F        C - N            │
│                                                                    │
│              Hour              Subject #                           │
│              9:00              1,8,15,22,29,36                     │
│              9:45              2,9,16,23,30,37                     │
│             10:30              3,10,17,24,31,38                    │
│             11:15              4,11,18,25,32,39                    │
│              1:00              5,12,19,26,33,40                    │
│              1:45              6,13,20,27,34,41                    │
│              2:30              7,14,21,28,35,42                    │
└──────────────────────────────────────────────────────────────────┘
```

FIGURE 6.17   Worksheet for Difference-from-control test. Example 6.12: analgesic cream.

See worksheet Figure 6.17. All subjects receive the labeled control first and the test sample second. Subjects are seated in individual booths which are well ventilated to reduce odor buildup and well lighted to permit visual cues to contribute to the assessment.

Conduct test — Weigh out samples within 15 min of each test. Label the two samples to be presented with a three-digit code. Using easily removed marks, trace the 10-cm^2 area on the backs of the hands of each subject. Instruct subjects to follow directions on the scoresheet (see Figure 6.18) carefully.

Analyze results — The results obtained are shown in Table 6.1, and an analysis of variance (ANOVA or AOV) procedure appropriate for a randomized (complete) block design is used to analyze the data. The 42 judges are the "blocks" in the design. The three samples are the "treatments" (or, more appropriately, are the three levels of the treatment). (See Chapter 13.IV, pp. 288–295 for a general discussion of ANOVA and block designs.)

Table 6.2 summarizes the statistical results of the test. The total variability is "partitioned" into three independent sources of variability, that is, variability due to the difference among the panelists (i.e., the block effect), variability due to the differences among the samples (i.e., the treatment

---

DIFFERENCE-FROM-CONTROL TEST

Name: _____   Date: _____

Type of Sample:_____

_____   Code of test sample_____

Instructions

1. You have received two samples, a control sample labeled C and a test sample labeled with a 3 - digit number.

2. Remove all of the control sample from the watch glass using your right index and middle fingers.

3. Using the index and middle fingers, spread the control product around the area traced on the back of your left hand.

4. Wipe finger tips with cloth on tray.

5. Pick up all of the test sample from the labeled watchglass using your left index and middle fingers.

6. Using the index and middle fingers, spread the product across the area traced on your right hand.

7. Indicate the size of the difference in skinfeel of the sample, relative to the control, on the scale below.

_____	0= no difference
_____	1 =
_____	2 =
_____	3 =
_____	4 =
_____	5 =
_____	6 =
_____	7 =
_____	8 =
_____	9 =
_____	10= extreme difference

REMEMBER THAT A DUPLICATE CONTROL IS THE SAMPLE SOME OF THE TIME.

COMMENTS : _____

_____

FIGURE 6.18   Worksheet for Difference-from-control test. Example 6.12: analgesic cream.

effect of interest), and the unexplained variability that remains after the other two sources of variability have been accounted for (i.e., the experimental error).

The F-statistic for samples is highly significant (Table T6); $F_{2,82} = 127.0$, $p < 0.0001$. The F-statistic is a ratio: the mean square for samples divided by the mean square for error. The appropriate degrees-of-freedom are those associated with the mean squares in the numerator and denominator of the F-statistic (2 and 82, respectively). A Dunnett's test (Dunnett, 1955, 1984) for multiple comparisons with a control was applied to the sample means and revealed that both of the test samples were significantly different from the blind control. It could also be concluded that Product N is significantly ($p < 0.05$) more different from the control than Product F based on an LSD multiple comparison (LSD = 0.4).

TABLE 6.1
Results from Example 6.12:
Difference-from-control Test — Analgesic Cream

Judge	Blind control	Product F	Product N	Judge	Blind control	Product F	Product N
1	1	4	5	22	3	6	7
2	4	6	6	23	3	5	6
3	1	4	6	24	4	6	6
4	4	8	7	25	0	3	3
5	2	4	3	26	2	5	1
6	1	4	5	27	2	5	5
7	3	3	6	28	2	6	4
8	0	2	4	29	3	5	6
9	6	8	9	30	1	4	7
10	7	7	9	31	4	6	7
11	0	1	2	32	1	4	5
12	1	5	6	33	3	5	5
13	4	5	7	34	1	4	4
14	1	6	5	35	4	6	5
15	4	7	6	36	2	3	6
16	2	2	5	37	3	4	6
17	2	6	7	38	0	4	4
18	4	5	7	39	4	8	7
19	0	3	4	40	0	5	6
20	5	4	5	41	1	5	5
21	2	3	3	42	3	4	4

TABLE 6.2
Analysis of Variance Table for Example 6.12:
Difference-from-control Test — Analgesic Cream

Source	Degrees of freedom	Sum of squares	Mean square	F	p
Total	125	545.78			
Judges	41	247.11	6.03	6.8	0.0001
Samples	2	225.78	112.89	127.00	0.0001
Error	82	72.89	0.89		

Sample Means with Dunnett's Multiple Comparisons

Sample	Blind control	Product F
Mean response	2.4a	4.8b
Sample	Blind control	Product N
Mean response	2.4a	5.5b

Note: Within a row, means not followed by the same letter are significantly different at the 95% confidence level. Dunnett's $d_{0.05} = 0.46$. Product N is significantly more different from the control than Product F ($LSD_{0.05} = 0.4$).

Interpretation — Significant differences were detected for both samples, and it is concluded that the two formulas are sufficiently different from the control to make it worthwhile to conduct attribute difference tests (see Chapter 15, Table 15.3, p. 341) or descriptive tests (see Chapter 11, pp. 184–186) for viscosity/thickness, skin heat, skin cool, and afterfeel.

G. EXAMPLE 6.13: FLAVORED PEANUT SNACKS

Problem/situation — The quality assurance manager of a large snack processing plant needs to monitor the sensory variation in a line of flavored peanut snacks and to set specifications for production of the snacks. The innate variations among batches of each of the added flavors (honey, spicy, barbecue, etc.) preclude the use of the Triangle, Duo-trio, or Same/Different tests. In most overall difference tests such as these, if subjects can detect variations within a batch, then this severely reduces the chances of a test detecting batch-to-batch differences. What is needed is a test which allows for separation of the variation within batches from the variation between batches.

Project objective — To develop a test method suitable for monitoring batch-to-batch variations in the production of flavored peanut snacks. Ultimately to set QA/QC sensory specifications.

Test objective — To measure the perceived difference within batches and between batches of flavored peanuts of known origin.

Test design — Samples from a recent control batch (normal production) are pulled from the warehouse. Jars from each of two lines are sampled and labeled Control A and Control B. These samples represent the variation within a batch. Samples are also pulled from a lot of production in which a different batch of peanuts served as the raw material. The sample is marked "test." A Difference-from-control test design is set up in which three pairs are tested:

- Control A vs. Control A (the blind control)
- Control A vs. Control B (the within batch measure)
- Control A vs. Test (the between batch measure)

Fifty subjects are scheduled to participate in three separate tests (C_A vs. C_A; C_A vs. C_B; C_A vs. Test) over a 3-day period. The pairs are randomized across subjects. In all pairs, C_A is given first as the control, and subjects rate the difference between the members of the pair on a scale of 0 to 10. The results are analyzed by the procedure of Aust et al. (1985), according to which the difference between the score for the blind control and that for the within batch measure is subtracted from the between batch measure in order to determine statistical significance for a difference.

Screen samples — The samples are prescreened for flavor, texture, and appearance by individuals from production, QA, marketing, and R & D who are familiar with the product, in order to determine that each sample is representative of the within and between batch variations for the product. Along with the sensory analyst the group decides that for the test, only whole peanuts will be sampled and tested.

Conduct test — Count out 15 whole peanuts for each sample and place in a labeled cup. Control A when in first position is labeled "control"; all other samples have three-digit codes:

Pair 1:	Control A vs. Control A
Labels:	"Control" vs. [three-digit code]
Pair 2:	Control A vs. Control B
Labels:	"Control" vs. [three-digit code]
Pair 3:	Control A vs. Test Sample
Labels:	"Control" vs. [three-digit code]

The scoresheet is shown in Figure 6.19.

DIFFERENCE-FROM-CONTROL TEST

Name: _____ Date: _8-7-98_ Test # _1103-6B_

Type of Sample: _____Flavored peanut snacks_____

Instructions

1. Taste the sample marked "Control" first.

2. Taste the sample marked with the three digit code.

3. Assess the overall sensory difference between the two samples using the scale below.

4. Mark the scale to indicate the size of the overall difference.

	SCALE	MARK TO INDICATE DIFFERENCE
No difference	0	_____
	1	_____
	2	_____
	3	_____
	4	_____
	5	_____
	6	_____
	7	_____
	8	_____
	9	_____
Extremely different	10	_____

REMEMBER THAT A DUPLICATE CONTROL IS THE SAMPLE SOME OF THE TIME.

COMMENTS : _____

FIGURE 6.19   Scoresheet for Difference-from-control test. Example 6.13: flavored peanut snacks.

Analyze results — The data from the evaluations (see Table 6.3) were analyzed according to the procedure described by Aust et al. (loc. cit.). This procedure tests whether the score for the test sample is significantly different from the average of the two control samples. The null and alternate hypotheses are

$$H_0: \mu_T = \left(\mu_{C_A} + \mu_{C_B}\right)/2 \text{ vs. } H_a: \mu_r > \left(\mu_{C_A} + \mu_{C_B}\right)/2$$

The error term used to test this hypothesis, called "pure error mean square" (1.13 in the analysis, Table 6.4) is calculated by summing the squared differences between the two control samples over all the panelists, then dividing by twice the number of panelists. The resulting ANOVA in Table 6.4 shows that the F-test ($F_{1,24} = MS_{T \text{ vs. } R}/MS_{\text{pure error}} = 326.54$) for differences between the test and control samples is highly significant.

TABLE 6.3
Results from Example 6.13:
Difference-from-control Test —
Flavored Peanut Snacks

Judge	Control A	Control B	Test
1	2	1	6
2	0	3	7
3	1	2	5
4	1	3	7
5	0	3	6
6	2	2	6
7	3	1	6
8	2	3	6
9	2	2	6
10	3	4	6
11	1	2	7
12	0	1	7
13	3	1	4
14	0	2	8
15	0	0	6
16	0	1	8
17	1	1	7
18	3	4	6
19	1	1	9
20	0	3	6
21	0	1	7
22	1	2	6
23	2	1	4
24	1	1	6

TABLE 6.4
Analysis of Variance Table According to the Difference-from-control Test of Aust et al. (1985) for the Data of Example 6.13: Flavored Peanut Snacks

Source	Degrees-of-freedom	Sum of squares	Mean square	F	p
Total	71	456.61			
Test vs. references	1	367.36	367.36	326.54	<0.0001
Pure error	24	27.00	1.13		
Residual	46	62.25			

Interpretation — The analyst concludes that, even in the presence of variability among the control samples, the test sample is significantly different from the average of the controls. He suggests, as a next step, to determine with consumers whether the test batch is different in *preference* or *acceptance*. Such determination allows the company to determine the degree to which the difference perceived by the panel is meaningful to consumers. Further study with the

Difference-from-control test paired with consumer tests permits the establishment of realistic specifications for QA.

IX. SEQUENTIAL TESTS

A. SCOPE AND APPLICATION

Sequential tests are a means to economize the number of evaluations required to draw a conclusion, for example, acceptance vs. rejection of a trainee on a panel or shipment vs. destruction of a lot of produced goods. Unlike the preceding tests in this chapter, where the size of the Type II error (β) is minimized for a fixed α and number of judgments, n, in sequential tests the values of α and β are decided upon beforehand, and n is determined by evaluating the outcome of each sensory evaluation as it occurs. Also, because α and β are determined beforehand, sequential tests provide a direct approach to simultaneously test for either the difference or the similarity (see Section II, p. 60) between the two samples.

Sequential tests are very practical and efficient because they take into consideration the possibility that the evidence derived from the first few evaluations may be quite sufficient (for fixed values of α and β) to draw a conclusion. Any further testing would be a waste of time and money. In fact, sequential tests can reduce the number of evaluations required by as much as 50%.

The sequential approach may be used with those existence-of-difference tests in which there is a correct and an incorrect answer, e.g., the Triangle, Two-out-of-five, and Duo-trio tests.

B. PRINCIPLE OF THE TEST

Conduct a sequence of evaluations according to the procedure appropriate for the chosen method and enter the results of each completed test into a graph such as Figure 6.20 in which three regions are identified: the acceptance region, the rejection region, and the continue-testing region. In Figure 6.20, the number of trials is plotted on the horizontal (x) axis and the total number of correct responses is plotted on the vertical (y) axis. Enter the result of the first test, if correct, as (x,y) = (1,1) and if incorrect, as (x,y) = (1,0). For each succeeding test, increase x by 1, and increase y by 1 for a correct reply and by 0 for an incorrect reply. Continue testing until a point touches or crosses one of the lines bordering the region of indecision. The indicated conclusion (i.e., accept or reject) is then drawn.

C. ANALYSIS AND INTERPRETATION OF RESULTS: PARAMETERS OF THE TEST

The version of the sequential test used here is that of the ISO (1983). The test itself is due to Wald (1947), and an alternative test is presented by Rao (1950). Both tests are clearly explained by Bradley (1953), who gives methods for calculating the expected number of evaluations needed to reach a decision, as well as rules for choosing the parameters associated with the method, as shown in Examples 6.14 and 6.15.

D. EXAMPLE 6.14: ACCEPTANCE VS. REJECTION OF TWO TRAINEES ON A PANEL

Project objective — To select or reject the trainees on the basis of their sensitivity to the differences in a series of test samples.

Test objective — To determine for each trainee whether his/her long-term proportion, p, of correct answers is sufficiently large for admittance onto the panel.

Test design — The sample pairs are submitted one at a time in the form of Triangle tests. Intervals between tests are kept long enough to avoid fatigue. As each triangle is completed,

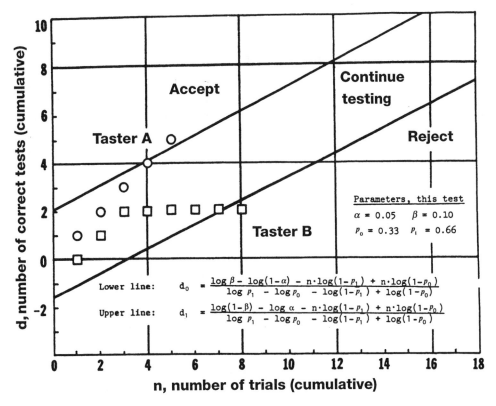

a is the probability of stating that a difference occurs when it does not
β is the probability of stating that no difference occurs when it does
p_0 is the expected proportion of correct decisions when the samples are identical
p_1 is the expected proportion of correct decisions when the odd sample is detec-
    ted (other than by guess) on half the total number of occasions

FIGURE 6.20 Example of sequential approach for selection of panel trainees by Triangle tests.

the result is entered in Figure 6.20. The tests series continue until the trainee is either accepted or rejected.

Analyze results. Test parameters — Values for four parameters are assigned by the panel leader:

- α is the probability of selecting an unacceptable trainee
- β is the probability of rejecting an acceptable trainee
- p_0 is the maximum unacceptable ability (measured as the proportion of correct answers)
- p_1 is the minimum acceptable ability (measured as the proportion of correct answers)

As can be seen in Figure 6.20, the equations for the lines dividing the graph into regions for acceptance, etc. depend on α, β, p_0, and p_1. In the present example, Trainee A is correct in all tests and is accepted after five triangles. Trainee B fails in the first triangle, succeeds in triangles two and three, but then fails on every subsequent triangle and is rejected after number eight.

Various values of the four parameters may be used. As p_o approaches p_1, the number of required trials increases. There are several methods for reducing the average number of trials required. First, using our Triangle tests example, the minimum acceptable probability of detecting a difference can be set higher, e.g., increased from 50% in our present example to 67% which would make $p_1 = 0.78$ [from $p_1 = 0.67 + (1 - 0.67)(\frac{1}{3})$].* Second, if many trainees are available, α and β could be assigned larger values (e.g., $\alpha > 0.05$ and/or $\beta > 0.10$).

E. EXAMPLE 6.15: SEQUENTIAL DUO-TRIO TESTS — WARMED-OVER FLAVOR IN BEEF PATTIES

Project objective — The routine QC panel at an Army Food Engineering station has detected warmed-over flavor (WOF) in beef patties refrigerated for 5 days and then reheated. The project leader, knowing that "an army marches on its stomach," wishes to set a realistic maximum for the number of days beef patties can be refrigerated.

Test objective — To determine, for samples stored 1 day, 3 days, and 5 days, whether a difference can be detected vs. a freshly grilled control.

Test design — Preliminary tests show that in Duo-trio tests, 5-day patties show strong WOF and 1-day patties none, hence a sequential test design is appropriate; a decision for these two samples could occur with few responses.

The three sample pairs (control vs. 1-day; control vs. 3-day; control vs. 5-day) are presented in separate Duo-trio tests, in which the control and storage samples are presented as the reference for every other subject. As each subject completes one test, the result is added to previous responses, and the cumulative results are plotted (see later). The test series continues until the storage sample is declared similar to or different from the control.

Analyze results — The results obtained are shown in Table 6.5. Here α is the probability of declaring a sample different from the control, when no difference exists; β is the probability of declaring a sample similar to the control, when it is really different.

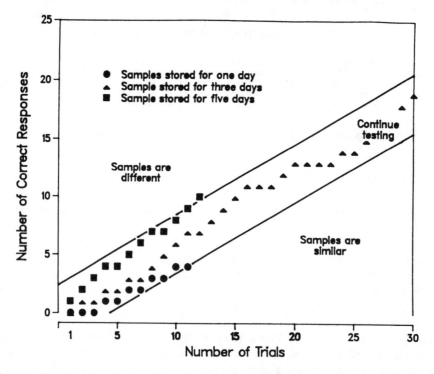

FIGURE 6.21 Test plot of results from Example 6.15: sequential Duo-trio tests, warmed-over flavor in beef patties.

---

* See Chapter 13, p. 275 for the derivation of this equation.

TABLE 6.5
Results Obtained in Example 6.15: Sequential Duo-trio Tests — Warmed-over Flavor in Beef Patties

Subject no.	Test A Control vs. 1 day		Test B Control vs. 3 day		Test C Control vs. 5 day	
1	I	0	I	0	C	1
2	I	0	C	1	C	2
3	I	0	I	1	C	3
4	C	1	C	2	C	4
5	I	1	I	2	I	4
6	C	2	C	3	C	5
7	I	2	I	3	C	6
8	C	3	C	4	C	7
9	I	3	C	5	I	7
10	C	4	C	6	C	8
11	I	4	C	7	C	9
12			I	7	C	10
13			C	8		
14			C	9		
15			C	10		
16			C	11		
17			I	11		
18			I	11		
19			C	12		
20			C	13		
21			I	13		
22			I	13		
23			I	13		
24			C	14		
25			I	14		
26			C	15		
27			C	16		
28			C	17		
29			C	18		
30			C	19		

Note: Column 1: I, incorrect; C, correct; column 2: cumulative correct.

The sensory analyst and the project leader decide to set both $\alpha = 0.10$ and $\beta = 0.10$. They set $p_0 = 0.50$, the null hypothesis p-value of a Duo-trio test. Further, they decide that the maximum proportion of the population that can distinguish the fresh and stored samples should not exceed 40%. Therefore, the value of p_1 is

$$p_1 = (0.40)(1.0) + (0.60)(0.50) = 0.70$$

(from:

$p = $ Pr[distinguisher]Pr[correct response given by a distinguisher]

+ Pr[nondistinguisher]Pr[correct response given by a nondistinguisher])

The equations of the two lines that form the boundaries of the acceptance, rejection, and continue-testing regions are

$$d_0 = -2.59 + 0.60n$$

$$d_1 = 2.59 + 0.60n$$

These lines are plotted in Figure 6.21 along with the cumulative number of correct duo-trio responses for each of the three stored samples (see Table 6.5). The sample stored 1 day is declared similar to the control. The sample stored for 5 days is declared significantly different from the control. The sample stored for 3 days had not been declared significantly similar to nor different from the control after 30 trials.

Interpret results — The project leader receives the decisive results for 1-day and 5-day samples and is informed that the result for the 3-day samples is indecisive after 30 tests. He can accept 3 days as the specification or choose to continue testing until a firm decision results.

REFERENCES

Aust, L.B., Gacula, M.C., Jr., Beard, S.A., and Washam, R.W., 1985. Degree of difference test method in sensory evaluation of heterogeneous product types. *J. Food Sci.* 50, 511.

Bradley, R.A., 1953. Some statistical methods in taste testing and quality evaluation. *Biometrics* 9, 22.

Conover, W.J., 1980. *Practical Nonparametric Statistics.* John Wiley & Sons, New York.

Dunnett, C.W., 1955. A multiple comparison procedure for comparing several treatments with a control. *J. Am. Stat. Assoc.* 50, 1096.

Dunnett, C.W., 1984. New tables for multiple comparisons with a control. *Biometrics* 20, 482.

Ferdinandus, A., Oosterom-Kleijngeld, I., and Runneboom, A.J.M., 1970. Taste testing. *Tech. Q. Master Brew. Assoc. Am.* 7, 210.

ISO, 1983. *Sensory Analysis — Methodology — "A" - "not A" Test.* International Organization for Standardization, ISO Standard 8588. Available from ISO, 1 rue Varembé, CH 1211 Génève 20, Switzerland, or from ANSI, New York, fax 212-302-1286.

ISO, 1999. *Sensory Analysis — Methodology — Triangle Test.* International Organization for Standardization, ISO Standard 4120. Available from ISO, 1 rue Varembé, CH 1211 Génève 20, Switzerland, or from ANSI, New York, fax 212-302-1286.

ISO, 1999. *Sensory — Analysis — Methodology — Sequential Tests.* International Organization for Standardization, ISO Draft Standard, under preparation. For availability, see above.

Macmillan, N.A. and Creelman, C.D., 1991. *Detection Theory, A User's Guide.* Cambridge University Press, 391 pp.

Rao, C.R., 1950. Sequential tests of null hypothesis. *Sankhya* 10, 361.

Wald, A., 1947. *Sequential Analysis.* John Wiley & Sons, New York.

7 Attribute Difference Tests: How Does Attribute X Differ Between Samples?

CONTENTS

I. INTRODUCTION: PAIRED COMPARISON DESIGNS

Attribute difference tests measure a single attribute, e.g., sweetness, comparing one sample with one or several others. Note that the lack of a difference between samples with regard to one attribute does *not* imply that no overall difference exists. Attribute difference tests involving two samples (Section II) are simple regarding test design and statistical treatment; the main difficulty is that of determining whether test situations are one-sided or two-sided (see next page and examples 7.1 and 7.2).

With more than two samples, some designs can be analyzed by the analysis of variance while others require specialized statistics. The degree of complexity increases rapidly with sample numbers, as does the economy of testing which is possible by improved test designs. A description of the various multiple pair tests follows; multisample tests and their designs are discussed in Section IV, p. 106.

In Sections III and IV we ask the subjects to compare each sample with every other sample. Such paired comparisons provide good measures for the intensity of the attribute of interest for each sample on a meaningful scale, and they have the advantage that a measure is obtained of the relative intensity of the attribute within each pair that can be formed. However, the number of possible pairs increases polynomially with the number of samples:

Number of samples, t	3	4	5	6	7	8	9
Number of possible pairs, $N = t(t-1)/2$	3	6	10	15	21	28	36

In Sections III and IV we ask the question: "Which sample is sweeter (fresher, preferred)?" This approach is based on rank data (e.g., the sweeter sample is assigned rank 2 and the other sample, rank 1) which introduces a degree of artificiality; no measure of the degree of difference is obtained directly from each respondent. In return, the statistics are simpler. With rating data, specialized statistics become necessary.

II. DIRECTIONAL DIFFERENCE TEST: COMPARING TWO SAMPLES

A. Scope and Application

Use this method when the test objective is to determine in which way a particular sensory characteristic differs between two samples (e.g., which sample is sweeter). In this mode, the method is also called the Paired Comparison test or the 2-AFC (2-Alternative Forced Choice) test. It is one of the simplest and most used sensory tests which is often used first to determine if other more sophisticated tests should be applied. Other forms of paired comparisons of two samples are the Same/Different test (see Chapter 6, p. 79) and the Paired Preference test (see Chapter 12, p. 242).

In using a Paired Comparison test it is necessary from the outset to distinguish between two-sided applications (bilateral, the most common) and one-sided applications (unilateral, when only one reply is of interest or only one reply is correct). (See Chapter 13, p. 277 and the Note on p. 102.)

The Unified Approach also applies to the Paired Comparison test. The number of respondents required for the test is affected by 1) whether the test is one-sided (use Table T9) or two-sided (use Table T11); and 2) by the values chosen for the test-sensitivity parameters α, β, and p_{max}. In Paired Comparison tests, the parameter p_{max} replaces the parameter p_d from the overall difference methods discussed in Chapter 6. p_{max} is the departure from equal intensity (i.e., a 50:50 split of opinion among respondents) that represents a meaningful difference to the researcher. For example, if the researcher considers a 60:40 split in the population of respondents to be a meaningfully large departure from equal intensity, then $p_{max} = 0.60$ and the researcher finds the number of respondents in that section of the appropriate table (T9 or T11) for the chosen values of α and β. As a rule of thumb:

- p_{max} <55% represents small departures from equal intensity;
- 55% $\leq p_{max} \leq$ 65% represents medium departures; and
- p_{max} >65% represents large departures.

B. Principle

Present to each subject two coded samples. Prepare equal numbers of the combinations AB and BA and allot them at random among the subjects. Ask the subject to taste the products from left to right and fill in the scoresheet. Clearly inform the subject whether "no difference" verdicts are permitted.

Only the "forced choice technique" is amenable to formal statistical analysis. However, in some cases subjects may object quite strenuously to inventing a difference when none is perceived. The sensory analyst must then decide whether: (1) to divide their scores evenly over the two samples or (2) to ignore them. Procedure (1) increases the probability of finding a difference while procedure (2) decreases it; hence, the analyst must face the duty/temptation to influence the results one way or the other. In practice, about one half of analysts prohibit "no difference" verdicts. The other half, having found that a happy panel is a better panel, most frequently use procedure (1).

C. Test Subjects

Because of the simplicity of the test, it can be conducted with subjects who have received a minimum of training: it is sufficient that subjects are completely familiar with the attribute under test. Or, if a test is of particular importance (e.g., an off-flavor in a product already on the market), highly trained subjects may be selected, who have shown special acuity for the attribute.

DIRECTIONAL DIFFERENCE TEST

Name: _____ Date: _____

Type of Sample: _____

_____

Characteristic studied: _____

Instructions:

Taste each pair from left to right and enter your verdict below.

If no difference is apparent, enter your best guess, however uncertain. "No difference" verdicts are permitted, but only as a last resort.

Test pairs   Which sample is more _____

_____ _____   _____
_____ _____   _____
_____ _____   _____

Comments: _____

_____

_____

FIGURE 7.1 Example of scoresheet for Directional Difference test. Presentation: paired comparisons. "No difference" verdicts permitted.

Because the chance of guessing is 50%, fairly large numbers of test subjects are required. Table T12 shows that, e.g., with 15 presentations, 13 must agree if a significance level of $\alpha = 0.01$ is to be obtained, while with 50 presentations, the same significance can be obtained with 35 agreeing verdicts.

D. Test Procedure

For test controls and product controls, see pp. 24 and 32. Offer samples simultaneously if possible, or else sequentially. Prepare equal numbers of the combinations A/B and B/A and allocate the sets at random among the subjects. Refer to pp. 61–62 for details of procedure. A typical scoresheet is shown in Figure 7.1. Note that the scoresheet is the same whether the test is one- or two-sided, but the scoresheet must show whether "no difference" verdicts are permitted (or the subjects must know this). Space for several successive paired comparisons may be provided on a single scoresheet, but do not add supplemental questions because these may introduce bias.

Count the number of responses of interest. In a one-sided test, count the number of *correct* responses, or the responses *in the direction of interest,* and refer to Table T10. In a two-sided test, count the number of *agreeing* responses citing one sample more frequently, and refer to Table T12.

E. Example 7.1: Directional Difference (Two-Sided) — Crystal Mix Lemonade

Problem/situation — Consumer research on lemonades indicates that consumers are most interested in a lemon/lemonade flavor most like "fresh-squeezed lemonade." The company has developed two promising flavor systems for a powdered mix. The developers wish to get some measure of whether one of these has more fresh-squeezed lemon character than the other.

Project objective — To develop a product which is high in fresh-squeezed lemon character.

Test objective — To measure the relative ability of the two flavor systems to deliver flavor more like that in fresh-squeezed lemonade.

Test design — As different people may have different ideas of fresh-squeezed flavor, a large panel is needed, but training is not a strong requirement. A Paired Comparison test with 40 subjects and an α-error of 5%, i.e., $\alpha = 0.05$, is deemed suitable. The null hypothesis is H_0: Freshness A = Freshness B. The alternative hypothesis is H_a: Freshness A $\neq$ Freshness B; either outcome is of interest, hence the test is two-sided. The samples are coded "691" and "812," and the scoresheet shown in Figure 7.1 is used to collect the data.

Screen samples — Taste samples in advance to confirm that the intensity of lemon flavors is similar in the two samples.

Analyze results — Sample 812 is chosen by 26 subjects as having more fresh-squeezed lemon flavor. Four subjects report "no difference" and are divided between the two samples. From Table T12, conclude that, with 28 out of 40 choosing 812, the number is sufficient to constitute a significant difference.

Interpret results — Suggest that formulation 812 is used in the future, as it has significantly more fresh-squeezed lemon character in the opinion of this test panel.

F. EXAMPLE 7.2: DIRECTIONAL DIFFERENCE (ONE-SIDED) — BEER BITTERNESS

Problem/situation — A brewer receives reports from the market that his beer "A" is deemed insufficiently bitter, and a test brew "B" is made using a higher hop rate.

Project objective — To produce a beer which is perceptibly more bitter, but not too much so.

Test objective — To compare beers A and B to determine whether a small but significant increase in bitterness has been obtained.

Test design — A paired-comparison/directional difference test is chosen, as the point of interest is the increase in bitterness, nothing else. The project leader opts for a high degree of certainty, i.e., $\alpha = 0.01$. The sensory analyst codes the beers "452" and "603" and offers them to a panel of 30 subjects of proven ability to detect small changes in bitterness. The scoresheet asks "Which sample is more bitter?" (and not: "Is 603 more bitter than 452?") so as not to bias the subjects.

Screen samples — The samples are tasted by a small panel of six to make certain that differences other than bitterness are minimal.

Analyze results — Sample B is selected by 22 subjects. The null hypothesis is H_0: Bitterness A = Bitterness B, but the alternate hypothesis is H_a: Bitterness B > Bitterness A, making the test one-sided. The analyst concludes from Table T10 that a difference in bitterness was perceived at $\alpha = 0.008$. The test brew was successful.

Note: The important point in deciding whether a paired-comparison test is one- or two-sided is whether the alternate hypothesis is one- or two-sided, not whether the question asked of the subjects has one or two replies. One-sided test situations occur mainly where the test objective is to confirm a definite "improvement" or treatment effect (see also Chapter 13, p. 277). Some examples of one- and two-sided test situations are

One-sided	Two-sided
Confirm that test brew is more bitter	Decide which test brew is more bitter
Confirm that test product is preferred (as we had prior reason to expect)	Decide which product is preferred
In training tasters: which sample is more fruity (doctored samples used)	Most other test situations … whenever the alternate hypothesis is that the samples are different, rather than "one is more than the other"

III. PAIRWISE RANKING TEST: FRIEDMAN ANALYSIS — COMPARING SEVERAL SAMPLES IN ALL POSSIBLE PAIRS

A. SCOPE AND APPLICATION

Use this method when the test objective is to compare several samples for a single attribute, e.g., sweetness, freshness, or preference. The test is particularly useful for sets of three to six samples which are to be evaluated by a relatively inexperienced panel. It arranges the samples on a scale of intensity of the chosen attribute and provides a numerical indication of the differences between samples and the significance of such differences.

B. PRINCIPLE OF THE TEST

Present to each subject one pair at a time in random order, with the question: "Which sample is sweeter?" (fresher, preferred, etc.). Continue until each subject has evaluated all possible pairs that can be formed from the samples. Evaluate the results by a Friedman-type statistical analysis.

C. TEST SUBJECTS

Select, train, and instruct subjects as described on p. 61. Use not less than 10 subjects; discrimination is much improved if 20 or more can be used. Ascertain that subjects can recognize the attribute of interest, e.g., by training with various pairs of known intensity difference in the attribute. Depending on the test objective, subjects may be required who have proven ability to detect small differences in the attribute.

D. TEST PROCEDURE

For test controls and product controls, see pp. 24 and 32. Offer samples simultaneously if possible or else sequentially. Refer to p. 62 for details of procedure. Make certain that the order of presentation is truly random: subjects must not be led to expect a regular pattern, as this will influence verdicts. Randomize presentation within pairs, between pairs, and among subjects. Ask only one question: "Which sample is more …?" Do not permit "no difference" verdicts; if they nevertheless occur, distribute the votes evenly among the samples.

E. EXAMPLE 7.3: MOUTHFEEL OF CORN SYRUP

Problem/situation — A manufacturer of blended table syrups wishes to market a product with low thickness at a given solids content. Four unflavored corn syrup blends, A, B, C, and D have been prepared for evaluation (Carr, 1985).

Project objective — To evaluate the suitability of the four syrup blends.

Test objective — To establish the positions of the four blends on a subjective scale of perceived mouthfeel thickness.

Test design — The Pairwise Ranking test with Friedman analysis is chosen: (1) because paired presentation is less affected by fatigue with these samples and (2) because this test establishes a meaningful scale. Twelve subjects of proven ability evaluate the six possible pairs AB, AC, AD, BC, BD, and CD. The worksheet and the scoresheet are shown in Figures 7.2 and 7.3, respectively.

Analyze results — The table below shows the number of times (out of 12) each "row" sample was chosen as being thicker than each "column" sample. For example, when Sample B was presented with Sample D it was perceived thicker by 2 of the 12 subjects.

		Column samples (thinner)			
		A	B	C	D
Row samples	A	—	0	1	0
(thicker)	B	12	—	6	2
	C	11	6	—	7
	D	12	10	5	—

Date 11-6-98	WORKSHEET	No. 78

CODE	SAMPLE	CODE	SAMPLE
A	Blend 4238	C	CCSA Blend III
B	Blend 133.8B	D	Test Sample 11.3A

Each panelist receives the six possible pairs in balanced random order. Each sample is coded with a random number.

Panelist No.	Order of presentation and serving code											
	1st		2nd		3rd		4th		5th		6th	
1	A 119	D 634	B 128	D 824	B 316	C 967	C 242	D 659	A 978	C 643	A 224	B 681
2	B 293	D 781	A 637	D 945	A 661	B 153	A 837	C 131	C 442	D 839	B 659	D 718
3	A 926	C 563	B 873	C 611	C 194	D 228	A 798	B 478	A 184	D 278	B 478	D 924
4	B 455	C 857	C 764	D 452	A 975	C 815	B 523	D 824	A 556	B 982	A 737	D 539
5	C 834	D 245	A 285	B 299	B 782	D 679	A 114	D 966	B 713	C 561	A 393	C 495
6	A 662	B 196	A 516	C 777	A 843	D 581	B 375	C 313	B 327	D 415	C 881	D 242
7	A 341	D 918	B 949	D 188	B 428	C 742	C 486	D 585	A 635	C 154	A 545	B 363
8	A 787	B 479	A 491	C 563	A 259	D 396	B 659	C 797	B 899	D 727	C 112	D 157
9	C 578	D 322	A 352	B 336	B 537	D 434	A 961	D 242	B 261	C 396	A 966	C 876
10	A 814	C 952	B 378	C 381	C 148	D 297	A 848	B 383	A 679	D 165	B 448	D 781
11	B 498	D 383	A 131	D 919	A 466	B 866	A 794	C 898	C 526	D 851	B 721	D 122
12	B 675	C 536	C 495	D 778	A 622	C 159	B 263	D 751	A 953	B 779	B 296	D 956

FIGURE 7.2 · Worksheet for Pairwise Ranking test—Friedman analysis. Example 7.3: mouthfeel of corn syrup.

The first step in the Friedman analysis (Friedman, 1937; Hollander and Wolfe, 1973) is to compute the rank sum for each sample. In the present example the rank of one is assigned to the "thicker" and the rank of two to the "thinner" sample. The rank sums are then obtained by adding the sum of the row frequencies to twice the sum of the column frequencies, e.g., for Sample B $(12 + 6 + 2) + 2(0 + 6 + 10) = 52$:

Sample	A	B	C	D
Rank sum	71	52	48	45

The test statistic, Friedman's T, is computed as follows:

$$T = (4/pt)\sum_{i=1}^{t} R^2 - (9p[t-1]^2)$$

$$= [4/(12)(4)][71^2 + 52^2 + 48^2 + 45^2] - [9(12)(3^2)] = 34.17$$

MULTIPLE PAIRED COMPARISONS TEST

Name: _____   Date: _____

Type of Sample: _Unflavored table syrup_

and Difference: _thickness (mouthfeel)_

Instructions:

1. Receive the sample tray and note each sample code below according to its position on the tray.

2. Taste the first sample pair from left to right and note which sample is thicker (more viscous). Indicate by placing an X next to the code.

3. Continue until all 6 pairs have been evaluated. Rinse with water as needed to clear your palate.

Pair no.	Left sample	Right sample	Remarks
6	_____	_____	_____
5	_____	_____	_____
4	_____	_____	_____
3	_____	_____	_____
2	_____	_____	_____
1	_____	_____	_____

If you perceive no difference, please make a best guess. Comments regarding reasons for your choice or the characteristics of the samples may be made under Remarks.

FIGURE 7.3 Scoresheet for Pairwise Ranking test—Friedman analysis. Example 7.2: mouthfeel of corn syrup.

where p = the number of times the basic design is repeated (here = 12); t = the number of treatments (here = 4); R_i = the rank sum for the i'th treatment; and $\sum R^2$ = sum of all R's squared, from R_1 to R_t.

Critical values of T have been tabulated (Skillings and Mack, 1981) for t = 3, 4, and 5 and small values of p; for experimental designs not in the tables, the value of T is compared to the critical value of χ^2 with $(t-1)$ degrees of freedom (see Table T5). In the present case, the critical Ts are

Level of significance, α	0.10	0.05	0.01
Critical T	6.25	7.81	11.3

The results can be shown on a rank sum scale of thick vs. thin:

On the same scale, the HSD value (see Chapter 13, p. 298) for comparing two rank sums (α = 0.05) looks like this:

$$HSD = q_{\alpha,t,\infty}\sqrt{pt/4} = 3.63\sqrt{(12)(4)(4)/4} = 12.6$$

(where the value $q_{\alpha t,\infty}$ is found in Table T4). The difference between A and B is much larger than 12.6, i.e., A is significantly thinner and thus more desirable than the group formed by B, C, and D.

IV. INTRODUCTION: MULTISAMPLE DIFFERENCE TESTS — BLOCK DESIGNS

The tests described in Sections I to III dealt with pairwise comparison of samples according to one selected attribute. The tests in the next four sections are based on groups of more than two samples, again compared according to one selected attribute (such as sweetness, freshness, or preference) and using the blocking designs discussed in Chapter 13, p. 287.

Complete block designs — The simplest design is to rank all of the samples simultaneously (see Method VI), but results are not as precise or actionable as those of more complex tests. The next simplest is to compare all samples together, using a rating scale. We can compare all samples in one complete block (Method VI, p. 111, Multisample Difference test, complete balanced design), or we can limit the load on the taste buds (or other sensory organs) and the short-term memory of the panelists, by splitting the comparison into several smaller blocks (balanced incomplete block [BIB] designs, Methods VII and VIII, pp. 116 and 119).

Balanced incomplete block (BIB) designs — In the complete block designs, the size of each block (row) equals the number of treatments being studied. A block in the present context is identified by the set of samples served to one panelist. Generally the panelist cannot evaluate more than four to six samples in a single setting. If the number of samples (treatments) to be compared is larger, for example, 6 to 12, a BIB design can be used. Instead of presenting all the t samples in one large block, the experimenter presents them in b smaller blocks, each of which contains $k < t$ samples. The k samples that form each block must be selected so that all the samples are evaluated an equal number of times and so that all pairs of samples appear together in the b blocks an equal number of times. Cochran and Cox (1957) present an extensive list of BIB designs that can be used in most test situations.

V. SIMPLE RANKING TEST: FRIEDMAN ANALYSIS — RANDOMIZED (COMPLETE) BLOCK DESIGN

A. Scope and Application

Use this method when the test objective is to compare several samples according to a single attribute, e.g., sweetness, freshness, preference. Ranking is the simplest way to perform such comparisons, but the data are merely ordinal, and no measure of the degree of difference is obtained from each respondent. Consecutive samples which differ widely, as well as those which differ slightly, will be separated by one rank unit. A good, detailed discussion of the virtues and limitations of rank data is given by Pangborn (1984). Ranking is less time-consuming than other methods and is particularly useful when samples are to be presorted or screened for later analysis.

B. Principle of the Test

Present the set of samples to each subject in balanced, random order. Ask subjects to rank them according to the attribute of interest. Calculate the rank sums and evaluate them statistically with the aid of Friedman's test, as described in Chapter 13, p. 292.

C. Test Subjects

Select, train, and instruct the subjects as described on p. 61. Use no fewer than 8 subjects; discrimination is much improved if 16 or more can be used. Subjects may require special instruction or training to enable them to recognize the attribute of interest reproducibly (see Chapter 9, p. 136).

Depending on the test objective, subjects may be selected on the basis of proven ability to detect small differences in the attribute.

D. Test Procedure

Test controls and product controls: see pp. 24 and 32. Offer samples simultaneously if possible, or else sequentially. The subject receives the set of t samples in balanced random order, and the task is to rearrange them in rank order. The set may be presented once or several times with different coding. Accuracy is much improved if the set can be presented two or more times. In preference tests, instruct subjects to assign rank 1 to the preferred sample, rank 2 to the next preferred, etc.* For intensity tests, instruct subjects to assign rank 1 to the lowest intensity, rank 2 to the next lowest, etc.

Recommend that subjects arrange the samples in a provisional order based upon a first trial of each and then verify or change the order based on further testing. Instruct subjects to make a "best guess" about adjacent samples, even if they appear to be the same; however, if a subject declines to guess, he or she should indicate under "comments" the samples considered identical. Assign the average rank to each of the identical samples for statistical analysis. For instance, in a four-sample test, if a panelist cannot differentiate the two middle samples, assign the average rank of 2.5 to each, i.e., $(2 + 3)/2$.

If a rank order for more than one attribute of the same set of samples is needed, carry out the procedure separately for each attribute, using new samples coded differently so that one evaluation does not affect the next. A scoresheet is shown in Figure 7.4. Space for several sets of samples may be provided, but note that a new set of codes is required for each set, and it is often simpler to supply one scoresheet for each set and subject.

E. Analysis and Interpretation of Results

Analysis by Friedman's test (Friedman, 1937; Hollander and Wolfe, 1973) is preferred to the use of Kramer's tables (Kramer et al., 1974), as the latter provides inaccurate evaluation of samples of intermediate rank. Tabulate the scores as shown in Example 7.3 and calculate the rank sums for each sample (column sums). Then use Equation 13.14 (p. 292) to calculate the value of the test statistic T. If the value of T exceeds the upper-α critical value of a χ^2 random variable with $(t - 1)$ degrees of freedom, then conclude that significant differences exist among the samples. Use the multiple comparison procedure appropriate for rank data, presented in Chapter 13, Equation 13.15, p. 292, and Equation 13.24, p. 299, to determine which samples are different.

F. Example 7.3: Comparison of Four Sweeteners for Persistence

Problem/situation — A laboratory of psychophysics wishes to compare four artificial sweeteners A, B, C, and D for degree of persistence of sweet taste.

Project/test objective — To determine whether there is a significant difference among the sweeteners in the persistence of sweetness in the mouth after swallowing.

Test design — The feeling of persistence may show large person-to-person variations so it is desirable to work with a large panel. The ranking test is suitable because it is simple to carry out and does not require much training. The four samples are tested with a panel of 48 students. Each subject receives the four samples coded with three-digit numbers and served in balanced, random order. The scoresheet is shown in Figure 7.4.

---

* Ranking is less confusing for the subjects if rank 1 can be assigned to the preferred sample; however, for those who must interpret the results, the least confusing is to have the "best" sample receive the highest score. A way out of this dilemma is for the analyst to enter the scores into the computer as "R minus the score," where R is one more than the number of samples ($R = t + 1$).

RANKING TEST

Name: _____   Date: _____

Type of Sample: __Artificial sweeteners_____

_____

Characteristic Studied: __Persistence of sweet taste_____

Instructions

1. Receive the sample tray and note each sample code below according to its position on the tray.

2. Taste the samples from left to right and note the
   degree of persistence of the sweetness _____.

   Wait at least 30 seconds between samples and rinse palate as required.

3. Write "1" in the box of the sample which you find
   least persistent _____.

   Write "2" for the next, "3" for the next, and "4" for the _most persistent_ _____.

   You may find it expedient to first arrange the samples in a provisional order, and then resolve the positions of adjacent samples by more careful tasting.

4. If two samples appear the same, make a "best guess" as to their rank order.

Code _____ _____ _____ _____

Rank ☐   ☐   ☐   ☐

Comments: _____

_____

_____

FIGURE 7.4   Scoresheet for Simple Ranking test. Example 7.3: comparison of four sweeteners for persistence.

Screen samples — This test requires very careful preparation to ensure that there are no other differences between the four compounds than those intended, i.e., those resulting from different chemical composition. Four experienced tasters evaluate and adjust the samples to ensure that they are equally sweet to the average observer and that any differences in temperature, viscosity, appearance (color, turbidity, and remains of foam, etc.) are absent or masked so as to preclude recognition by means other than taste and smell.

Analyze results — Table 7.1 shows how the results are compiled and the rank sums calculated. The value of the test statistic T in Equation 13.14, p. 292, is

$$T = ([12/(48)(4)(5)][135^2 + 103^2 + 137^2 + 105^2]) - 3(48)(5)$$
$$= 12.85$$

Use Table T5 to find that the upper 5% critical value of a χ^2 with 3 degrees-of-freedom is 7.81. As the value of $T = 12.85$ is greater than 7.81, conclude that the samples are significantly different at the 5% level in their persistence of sweet taste. To determine which samples are significantly different, calculate the critical value of the multiple comparison in Equation 13.15, p. 292, as:

$$\text{LSD}_{\text{rank}} = 1.96\sqrt{48(4)(5)/6}$$

$$= 24.8$$

TABLE 7.1
Table of Results for Example 7.3:
Comparison of Four Sweeteners for Persistence

Subject no.	Sample A	Sample B	Sample C	Sample D
1	3	1	4	2
2	3	2	4	1
3	3	1	2	4
4	3	1	4	2
5	1	3	2	4
—	—	—	—	—
—	—	—	—	—
—	—	—	—	—
44	4	2	3	1
45	3	1	4	2
46	3	4	1	2
47	4	1	2	3
48	4	2	3	1
Rank sum	135	103	137	105

Any two samples whose rank sums differ by more than $LSD_{rank} = 24.8$ are significantly different at the 5% level. Therefore, conclude that Samples B and D both show significantly less persistence of sweet taste than both Samples A and C. Sample B is not significantly different from D, nor A from C.

G. EXAMPLE 7.4: BITTERNESS IN BEER NOT AGREEING WITH ANALYSIS

Problem/situation — A manager of quality control at a brewery knows that the company's brand P reads the same as the competition's by the standard analysis method for hop bitter substances, yet he hears reports that it tastes more bitter. Before commencing an investigation into possible contamination by nonhop bitter substances, he wishes to confirm that there is a difference in perceivable bitterness.

Project/test objective — To taste beer P for bitterness against the competitive brands A, B, and C.

Test design — The four samples are ranked by 12 subjects of proven ability to detect small differences in bitterness. The null hypothesis is H_0: Bitterness P = Bitterness A, B, or C; and the alternative hypothesis is H_a: Bitterness P ≠ Bitterness A, B, or C, there being no advance information about any systematic difference between A, B, and C. The scoresheet used is patterned on Figure 7.4.

Analyze results — See Table 7.2. Note that the experienced panelists were permitted to assign equal ranks or "ties" to the samples.* The alternate form of the test statistic T' in Equation 13.16, p. 292, must be used when ties are present in the data. To calculate the value of T', we must first determine the number of tied groups (g_i) in each block (i) and the size of each tied group $(t_{i,j})$. (Each nontied sample is considered as a separate group of size $t_{i,j} = 1$.) Only blocks in which ties

---

* Inexperienced panelists should not be allowed to assign equal ranks or "ties" for two reasons. First, allowing ties provides an easy way out for lazy panelists who, when forced to choose, could differentiate the samples. Second, ties decrease the sensitivity of the statistical test for differences.

TABLE 7.2
Table of Results for Example 7.4:
Bitterness of Beer Not Agreeing with Analysis

Subject no.	Sample A	Sample B	Sample C	Sample P
1	1	2.5	2.5	4
2	2	1	4	3
3	1	3	3	3
4	2	1	3	4
5	2	3	1	4
6	2	1	4	3
7	3	1	2	4
8	1	2	3.5	3.5
9	2	3	4	1
10	2	1	3.5	3.5
11	2	3	1	4
12	2	1	4	3
Rank sum	22	22.5	35.5	40

occur need to be considered because only these blocks affect the calculation of T'. From Table 7.2 we see that ties occur in blocks 1, 3, 8, and 10. The values of g_i and $t_{i,j}$ for these blocks are

$$g_1 = 3, t_{1,1} = 1 \qquad g_3 = 2, t_{3,1} = 1 \qquad g_8 = 3, t_{8,1} = 1 \qquad g_{10} = 3, t_{10,1} = 1$$
$$t_{1,2} = 2 \qquad\qquad t_{1,2} = 3 \qquad\qquad t_{8,2} = 1 \qquad\qquad t_{10,2} = 1$$
$$t_{1,3} = 1 \qquad\qquad\qquad\qquad\qquad t_{8,3} = 2 \qquad\qquad t_{10,3} = 2$$

These values are used to calculate the second term in the denominator of T' in Chapter 13, Equation 13.16 as:

$$T' = \left[12\sum_{j=1}^{t}\left(X_j - G/t\right)^2 \right] \bigg/ \left[bt(t+1) - \left(1/(t-1)\right)\sum_{i=1}^{b}\left(\left(\sum_{j=1}^{g_i} t_{i,j}^{3}\right) - t\right) \right]$$

$$= \frac{12\left[(22-30)^2 + (22.5-30)^2 + (35.5-30)^2 + (40-30)^2\right]}{(12)(4)(5) - (1/3)(6+24+6+6)}$$

$$= 13.3$$

The value of $T' = 13.3$ exceeds the upper 5% critical value of a χ^2 with 3 degrees of freedom ($\chi^2_{0.05,3} = 7.81$). Conclude that significant differences exist among the samples.

Only comparisons of Samples A, B, and C vs. Sample P are of interest. Therefore, the multiple comparison procedure for comparing test samples to a con trol or standard sample, appropriate for rank data, is used (see Hollander and Wolfe, 1973). The upper 5% (one-sided) critical value of the multiple comparison is 13.1. The rank sum of Sample P is more than 13.1 units higher than the rank sums of Samples A and B.

Test report — The QA manager concludes that the company's Sample P is significantly more bitter than the competition's beers A and B, and he therefore commences an investigation of possible contamination of P with extraneous bitter-tasting substances.

VI. MULTISAMPLE DIFFERENCE TEST: RATING APPROACH — EVALUATION BY ANALYSIS OF VARIANCE (ANOVA)

A. SCOPE AND APPLICATION

Use this method when the test objective is to determine in which way a particular sensory attribute varies over a number of t samples, where t may vary from 3 to 6 or at most 8, and it is possible to compare all t samples as one large set.

Note: In descriptive analysis (see Chapter 10), when several samples are compared, the present method may be applied to each attribute.

B. PRINCIPLE OF THE TEST

Subjects rate the intensity of the selected attribute on a numerical intensity scale, e.g., a category scale (see pp. 52–56). Specify the scale to be used. Evaluate the results by the analysis of variance.

C. TEST SUBJECTS

Select, train, and instruct the subjects as described on p. 61. Use no fewer than 8 subjects; discrimination is much improved if 16 or more can be used. Subjects may require special instruction to enable them to recognize the attribute of interest reproducibly (see Chapter 9, pp. 136ff.). Depending on the test objective, subjects may be selected who show high discriminating ability in the attribute.

D. TEST PROCEDURE

For test controls and product controls, see pp. 24 and 32. Offer samples simultaneously if possible, or else sequentially. The subject receives the set of t samples in balanced randomized order, and the task is to rate each sample using the specified scale. The set may be presented once only, or several times with different coding. Accuracy is much improved if the set can be presented two or more times.

If more than one attribute is to be rated, theoretically the sample should be presented separately for each attribute. In practical descriptive analysis this can become impossible because of the number of attributes to be rated in a given sample (typically from 6 to 25). *In dispensing with the requirement to rate each attribute separately, the sensory analyst accepts that there will be some interdependence between the attributes.* For example, if in a shelf-life study the product can go stale microbiologically (e.g., sourness) or oxidatively (e.g., rancidity), high ratings on one will raise the rating on the other, even if it is absent. The effect must be counteracted by making subjects aware of it and by vigorous training enabling them to recognize each attribute independently.

E. ANALYSIS AND INTERPRETATION OF RESULTS

The results are analyzed by the analysis of variance, see Chapter 13, p. 291, and the examples.

F. EXAMPLE 7.5: POPULARITY OF COURSE IN SENSORY ANALYSIS. RANDOMIZED COMPLETE BLOCK DESIGN

Problem/test objective — A department of food science routinely asks the students at the end of each semester to rate the courses they have taken on a scale of –3 to +3, where –3 = very poor, 0 = indifferent, and +3 = excellent. Thirty students complete the scoresheet with the results shown in Table 7.3. The objective of the evaluation is to identify courses that require improvement.

TABLE 7.3
Results Obtained in Example 7.5:
Multisample Difference Test (Rating) —
Popularity of Courses in Food Science

Courses evaluated Student no.	Biology	Nutrition	Sensory analysis	Statistics
1	2	-2	1	1
2	3	0	2	1
3	1	-3	0	0
4	2	0	1	0
5	0	1	0	0
6	-3	-3	-3	-3
7	1	3	1	1
8	-1	-1	-1	-1
9	2	-2	1	1
10	0	-3	-1	-1
11	2	0	2	2
12	-1	-2	0	1
13	3	-3	3	3
14	0	0	0	0
15	-2	2	-1	-1
16	2	-2	1	1
17	1	-1	0	0
18	0	-1	0	-1
19	3	3	3	3
20	1	-2	1	0
21	-2	-2	-2	-2
22	2	-1	1	1
23	1	0	1	1
24	3	-3	3	3
25	1	1	1	1
26	0	-1	1	-1
27	1	0	2	-1
28	2	-2	0	0
29	-2	-3	-1	-2
30	2	2	2	2

Note: Scale used: -3 to +3, where -3 = very poor, 0 = indifferent, +3 = excellent.

Analyze results — The data lend themselves to analysis of variance for a randomized (complete) block design. The students are treated as "blocks"; the courses evaluated are the "treatments." The F-statistic for "courses evaluated" in Table 7.4 is highly significant ($F_{3,87} = 12.91, p < 0.0001$). Therefore, the course evaluator concludes that there are differences among the average responses for the courses. The course evaluator performs an LSD multiple comparison procedure to determine which of the course means are significantly different from each other (see Table 7.4, bottom). The results of the LSD procedure reveal that the Nutrition course has a significantly lower (poorer) average rating than the other three. There are no other significant differences among the mean ratings of the other three courses. The course evaluator communicates these results to the professor and the department for further action.

TABLE 7.4
Randomized (Complete) Block ANOVA of Results in Table 7.3:
Popularity of Courses in Food Science

Source of variation	Degrees of freedom	Sum of squares	Mean square	F	p
Total	119	344.37			
Students (blocks)	29	188.87			
Courses evaluated	3	47.90	15.97	12.91	<0.0001
Error	87	107.60	1.24		

Average ratings for the items evaluated with the 95% LSD multiple-comparison results:

Courses evaluated	Biology	Nutrition	Sensory analysis	Statistics
Mean rating	0.80a	−0.83b	0.60a	0.30a

Note: Mean ratings not followed by the same letter are significantly different at the 95% confidence level — $LSD_{95\%} = 0.57$.

G. EXAMPLE 7.6: HOP CHARACTER IN FIVE BEERS. SPLIT PLOT DESIGN

Problem/situation — A brewer is producing a new brand of beer which is to have a high level of hop character. He is brewing with five alternative lots of hops which cost $1.00, $1.20, $1.40, $1.60, and $1.80/lb.

Project objective — To choose the lot which gives the most hop character for money.

Test objective — To compare the resulting five beers for degree of hop character. To obtain a measure of the reliability of the results.

Test design — The logical way is to line up the five beers in front of a large enough number of capable tasters, so this is a typical Multisample Difference test; 20 subjects evaluate the samples on a scale of 0 to 9, using the scoresheet in Figure 7.5. The order of presentation is randomized, and the samples are presented on three separate occasions with different coding.

Screen samples — Two experienced tasters evaluate the samples to make certain that they are representative of the type of beer to be produced and that there are no disturbing sensory differences in attributes other than hop character.

Analyze results — The results of the evaluations are shown in Table 7.5 and the corresponding split-plot ANOVA in Table 7.6. The subject-by-sample interaction was not significant:

$$F_{\text{interaction}} = 0.97, \ \Pr\left[F_{76,190} \geq 0.97\right] = 0.56 > 0.05$$

The sample effect and the subject effect were both highly significant:

$$F_{\text{Sample}} = 41.88 \ \Pr\left[F_{4,8} \geq 41.88\right] < 0.01$$

$$F_{\text{Subject}} = 17.79 \ \Pr\left[F_{19,190} \geq 17.79\right] < 0.01$$

MULTISAMPLE COMPARISONS TEST

Name: _____   Date: _____

Type of Sample: Beer _____

_____

Characteristic Studied: Hop character _____

Instructions

Taste the samples from left to right and note the intensity of the characteristic studied. Rate each sample on the following scale:

0 1	Imperceptible
2 3	Slightly perceptible
4 5	Moderately perceptible
6 7	Strongly perceptible
8 9	Extremely perceptible

Sample Code: _____ _____ _____ _____

Rating: _____ _____ _____ _____

Comments: _____

_____

FIGURE 7.5 Scoresheet for Multisample Difference test (rating). Example 7.6: hop character in five beers.

As the interaction was not significant, it may be assumed that the subjects were consistent in their ratings of the samples. However, the significance of the subject effect suggests that the subjects used different parts of the scale to express their perceptions. This is not uncommon. Further, when there is no interaction, subject-to-subject differences are normally of secondary interest. The differences among the samples are of primary concern. To determine which samples differ significantly in average hop character, compare the sample means using an HSD multiple comparison procedure, as shown below:

Sample	4	2	5	1	3
Mean	3.9	3.0	2.9	2.1	1.4

Note: Means not connected by a common underscore are significantly different at the 5% significance level. $\text{HSD}_{5\%} = q_{0.05,5,8} \sqrt{MS_{Error(A)}/n} = 4.89\sqrt{1.32/60} = 0.7$($q$-value from Table T4).

Sample 4 had a significantly greater average rating than all of the other samples. Samples 2 and 5, with nearly identical average ratings, had significantly less hop character than Sample 4 and significantly more than Samples 1 and 3. Samples 1 and 3 showed significantly less hop character than Samples 2, 4, and 5.

TABLE 7.5
Results Obtained in Example 7.6:
Multisample Difference Test (Rating) —
Hop Character in Five Beers

Sample no.	1	2	3	4	5
1	2,2,1	3,4,5	1,0,2	5,4,3	3,2,4
2	0,0,1	1,2,1	0,0,0	2,1,2	2,1,1
3	0,2,1	2,0,2	0,2,0	2,3,2	0,2,2
4	3,3,3	4,5,6	2,3,1	5,8,4	5,6,4
5	2,4,3	4,3,1	3,0,3	3,5,6	1,4,3
6	2,4,1	3,2,4	3,2,1	4,6,7	3,4,2
7	0,0,1	1,2,1	0,0,0	0,2,1	2,1,1
8	6,4,3	4,6,3	3,4,6	4,6,3	3,4,6
9	2,2,2	3,3,5	0,1,1	4,6,5	3,5,3
10	1,4,3	2,5,3	2,0,2	5,4,5	5,2,3
11	3,4,2	1,3,4	3,0,3	6,5,3	3,4,1
12	1,0,0	1,2,1	0,0,0	1,2,1	1,1,2
13	1,0,0	1,2,1	0,0,0	2,1,2	1,1,2
14	3,3,3	6,5,4	1,3,2	4,8,5	4,6,5
15	2,2,2	5,3,3	1,1,0	5,6,4	3,5,3
16	1,4,2	4,2,3	1,2,3	7,6,4	2,4,3
17	3,4,1	3,5,2	2,0,2	5,4,5	3,2,5
18	1,2,0	2,0,2	0,2,0	2,3,2	2,2,0
19	1,2,2	5,4,3	2,0,1	3,4,5	4,2,3
20	3,4,6	3,6,4	6,4,3	3,6,4	6,4,3

Explanation: e.g., Subject no. 20 rated Sample no. 1 a "3" the first time, a "4" the second time, and a "6" the third time.

TABLE 7.6
Split-Plot ANOVA of Results in Table 7.5:
Hop Character in Five Beers

Source of variation	Degrees of freedom	Sum of squares	Mean squares	F
Total	299	975.64		
Replications	2	8.89		
Samples	4	221.52	55.38	41.88[a]
Error(A)	8	10.58	1.32	
Subjects	19	412.30	21.70	17.79[a]
Sample × subject	76	89.81	1.18	0.97
Error(B)	190	232.53	1.22	

Note: Error(A) is calculated as would be the Rep × Sample interaction. Error(B) is calculated by subtraction.

[a] Significant at the 1% level.

Interpret and report results — The sensory analyst's report to the brewer contains the table of sample means and the ANOVA table and concludes that, of the five samples tested, Sample 4 produced a significantly higher level of hop character. Sample 2, of a less expensive variety, also merits consideration.

VII. MULTISAMPLE DIFFERENCE TEST: BIB RANKING TEST (BALANCED INCOMPLETE BLOCK DESIGN) — FRIEDMAN ANALYSIS

A. Scope and Application

Use this method when the test objective is to determine in which way a particular sensory attribute varies over a number of samples, and there are too many samples to evaluate at any one time. Typically, the method is used when the number of samples to be compared is from 6 to 12 or at most 16.

Choose the present method (ranking) when the panelists are relatively untrained for the type of sample and/or a relatively simple statistical analysis is preferred. Use Method VIII (rating) when panelists trained to use a rating scale are available.

B. Principle of the Test

Instead of presenting all t samples as one large block, present them in a number of smaller blocks according to one of the designs of Cochran and Cox (1957). Ask subjects to rank the samples according to the attribute of interest.

C. Test Subjects

Select, train, and instruct the subjects as described on p. 61. Ascertain that subjects can recognize the attribute of interest, e.g., by training with sets of known intensity levels in the attribute, see p. 136ff. and pp. 195–208.

D. Test Procedure

For test controls and product controls: see pp. 24 and 32. Offer samples simultaneously if possible, or else sequentially. Refer to p. 62 for details of procedure. Make certain that order of presentation is truly random: subjects must not be led to suspect a regular pattern, as this will influence verdicts. For example, state only to "rank the samples according to sweetness, giving rank 1 to the sample of lowest sweetness, rank 2 to the next lowest, etc."

E. Example 7.7: Species of Fish

Problem/situation — Field Ration XPQ-6, Fish Fingers in Aspic, has been prepared in the past from 15 different species of fish. Serious complaints of "fishy" flavor have been traced to the use of certain of these species. The Strategic Command wants to specify a limited number of species so as to be able to weigh availability and price against the probability of food riots.

Project objective — To compare the 15 species in such a way that quantitative information on the degree of fishy flavor is obtained which can be applied to the problem at hand.

Test objective — To compare Fish Fingers produced from the 15 species for degree of fishy flavor.

Test design — The Multisample Difference test with balanced incomplete design is chosen because it permits comparison of the 15 test products in groups of three. A randomly selected group of 105 enlisted personnel are randomly divided into 35 groups of three subjects each. Each group of three subjects is randomly assigned one of the 35 groups of three samples according to the design in Table 7.7. The scoresheet asks the subject to rank his three samples according to fishy flavor, from least (= 1) to most (= 3).

TABLE 7.7
Multisample Difference Test: BIB Design for Example 7.7 — Fish Fingers in Aspic

$t = 15$, $k = 3$, $r = 7$, $b = 35$, $\lambda = 1$, $E = 0.71$

Block

(1) 1	2	3	(6) 1	4	5	(11) 1	6	7	(16) 1	8	9
(2) 4	8	12	(7) 2	8	10	(12) 2	9	11	(17) 2	13	15
(3) 5	10	15	(8) 3	13	14	(13) 3	12	15	(18) 3	4	7
(4) 6	11	13	(9) 6	9	15	(14) 4	10	14	(19) 5	11	14
(5) 7	9	14	(10) 7	11	12	(15) 5	8	13	(20) 6	10	12

(21) 1	10	11	(26) 1	12	13	(31) 1	14	15
(22) 2	12	14	(27) 2	5	7	(32) 2	4	6
(23) 3	5	6	(28) 3	9	10	(33) 3	8	11
(24) 4	9	13	(29) 4	11	15	(34) 5	9	12
(25) 7	8	15	(30) 6	8	14	(35) 7	10	13

From Cochran, W. G. and Cox, G. M. *Experimental Designs*, John Wiley & Sons, New York, 1957, 469. With permission.

TABLE 7.8
Results Obtained in Example 7.7, Multisample Difference Test:
BIB Design with Rank Data — Fish Fingers in Aspic

Block/subject	

Block/subject	1	2	3	4	5	6	7	8	9	10	11	12	13	14	15
1	1	2	3												
2				3			1			2					
3					1					3					2
4						3					2		1		
5							3		1					2	
6	3			2	1										
7		2						3		1					
8			3										2	1	
9						3			2						1
10							2				1	3			
—	—	—	—	—	—	—	—	—	—	—	—	—	—	—	—
—	—	—	—	—	—	—	—	—	—	—	—	—	—	—	—
—	—	—	—	—	—	—	—	—	—	—	—	—	—	—	—
101	2													3	1
102		1		3		2									
103			1					2			3				
104					1				2			3			
105							3			2			1		
Rank sum	35	45	54	43	28	37	55	42	37	50	49	50	34	42	29

Note: Response: 1, least fishy; 2, intermediate; 3, most fishy.

TABLE 7.9
Summary of Results and Statistical
Analysis of the Data in Table 7.8:
Fish Fingers in Aspic

Sample/species	Rank sum					
5	28	a				
15	29	a				
13	34	a	b			
1	35	a	b	c		
6	37	a	b	c		
9	37	a	b	c		
14	42		b	c	d	
8	42		b	c	d	
4	43		b	c	d	
2	45			c	d	e
11	49				d	e
10	50				d	e
12	50				d	e
3	54					e
7	55					e

Note: Means followed by the same letter are not significantly different at the 5% significance level ($LSD_{rank} = 10.74$).

Screen samples — The help of the cook is enlisted in preparing samples as uniformly as possible regarding texture, appearance, and flavor, minimizing the differences attributable to species by suitable changes in cooking methods and secondary ingredients. The pieces prepared for each serving are screened for appearance, and any that contain coarse fragments or show other visible deviations are discarded.

Analyze results — To make the results easier to analyze, the rank data from the study are arranged as shown in Table 7.8. The rank sum for a given species of fish is simply the sum of all the numbers in the column corresponding to that species. The value of Friedman's test statistic T (see Equation 13.18, p. 295) is computed to determine if there are any differences among the species in the intensity of fishy flavor. The value of $T = 68.53$ exceeds the upper 5% critical value of a χ^2 with $(t-1) = 14$ degrees of freedom ($\chi^2_{14,0.05} = 23.69$), and it is concluded that there are indeed significant differences in the data set. Next, Equation 13.19, p. 295, is used to calculate the value of a 95% LSD multiple comparison to determine which of the species are significantly different. (See Table 7.9.)

Interpret and report results — The Strategic Command concludes from Table 7.9 that the species identified as Samples 5, 15, 13, 1, 6, and 9 should be retained for price and availability consideration, as these produce the least degree of fishy flavor and are not significantly different from each other. The species denoted as Samples 14, 8, and 4 are provisionally retained in case too many of the species in the first group are eliminated because of high cost or unavailability. This is done in recognition of the fact that Samples 14, 8, and 4 have rank sums for the intensity of fishy flavor that are significantly greater than only Samples 5 and 15 and are not significantly different from the remaining samples in the first group. The remaining species in Table 7.9 (2, 11, 10, 12, 3, and 7) are eliminated from use in Field Ration XPQ-6.

VIII. MULTISAMPLE DIFFERENCE TEST: BIB RATING TEST (BALANCED INCOMPLETE BLOCK DESIGN) — EVALUATION BY ANALYSIS OF VARIANCE

A. SCOPE AND APPLICATION

Use this method when the test objective is to determine in which way a particular sensory attribute varies over a number of samples, and there are too many samples to evaluate at any one time. Typically, the method is used when the number of samples to be compared is from 6 to 12 or at most 16.

Choose the present method (rating) when panelists trained to use a rating scale are available and results need to be as precise and actionable as possible. Use Method VII (ranking) when panelists have less training and/or the ranking test gives sufficient information.

Note: In descriptive analysis (see Chapter 10), when the number of samples to be compared is large, the present method may be applied to each attribute.

B. PRINCIPLE OF THE TEST

Instead of presenting all *t* samples as one large block, present them in a number of smaller blocks according to one of the designs of Cochran and Cox (1957). Ask subjects to rate the intensity of the attribute of interest on a numerical intensity scale (see Chapter 5, pp. 52–56). Specify the scale to be used. Evaluate the results by the analysis of variance.

C. TEST SUBJECTS

Select, train, and instruct the subjects as described on p. 61. Ascertain that subjects can recognize the attribute of interest, e.g., by training with sets of known intensity levels in the attribute. Use no fewer than 8 subjects; discrimination is much improved if 16 or more can be used.

Subjects may require special instruction to enable them to recognize the attributes of interest reproducibly (see Chapter 9, pp. 136ff.). Depending on the test objective, subjects may be selected who show high discriminating ability in the attribute(s) of interest.

D. TEST PROCEDURE

For test controls and product controls: see pp. 24 and 32. Offer samples simultaneously if possible, or else sequentially. Refer to pp. 61–62 for details of procedure. Make certain that order of presentation is truly random: subjects must not be led to suspect a regular pattern, as this will influence verdicts.

Note: If more than one attribute is to be rated, unavoidably there will be some interdependence in the resulting ratings (see Section IV.D, p. 111).

E. ANALYSIS AND INTERPRETATION OF RESULTS

The results are analyzed by the analysis of variance, see Chapter 13, p. 294, and Example 7.8 below.

F. EXAMPLE 7.8: REFERENCE SAMPLES OF ICE CREAM

Problem/situation — As part of an ongoing program, the QC manager of an ice cream plant routinely screens samples of finished product in order to select lots that will be added to the pool of quality reference samples for use in the main QC testing program. New reference samples are needed at regular intervals, as the older samples will have changed with time and are no longer appropriate. The procedure is also used to eliminate from the pool any current reference samples that may have deteriorated.

TABLE 7.10
BIB Design for Example 7.8:
Reference Samples of Ice Cream

$t = 6$, $k = 4$, $r = 10$, $b = 15$, $\lambda = 6$, $E = 0.90$

Block

(1) 1 2 3 4	(4) 1 2 3 5	(7) 1 2 3 6
(2) 1 4 5 6	(5) 1 2 4 6	(8) 1 3 4 5
(3) 2 3 5 6	(6) 3 4 5 6	(9) 2 4 5 6

(10) 1 2 4 5	(13) 1 2 5 6
(11) 1 3 5 6	(14) 1 3 4 6
(12) 2 3 4 6	(15) 2 3 4 5

From Cochran. W. G. and Cox, G.M. *Experimental Designs,* John Wiley & Sons, New York, 1957, 469. With permission.

TABLE 7.11
Table of Results for Example 7.8:
Reference Samples of Ice Cream

Block/subject	Sample					
	1	2	3	4	5	6
1	6	1	1	2		
2	6			1	3	3
3		4	2		5	2
4	7	2	3		2	
5	3	5		1		1
6		1	1		3	2
7	7	4	4			3
8	2		1	1	1	
9		2		2	2	3
10	4	2		2	5	
11	5		3		1	1
12		3	2	1		2
13	4	2			1	1
14	5		2	2	1	
15		2	4	5	3	
Adjusted means	5.0	2.5	2.2	2.0	2.6	1.9

Note 1: BIB design with rating.

Note 2: Response — 10-point category scale with 0 = no off-flavor, 9 = extreme off-flavor.

Note 3: Adjusted means that are not connected by a common underscore are significantly different at the 5% significance level (LSD$_{5\%}$ = 1.1).

TABLE 7.12
Balanced Incomplete Block ANOVA Table for Example 7.8:
Reference Samples of Ice Cream

Source of variation	Degrees of freedom	Sum of squares	Mean square	F	p
Total	59	150.98			
Judges (blocks)	14	39.73			
Samples (treatments, adjusted for blocks)	5	59.89	11.98	9.33	<0.0001
Error	40	51.36	1.28		

Project objective — To maintain a sufficient inventory of reference samples of finished ice cream for QC testing purposes.

Test objective — To rate the inventory of six lots each day for overall off-flavor and discard any lot that may not be suitable as a reference.

Test design — Samples of the six lots are evaluated for overall off flavor by 15 well-trained panelists who use a 10-point category scale from 0 = no off-flavor to 9 = extreme off-flavor. The panelists cannot evaluate more than four samples in one sitting, so the sensory analyst chooses a BIB design from Cochran and Cox (1957) (see Table 7.10). Each of the 15 panelists is randomly assigned one block of four samples from the design. The order of presentation of the samples within each block is randomized.

Analyze results — The ratings data for the overall off-taste attribute are presented in Table 7.11. The data are analyzed by a computer program capable of performing a balanced-incomplete-block ANOVA (see Chapter 13, p. 293). The resulting BIB ANOVA table is presented in Table 7.12. The F-statistic for "treatments" (i.e., samples of ice cream), when compared to the upper 5% critical value of an F-distribution with $(t-1) = 5$ and $(tpr - t - pb + 1) = 40$ degrees of freedom, is found to be significant ($F = 9.33 > F_{0.05;5,40} = 2.45$). An LSD multiple comparison procedure is applied to the average ratings of the samples to determine which samples have significantly different overall off-flavor (see *Note 3* at the foot of Table 7.11).

Interpret and report results — The average off-taste rating of Sample 1 is significantly greater than the average ratings of the remaining samples. There are no other significant differences among the mean ratings of the other samples. The sensory analyst reports the results to the QC manager with the recommendation that the lot from which Sample 1 was taken be discarded from the pool of reference samples.

REFERENCES

Carr, B.T., 1985. Statistical models for paired-comparison data. *Am. Soc. Qual. Contr. 39th Congr. Trans.*, Baltimore, MD, 295.

Cochran, W.G. and Cox, G.M., 1957. *Experimental Designs.* John Wiley & Sons, New York, p. 469.

Friedman, M., 1937. The use of ranks to avoid the assumption of normality implicit in the Analysis of Variance. *J. Am. Stat. Assoc.* 32, 675.

Hollander, M. and Wolfe, D.A., 1973. *Nonparametric Statistical Methods.* John Wiley & Sons, New York.

Kramer, A., Kahan, G., Cooper, D., and Papavasiliou, A., 1974. A non-parametric method for the statistical evaluation of sensory data. *Chem. Senses Flavor* 1, 121.

Pangborn, R.M., 1984. Sensory techniques of food analysis. In: *Food Analysis. Principles and Techniques,* Vol. 1, Gruenwedel, D.W. and Whitaker, J.R., Eds., Marcel Dekker, New York, p. 59.

Skillings, J.H. and Mack, G.A., 1981. On the use of a Friedman-type statistic in balanced and unbalanced block designs. *Technometrics* 23, 171.

8 Determining Thresholds

CONTENTS

I. INTRODUCTION

Sensory thresholds are ill-defined in theory (Lawless and Heymann, 1998; Morrison, 1982). A good determination requires hundreds of comparisons with a control, and results do not reproduce well at all (Brown et al., 1978; Stevens et al., 1988; Marin et al., 1988). Published group thresholds (Fazzalari, 1978; Van Gemert et al., 1984; Devos et al., 1990) vary by a factor of 100 for quinine sulfate in water and by much more in complex systems. Swets (1964) doubts even the existence of a sensory threshold. One's first reaction is that it is futile to invest time and money in threshold studies. However, in situations such as those described in the next paragraph, the threshold approach is still the best available.

Thresholds in air, determined by automated flow olfactometry, are used to determine degrees of air pollution (CEN, 1997) and to set legal limits for polluters. Thresholds of added substances are used with water supplies, foods, beverages, cosmetics, paints, solvents, etc. to determine the point at which known contaminants begin to reduce acceptability. These are the most important uses, and testing may be done with hundreds of panelists in order to map the distribution of relative sensitivity in the population. Thresholds may also be used as a means of selecting or testing panelists, but this should not be the principal basis for selection (see Chapter 9) unless the test objective requires detection of the stimulus at very low levels. The threshold of added desirable substances may be used as a research tool in the formulation of foods, beverages, etc.

The concepts of the *Odor Unit* (Guadagni et al., 1966) or *Flavor Unit* (Meilgaard, 1975) use the threshold as a measure of flavor intensity. For example, if H_2S escapes from a leaking bottle into a room, when the level reaches the threshold of detection we say that the odor intensity is at 1 O.U., and at double that level of H_2S, the intensity is at 2 O.U., and so on. This use of thresholds requires much caution and is not applicable above 3 to 6 O.U. (see Chapter 2, pp. 19–21). (Procedures for estimating sensory intensity at levels well above threshold are discussed in Chapter 5, pp. 46–50.)

The methods by which olfactory thresholds are determined can have a profound influence on the results. Hangartner and Paduch (1988) show that odorant flows below the usual sniffing volume of 1 to 2 L/sec will give rise to thresholds severalfold too low. Doty et al. (1986) found that use of a larger sniff bottle resulted in 10- to 20-fold lower thresholds. Training can lower thresholds as much as 1000-fold (see Powers and Shinholser, 1988).

Experience shows that with practice and training (Brown et al., 1978) it is possible to obtain reproducibility levels of ±20% for a given panel and ±50% between one large panel (>25) and another. The important factors, in addition to repeated training with the actual substance under test, are those described in Chapter 4: subjects will pride themselves and hope to please the experimenter by finding the lowest threshold, and this must be counteracted by meticulous attention to the details of sample preparation and sample presentation so as not to leave clues to their identity.

II. DEFINITIONS

Thresholds are the limits of sensory capacities. It is convenient to distinguish between the absolute threshold, the recognition threshold, the difference threshold, and the terminal threshold.

The *absolute threshold* (detection threshold) is the lowest stimulus capable of producing a sensation — the dimmest light, the softest sound, the lightest weight, the weakest taste.

The *recognition threshold* is the level of a stimulus at which the specific stimulus can be recognized and identified. The recognition threshold is usually higher than the absolute threshold. If a person tastes water containing increasing levels of added sucrose, at some point a transition will occur in sensation from "water taste or pure water" to "a very mild taste." As the concentration of sucrose increases, a further transition will occur from "a very mild taste" to "mild sweet." The level at which this second transition occurs is called the recognition threshold.

The *difference threshold* is the extent of change in the stimulus necessary to produce a noticeable difference. It is usually determined by presenting a standard stimulus which is then compared to a variable stimulus. The term *just noticeable difference* (*JND*) is used when the difference threshold is determined by changing the variable stimulus by small amounts above and below the standard until the subject notices a difference (see also Chapter 5, pp. 46–47).

The *terminal threshold* is that magnitude of a stimulus above which there is no increase in the perceived intensity of the appropriate quality for that stimulus; above this level, pain often occurs.

JNDs increase as one proceeds up the scale of concentration, and they have been used as scale steps of sensory intensity. Hainer et al. (1988) calculated that their subjects could distinguish some 29 JNDs between the absolute and the terminal thresholds. However, thresholds vary too much from person to person and from group to group for the JND to have gained practical application as a measure of perceived intensity.

The conventional notion of a threshold (say, for diesel exhaust in air) is that shown in Figure 8.1. Above 5 ppm the exhaust can be detected, and below 5 ppm it cannot. However, an observer making repeated tests using a dilution olfactometer will produce a set of results such as Figure 8.2. His sensitivity will vary with chance air currents over the olfactor membrane and with momentary or biorhythmic variations in the sensitivity of his nervous system. The ticking of a watch held at a certain distance can be heard one moment and is inaudible the next, then audible again, etc. The threshold is not a fixed point but rather a value on a stimulus continuum. By convention, the observer's personal threshold is that concentration he can detect *50% of the time,* and *not* the concentration he can detect at X% significance, an error frequently committed (Laing, 1987). As a rule, one finds a typical Gaussian dose–response curve from which the 50%-point can be accurately determined after transformation of the experimental percentage points by one of the methods described in Example 8.2.

To get from a collection of personal thresholds to a group threshold, we note that the frequency distribution tends to be bell shaped for the majority (Meilgaard, 1993). However, the right-hand tail of the curve tends to be longer than the left (see Figure 8.3) because most groups contain a proportion of individuals who show very low sensitivity to the stimulus in question. The measure of central tendency which makes most sense for such a group of observers may be the geometric mean, as it gives less weight to the highest thresholds. For a discussion of other measures of central tendency (e.g., the median) which may be appropriate with certain distributions, see ASTM, 1990. A rank probability graph (such as Figure 8.5, p.131) is a useful tool for testing whether a set of individual thresholds are normally distributed, which they are if a good straight line can be drawn through the

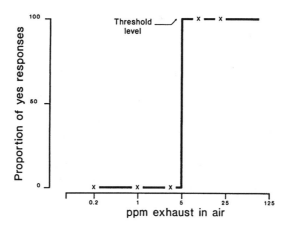

FIGURE 8.1 Conventional notion of the absolute threshold (for diesel exhaust in air).

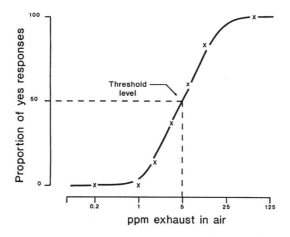

FIGURE 8.2 Typical data from determination of personal threshold (for diesel exhaust in air).

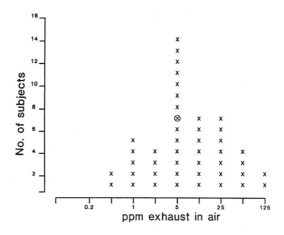

FIGURE 8.3 Typical histogram of threshold for a group of 45 subjects; ⊗ = subject from Figure 8.2.

points. In this case, the graph can serve to locate not only the group threshold as the 50%-point, but also the concentrations that, for example, 5% or 90% of the corresponding population can detect.

III. APPLICATIONS OF THRESHOLD DETERMINATIONS

Thresholds can be measured by a variety of the classical psychophysical designs based, e.g., on the Method of Limits, the Method of Average Error, or the Frequency Method (Kling and Riggs, 1971). In recent years, the tendency among psychophysicists has been to choose a different route by applying the Signal Detection Theory, SDT (Swets, 1964; Macmillan and Creelman, 1991). SDT is a system of methods based on the idea that the point of interest is not the threshold as such, but rather "the size of the psychological difference between the two stimuli," which has the name d'. The advantage of SDT is that the subject's decision process becomes more explicit and can be modeled statistically. However, SDT procedures are more time-consuming than the classical threshold designs, and it has been shown (Frijters, 1980) that for forced-choice methods of sample presentation, there is a 1:1 relationship between d' and the classical threshold.

For these reasons, both the ASTM (1976, 1990) and the International Standards Organization (1999) have decided to stick with the Method of Limits and what is known as the 3-Alternative Forced Choice (3-AFC) method of sample presentation in which three samples are presented: two are controls and one contains the substance under test. The ASTM's rapid method (E679, see Example 8.1) aims at determining a practical value close to the threshold, based on a minimum of testing effort (e.g., 50 to 150 3-AFC presentations). It makes a very approximate "best estimate" determination of each panelist's threshold. In return, the panel can be larger, and the resulting group threshold and distribution become more reliable by virtue of the fact that the variation between individuals is much greater (up to 100-fold) than the variation between tests by a single individual (up to 5-fold). The result is slightly biased at best and can be very biased if subjects falling on the upper or lower limits of the range under test are not reexamined (see Example 8.1).

The ASTM's intermediate method (E1432) proceeds to determine individual thresholds according to Figure 8.2 and then, in a second step, the group threshold according to Figure 8.3. For this, it requires approximately 5× as many sample presentations per panelist as the rapid method. In return, the group threshold and distribution of individual thresholds are both bias free.

The ISO Standard 13301 (ISO, loc. cit.) is in effect a combination of both of the above. For the curve-fitting step, the intermediate method uses nonlinear least squares regression (see Example 8.2). (The ISO procedure permits logistic regression and a maximum likelihood procedure for which a procedure of calculation using computer spreadsheets has been introduced.) If results more precise than can be expected with these methods are expected, one enters the field of research projects as such, and any of a number of designs may be appropriate, e.g., Powers' Multiple Pairs test (Powers and Ware, 1976; Kelly and Heymann, 1989) or Signal Detection Theory (Macmillan and Creelman, loc. cit.). Bi and Ennis (1998) provide a review of these methods and propose an additional procedure for population thresholds, based on the Beta-Binomial distribution, which takes account of the fact that data for one individual tend to have a much narrower distribution than those for a group of individuals.

A. EXAMPLE 8.1: THRESHOLD OF SUNSTRUCK FLAVOR COMPOUND ADDED TO BEER

Problem/situation — A brewer, aware that beer exposed to UV light develops sunstruck flavor (3-methyl-2-butene-l-thiol, a compound not otherwise present), wishes to test the protection offered by various types of packaging material.

Project objective — To choose packaging which offers acceptable protection at minimum cost, using as criterion the amount of the sunstruck compound formed during irradiation, compared with the threshold amount.

Test objective — To determine the threshold of purified 3-methyl-2-butene-1-thiol added to the company's beer.

TABLE 8.1
Sensory Threshold of the Sunstruck Flavor Compound Added to Beer

Procedure: ASTM E679 Ascending Concentration Series Method of Limits
Equipment: Colorless beer glasses, 250 mL; 50 mL beer "A" per glass
Sample: 3-Methyl-2-butene-1-thiol (Aldrich)
Purification: By preparative gas chromatography on two columns
Number of scale steps: 6   Concentration factor per step: 3.0
Number of subjects: 25   "High" and "low" results confirmed: Yes

Panelist	Concentrations presented (ppb)								Best estimate threshold	
	0.27	0.80	2.41	7.28	21.7	65.2	195	Over	ppb	Log (10)
01		0	0	+	+	+	+		4.19	0.622
02	0	+	+	+	+	+	+		0.46	−0.337
03		0	+	+	+	+	+		1.39	0.143
04	0	+	+	+	+	+	+		0.46	−0.337
05		0	+	0	+	+	+		12.6	1.100
06		0	+	+	+	+	+		1.39	0.143
07		+	0	+	+	+	+		4.19	0.622
08	0	+	+	+	+	+	+		0.46	−0.337
09	0	+	+	+	+	+	+		0.46	−0.337
10		0	+	0	0	+	0	+	338	2.529
11		0	+	+	+	+	+		1.39	0.143
12		0	+	+	+	+	+		1.39	0.143
13		+	0	+	+	+	+		4.19	0.622
14		0	0	+	+	+	+		4.19	0.622
15	0	+	+	+	+	+	+		0.46	−0.337
16		0	+	0	+	+	+		12.6	1.100
17		0	+	+	+	+	+		1.39	0.143
18		+	+	0	0	+	+		37.7	1.576
19	0	+	+	+	+	+	+		0.46	−0.337
20		+	0	+	+	+	+		4.19	0.622
21		0	+	+	+	+	+		1.39	0.143
22	0	+	+	+	+	+	+		0.46	−0.337
23		+	0	0	+	+	+		12.6	1.100
24	0	+	+	+	+	+	+		0.46	−0.337
25		0	0	+	+	+	+		4.19	0.622
									Sum →	9.299
			Group BET, geometric mean, ppb						2.35 ←	0.3720
			Log standard deviation =							0.719

Histogram of Individual BE Thresholds
G.M. = 2.35 ppb

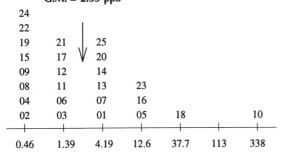

3-Methyl-2-butene-1-thiol, ppb in beer "A"

Test design — The E-679 rapid test is suitable as the need is for good coverage of the variability among people; 25 panelists each receive six 3-AFC tests with concentrations spaced by a factor of 3. Limit bias by: (1) choosing the range of concentrations offered with the aid of a preliminary test using five panelists, and (2) retesting those panelists who are correct at the lowest or fail at the highest level.

Screen samples — In the preliminary test, ascertain that the base beer is free of sunstruck flavor and that the 3-methyl-2-butene-1-thiol confers a pure sunstruck character at the chosen test concentrations.

Conduct the test — Test each panelist at the six concentrations. Test any panelist who is correct at the lowest level once or twice more at that level and include sets at one or two lower levels. Likewise, test any panelist who fails at the highest level twice more at that level and at one or two higher levels. Record and analyze results, as shown in Table 8.1. The best estimate threshold (BET) for each subject is the geometric mean of the highest concentration missed and the next higher concentration. The group BET is the geometric mean of the individual BETs. Repeat the test series at least once on a different day, using the same observers. Note that thresholds often decrease as panelists become accustomed to the flavor of the substance and the mechanics of the test. If the threshold decreases more than 20%, repeat the test series until the values stabilize.

Test report — Include the complete Table 8.1 and give demographics of the panelists.

B. EXAMPLE 8.2: THRESHOLD OF ISOVALERIC ACID IN AIR

Problem/situation — A rendering plant produces air emissions containing isovaleric acid as the most flavor-active component. The neighbors complain and an ordinance is passed requiring a reduction below threshold.

Project objective — To choose between various process alternatives and a higher chimney.

Test objective — To determine the threshold of isovaleric acid in air.

Test design — A fairly thorough method, such as the ASTM Intermediate Method (E1432) or the second example of ISO Standard 13301, is suitable because of the economic consequences of the issue. Use a dynamic olfactometer (CEN, 1997) and 20 panelists. Give each panelist 3-AFC tests six times at each of five or more concentrations spaced twofold apart and chosen in advance (see below). The apparatus contains three sniff ports; the panelist knows that two produce odor-free air and must choose the one which he believes to contain added isovaleric acid. The added concentration is at the lowest level in the first test and increases by a constant factor in each subsequent test. From the percentage of correct results at each concentration, calculate each panelist's threshold, and from these, the group threshold.

Screen samples — Ascertain that the air supply is free from detectable odors and that the isovaleric acid is of sensory purity and free from foreign odors. Check the reliability of the olfactometer by chemical analysis.

Conduct the test — Test each panelist in turn at the chosen concentrations. Make this choice in advance, by a single test (or a few tests) at each of a set of widely spaced concentrations (e.g., 2.5, 10, 40, 160, and 640 ppb). In the test itself, if a panelist should score 100% correct at the lowest concentration, reschedule the concentration series with this as the highest. Likewise, if a panelist scores less than 80% correct at the highest concentration, continue the series by presenting higher concentrations until this no longer happens. Calculate the individual thresholds by one of the six curve-fitting methods allowed by ASTM E1432 or ISO 13301 (e.g., by logistic regression using a computer package, as shown in Figure 8.4). Plot the individual thresholds in a rank/probability graph, as shown in Figure 8.5, and obtain the group threshold as the 50% point. If a straight line can be drawn through the points, conclude that the panelists represent a normal distribution and that other points of interest can be read from the graph, e.g., that concentration which 10% of a population similar to the panel can detect.

Test report — Include the information in Table 8.2 and Figure 8.5 and give demographics of the panelists.

TABLE 8.2
Determination of Olfactory Thresholds to Isovaleric Acid in Air by ASTM Intermediate Method E1432

Procedure: ASTM Intermediate Method E1432
Equipment: Dynamic Triangle Olfactometer after A. Dravnieks
Sample: Isovaleric acid (sigma)
Purification: Recrystallization as calcium salt
Number of panelists: 20   Number of scale steps presented to each: min 5
Concentration factor per scale step: 2.0

Example of results, showing six panelists: Number of correct tests (out of six):

Concentrations presented (ppb)	Panelist					
	2	11	13	15	18	19
640				6	6	5
320			6	5	6	4
160	6	6	5	4	2	3
80	5	4	4	2	2	2
40	4	4	2	3	3	1
20	6	0	2	2	1	2
10	5	3				
5	3					
2.5	2					

Converted to Proportion Correct = C/N where C = above number; N = 6:

Concentrations presented (ppb)	Panelist					
	2	11	13	15	18	19
640				1.000	1.000	.833
320			1.000	.833	1.000	.667
160	1.000	1.000	.833	.667	.333	.500
80	.833	.667	.667	.333	.333	.333
40	.667	.667	.333	.500	.500	.167
20	1.000	.000	.333	.333	.167	.333
10	.833	.500				
5	.500					
2.5	.333					

Using logistic regression (computer package SAS® PROC NLIN, see Figure 8.4) the individual thresholds are

Panelist No.	1	2	3	4	5	6	7	8	9	10
Log(threshold)	0.84	0.84	1.04	1.20	1.26	1.32	1.43	1.58	1.67	1.67
Threshold(ppb)	7	7	11	16	18	21	27	38	47	47

Panelist No.	11	12	13	14	15	16	17	18	19	20
Log(threshold)	1.81	1.91	1.95	2.07	2.19	2.25	2.29	2.40	2.52	2.82
Threshold(ppb)	64	81	90	118	154	178	196	249	330	665

```
** PURPOSE:  Fit logistic models P = (1/3 + EXP[K])/(1 + EXP[K]),
**              where K = B(T - LOG[X]),
**              P is the proportion of correct identifications,
**              B is the slope,
**              X is the actual concentration (ppb) of Isovaleric
**                  Acid in air,
**          and  T is the threshold value in log(ppb).

PROC NLIN   Method=DUD    Data=Input;    by panelist;
               PARMS B=-4  T=2
                 K   =  B*(T - LOG10(K));
                 E   =  EXP(K);
                 N   =  (1/3 + E);
                 D   =  (1 + E);
               MODEL  P = N/D
               TITLE2  'Logistic Regression Models';
RUN;
```

OUTPUT FOR PANELIST 13:

```
        Logistic Regression of Threshold Data Using SAS PROC NLIN
                        Logistic Regression Models
                    NON-LINEAR LEAST SQUARES ITERATIVE PHASE
               DEPENDENT VARIABLE: P                METHOD:  DUD
  ITERATION               B                T            RESIDUAL  SS
     -3                  -4         2.000000000        0.025885700365
     -2                  -4.4       2.000000000        0.02054415598
     -1                  -4         2.200000000        0.084958944779
      0                  -4.4       2.000000000        0.02054415598
      1              -5.852958      1.961443385        0.010812277188
      2              -6.259745      1.967823308        0.010766524899
      3              -6.189164      1.951938036        0.010504941622
      4              -6.283542      1.954261395        0.010481402394
      5              -6.280162      1.954257276        0.010481361251
      6              -6.281544      1.954068199        0.010481219887
      7              -6.277816      1.953905805        0.010481193047
      8              -6.280506      1.953919346        0.010481176612
      9              -6.281737      1.953896400        0.010481176219
     10              -6.281715      1.953899496        0.010481176193
  CONVERGENCE CRITERION MET.
```

```
NON-LINEAR LEAST SQUARES SUMMARY STATISTICS   DEPENDENT VARIABLE P
   SOURCE            DF     SUM OF SQUARES       MEAN SQUARE
   REGRESSION         2      2.3500748238       1.1750374119
   RESIDUAL           3      0.0104811762       0.0034937254
   UNCORRECTED TOTAL  5      2.3605560000
   (CORRECTED TOTAL)  4      0.3558448000

   PARAMETER   ESTIMATE       ASYMPTOTIC        ASYMPTOTIC 95%
                              STD. ERROR      CONFIDENCE INTERVAL
                                             LOWER          UPPER
   B         -6.281714751   1.6824126163   -11.635992903  -0.9274366000
   T          1.953899496   0.0473965533     1.803059965   2.1047390264
```

PREDICTED VALUES

P	C	LOG[X]	X
.967	.95	2.4226	265
.933	.90	2.3037	201
.833	.75	2.1288	135
.667	.50	1.9539	90
.500	.25	1.7790	60
.400	.10	1.6041	40
.367	.05	1.4770	30

FIGURE 8.4 Fitting of a dose-response curve to the data in Table 8.2 using "SAS® PROC NLIN" and the logistic method. The estimated value of $T = 1.954$ is the threshold concentration in log (ppb) for Panelist 13.

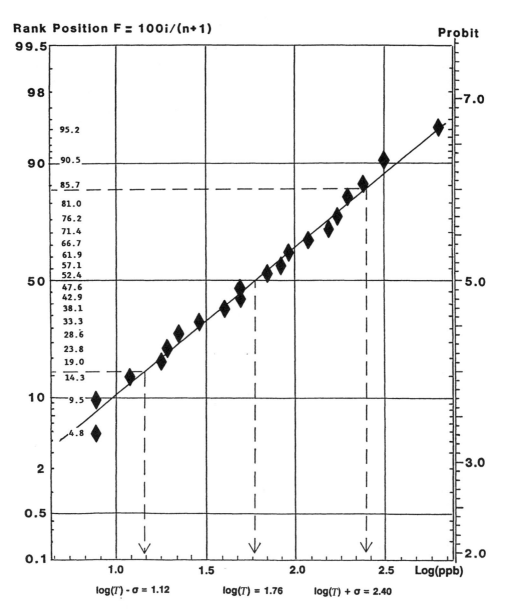

FIGURE 8.5   Rank probability for the 20 panelists in Table 8.2. Result: a straight line can be drawn through the points; consequently, the panelists are normally distributed with $\log(T) = 1.76$ ($T = 58$ ppb); $\log(T) - \sigma = 1.12$ (13 ppb); $\log(T) + \sigma = 2.40$ (255 ppb).

Group threshold.   Obtain the group threshold T by Rank Probability graph as shown in Figure 8.5. The result is $\log(T) = 1.76$; $T = 58$ ppb. Alternatively, calculate T as the geometric mean of the individual thresholds:

$$\log(T) = (0.84 + 0.84 + 1.04 \ldots\ldots 2.40 + 2.52 + 2.82)/20$$
$$= 36.06/20 = 1.753; \, T = 56.6 \text{ ppb.}$$

REFERENCES

ASTM, 1976. Standard Practice E679, *Determination of Odor and Taste Thresholds by a Forced-Choice Ascending Concentration Series Method of Limits*. American Society for Testing and Materials, Philadelphia.

ASTM, 1990. Standard Practice E1432, *Defining and Calculating Sensory Thresholds from Forced-Choice Data Sets of Intermediate Size*. American Society for Testing and Materials, Philadelphia.

Bi, J. and Ennis, D.M., 1998. Sensory thresholds: Concepts and methods. *J. Sensory Stud.* 13, 133–148.

Brown, D.G.W., Clapperton, J.F., Meilgaard, M.C., and Moll, M., 1978. Flavor thresholds of added substances. *J. Am. Soc. Brew. Chem.* 36, 73–80.

CEN, 1997. *Air Quality. Determination of Odour Concentration by Dynamic Olfactometry*. European Committee for Standardisation, 36 rue de Strassart, B-1050 Brussels, Belgium.

Devos, M., Patte, F., Rouault, J., Laffort, P., and Van Gemert, L.J., 1990. *Standardized Human Olfactory Thresholds*. IRL Press, Oxford.

Doty, R.L., Gregor, T.P., and Settle, R.G., 1986. Influence of intertrial interval and sniff-bottle volume on phenyl alcohol detection thresholds. *Chem. Senses* 11(2), 259–264.

Fazzalari, F.A., Ed., 1978. *Compilation of Odor and Taste Threshold Values Data*. American Society for Testing and Materials, Philadelphia.

Frijters, J.E.R., 1980. Three-stimulus procedures in olfactory psychophysics. An experimental comparison of Thurstone-Ura and Three-Alternative Forced-Choice models of Signal Detection Theory. *Percep. Psychophys.* 28(5), 390–397.

Guadagni, D.G., Okano, S., Buttery, R.G., and Burr, H.K., 1966. Correlation of sensory and gas-liquid chromatographic measurement of apple volatiles. *Food Technol.* 30, 518.

Hainer, R.M., Emslie, A.G., and Jacobson, A., 1954. In: *Basic Odor Research Correlation. Ann. N.Y. Acad. Sci.*, 58, 158.

Hangartner, M. and Paduch, M., 1988. Interface human nose — olfactometer. In: *Measurement of Odor Emissions*, Proceedings of workshop, Zürich. Commission of the EEC, Annex V, p. 53.

ISO, 1999. *Sensory Analysis — Methodology — General Guidance for Measuring Odour, Flavour, and Taste Detection Thresholds by a Three-Alternative Forced-Choice (3-AFC) Procedure*. International Organization for Standardization, Draft International Standard ISO/DIS 13301.2. ISO, Case Postale 56, 1 rue Varembé, CH1211 Génève 2, Switzerland.

Kelly, F.B. and Heymann, H., 1989. Contrasting the effect of ingestion and expectoration in sensory difference tests. *J. Sensory Stud.* 3(4), 249–255.

Kling, J.W. and Riggs, L.A., Eds., 1971. *Woodworth & Schlosberg's Experimental Psychology*, 3rd ed. Holt, Rinehart & Winston, New York, Chapter 2.

Laing, G.G., 1987. *Optimum Perception of Odours by Humans*. Report, CSIRO Division of Food Research, North Ryde, N.S.W., Australia, p. 8.

Lawless, H.T. and Heymann, H., 1998. *Sensory Evaluation of Food. Principles and Practices*. Chapman & Hall, New York, Chapter 6.

Macmillan, N.A. and Creelman, C.D., 1991. *Detection Theory, A User's Guide*. Cambridge University Press, 391 pp.

Marin, A.B., Acree, T.E., and Barnard, J., 1988. Variation in odor detection thresholds determined by Charm Analysis. *Chem. Senses.* 13(3), 435–444.

Meilgaard, M.C., 1975. Flavor Chemistry of Beer. Part I. Flavor interaction between principal volatiles. *Tech. Q. Master Brew. Assoc. Am.* 12, 107–117.

Meilgaard, M.C., 1993. Individual differences in sensory threshold for aroma chemicals added to beer. *Food Quality and Preference* 4, 153–167.

Morrison, G.R., 1982. Measurement of flavor threshold. *J. Inst. Brewing* 88, 170–174.

Powers, J.J. and Ware, G.O., 1976. Comparison of Sigmplot, Probit and Extreme-Value methods for the analysis of threshold data. *Chem. Senses and Flavor* 2(2), 241–253.

Powers, J.J. and Shinholser, K., 1988. Flavor thresholds for vanillin and predictions of higher or lower thresholds. *J. Sensory Stud.* 3(1), 49–61.

Stevens, J.C., Cain, W.W., and Burke, R.J., 1988. Variability of olfactory thresholds. *Chem. Senses* 13(4), 643–653.

Swets, J.A., 1964. Is there a sensory threshold? In: *Signal Detection and Recognition by Human Observers*, Swets, J.A., Ed. John Wiley & Sons, New York.

Van Gemert, L.J. and Nettenbreijer, A.H., 1984. *Compilation of Odour Threshold Values in Air and Water*. Central Institute for Nutrition and Food Research TNO, Zeist, Netherlands. Supplement V.

9 Selection and Training of Panel Members

CONTENTS

I. INTRODUCTION

This section is partly based on ASTM Special Technical Publication 758, "Guidelines for the Selection and Training of Sensory Panel Members" (1981) and on the ISO "Guide for Selection and Training of Assessors" (1993). The development of a sensory panel deserves thought and planning with respect to the inherent need for the panel, the support from the organization and its management, the availability and interest of panel candidates, the need for screening of training samples and references, and the availability and condition of the panel room and booths. In the food, fragrance, and cosmetic industries, the sensory panel is the company's single most important tool in research and development and in quality control. The success or failure of the panel development process depends on the strict criteria and procedures used to select and train the panel.

The project objective of any given sensory problem or situation determines the criteria for selection and training of the subjects. Too often in the past (ISO, 1979), the sole criterion was a low threshold for one or more of the basic tastes. Today sensory analysts use a wide selection of tests, specifically selected to correspond to the proposed training regimen and end use of the panel. Taste acuity is only one aspect; much more important is the ability to discern and describe a particular sensory characteristic in a "sea" or "fog" of other sensory impressions.

This chapter describes specific procedures for the decision to establish a panel, the selection and training of both difference and descriptive panels, and ways to monitor and motivate panels. This chapter does not apply to consumer testing (see Chapter 12) which uses naive subjects representative of the consuming population. Although the text uses the language of a commercial organization which exists to develop, manufacture, and sell a "product" and has its "upper" and "middle management" and reward structure, the system described can be easily modified to fit the needs of other types of organizations such as universities, hospitals, civil or military service organizations, etc.

II. PANEL DEVELOPMENT

Before a panel can be selected and trained, the sensory analyst must establish that a need exists in the organization and that commitment can be obtained to expend the required time and money to develop a sensory tool. Upper management and the project group (R&D or QA/QC) must see the *need* to make decisions based on sound sensory data with respect to overall differences and attribute differences (difference panels) or full descriptions of product standards, product changes over time, or ingredient and processing manipulation, and for construction and interpretation of consumer questionnaires (descriptive panels). The sensory analyst must also define the resources required to develop and maintain a sensory panel system.

Personnel — Heading the list of resources required is (1) a large enough pool of available candidates from which the panel can be selected; (2) a sensory staff to implement the selection, training, and maintenance procedures; and (3) a qualified person to conduct the training process. Panelists most often come from within the organization, as they are located at the site where the samples are prepared (e.g., R&D facility or plant).

Some companies choose to test products at a different site, which may be another company facility. With reduced laboratory staffing, many companies have opted to use residents recruited from the local community as panelists rather than bench chemists and support staff from the labs. Outside panelists may be available for more hours per week, and they may be cheaper and more focused for longer panel sessions. The primary drawbacks of external panelists are that they often require more time and effort to train in the technical aspects of panel work, and that they do not provide the inherent proprietary security of internal employees.

Panel candidates and management must understand in advance the amount of time required (personnel hours) for the selection and training of the particular panel in question. An assessment of the number of hours needed for panelists, technicians, and panel leader should be presented and accepted before the development process is initiated. The individual designated to select and train the panel is often a member of the sensory staff who is experienced and trained in the specific selection and training techniques needed for the problem at hand.

Facilities — The physical area for the selection, training, and ongoing work of a panel must be defined before development of the panel begins. A training room and panel testing facilities (booths and/or round table/conference room) must have the proper environmental controls (see Chapter 3), be of sufficient size to handle all of the panelists and products projected, and be located near to the product preparation area and panelist pool.

Data collection and handling — This is another resource to be defined: the personnel, hardware, and software required to collect and treat the data to be generated by the panel. Topics such as the use of personal computers with PC software vs. the company's mainframe should be addressed before the data begin to accumulate on the sensory analyst's desk. The specific ways in which the data are generated and used (that is, frequency data, scalar data — category, linear, magnitude estimation), the number of attributes, the number of replications, and the need for statistical analysis all contribute to the requirements for data collection and handling.

Projected costs — Once upper management and the project group understand the need to have a panel and the time and costs required for its development and use, the costs and benefits can be assessed from a business and investment perspective. This phase is essential so that the support from management is based on a full understanding of the panel development process. Once management and the project team are "on board," the sensory analyst can expect the support which is needed to satisfy the requirements for personnel, both panelists and staff, facilities, and data handling. Management can then, through circulars, letters, and/or seminars, communicate its support for the development of and participation in sensory testing. As the reader will have gathered by now, public recognition by management of the importance of the sensory program and of the involvement of employees as panelists are essential for the operation of the system. If participation in sensory tests is not seen by upper and middle management as a worthwhile expenditure of time, the sensory analyst will find the recruiting task to be difficult if not impossible, and test participation will dry up more quickly than new recruits can be enrolled.

Once management support has been communicated through the organization and has been demonstrated in terms of facilities and personnel for the panel, the sensory analyst can use presentations, questionnaires, and personal contacts to reach potential panel members. The time commitment and qualifications must be clearly iterated so that candidates understand what is required of them. General requirements include: interest in the test program, availability (about 80% of the time), promptness, general good health (no allergies or health problems affecting participation), articulateness, and absence of aversions to the product class involved. Other specific criteria are listed for individual tasks in Sections III and IV of this chapter.

III. SELECTION AND TRAINING FOR DIFFERENCE TESTS

A. Selection

Assume that the early recruitment procedure has provided a group of candidates free of obvious drawbacks, such as heavy travel or work schedules or health problems, which would make participation impossible or sporadic. The sensory analyst must now devise a set of screening tests which teach the candidates the test process while weeding out unsuitable nondiscriminators as early as possible. Such screening tests should *use the products to be studied* and *the sensory methods to be used in the study.* It follows that they should be patterned on those described below, rather than using them directly. The screening tests aim to determine differences among candidates in the ability to: (1) discriminate (and describe, if attribute difference tests are to be used) character differences among products, and (2) discriminate (and describe with a scale for attribute difference tests) differences in the intensity or strength of the characteristic.

Suggested rules for evaluating the results are given at the end of each section. Bear in mind that while candidates with high success rates may on the whole be satisfactory, the best panel will result if selection can be based on potential rather than on current performance.

TABLE 9.1
Suggested Samples for Matching Tests

Tastes, Chemical Feeling Factors

Flavor	Stimulus	Concentration (g/L[a])
Sweet	Sucrose	20
Sour	Tartaric acid	0.5
Bitter	Caffeine	1.0
Salty	Sodium chloride	2.0
Astringent	Alum	10

Aroma, Fragrances, Odorants[b]

Aroma descriptors	Stimulus
Peppermint, minty	Peppermint oil
Anise, anethole. licorice	Anise oil
Almond, cherry, Amaretto	Benzaldehyde, oil of bitter almond
Orange, orange peel	Orange oil
Floral	Linalool
Ginger	Ginger oil
Jasmine	Jasmine-74-D-10%
Green	cis-3-Hexenol
Vanilla	Vanilla extract
Cinnamon	Cinnamaldehyde, cassia oil
Clove, dentist's office	Eugenol, oil of clove
Wintergreen, BenGay	Methyl salicylate, oil of wintergreen

[a] In tasteless and odorless water at room temperature.
[b] Perfume blotters dipped in odorant, dried in hood 30 min, placed in widemouthed jar with tight cap.

---

1. Matching Tests

Matching tests are used to determine a candidate's ability to discriminate (and describe, if asked in addition) differences among several stimuli presented at intensities well above threshold level. Familiarize candidates with an initial set of four to six coded, but unidentified, products. Then present a randomly numbered set of eight to ten samples, of which a subset is identical to the initial set. Ask candidates to identify on the scoresheet the familiar samples in the second set and to label them with the corresponding codes from the first set.

Table 9.1 contains a selection of samples suitable for matching tests. These may be common flavor substances in water, common fragrances, lotions with different fat/oil systems, products made with pigments of different colors, fabrics of similar composition but differing in basis weight, etc. Care should be taken to avoid carryover effects, e.g., samples must not be too strong. Table 9.2 shows an example of a scoresheet for matching fragrances at above threshold levels in a nonodorous diluent.

2. Detection/Discrimination Tests

This type of selection test is used to determine a candidate's ability to detect differences among similar products with ingredient or processing variables. Present candidates with a series of three

---

TABLE 9.2
Scoresheet for Fragrance Matching Test

Instructions:	Sniff the first set of fragrances; allow time to rest after each sample. Sniff the second set of fragrances and determine which samples in the second set correspond to each sample in the first set. Write down the code of the fragrance in the second set next to its match from the first set.
Optional:	Determine which descriptor from the list below best describes the fragrance pair.

First set	Second set match	Descriptor[a]
079	_____	_____
318	_____	_____
992	_____	_____
467	_____	_____
134	_____	_____
723	_____	_____

[a] A list of descriptors, similar to the one given below, may be given at the bottom of the scoresheet. The ability to select and use descriptors should be determined if the candidates will be participating in attribute difference tests.

Floral	Peppermint	Vanilla	Wintergreen
Green	Cinnamon	Ginger	Clove
Jasmine	Orange	Cherry, almond	Anise/licorice

or more Triangle tests (Rainey, 1976; Zook and Wessman, 1977) with differences ranging from easy to moderately difficult (see, for example, Bressan and Behling, 1977). Duo-trio tests (Chapter 6.III, p. 68) may also be used. Table 9.3 lists some common flavor standards and the levels at which they may be used. "Doctored" samples, such as beers spiked (Meilgaard et al., 1982) with substances imitating common flavors and off-notes, may also be used. Arrange preliminary tests with experienced tasters to determine the optimal order of the test series and to control that stimulus levels are appropriate and detectable, but not overpowering. Use standard Triangle or Duo-trio

---

TABLE 9.3
Suggested Materials
for Detection Tests

Substance	Concentration (g/L[a])	
Caffeine	0.2[b]	0.4[c]
Tartaric acid	0.4[b]	0.8[c]
Sucrose	7.0b	14.0[c]
γ-Decalactone	0.002[b]	0.004[c]

[a] Amounts of substances added to tasteless and odorless water.
[b] 3 × threshold level.
[c] 6 × threshold level.

TABLE 9.4
Suggested Materials for Ranking/Rating Tests

		Sensory stimuli			
Taste					
Sour	Citric acid/water, g/L	0.25	0.5	1.0	1.5
Sweet	Sucrose/water, g/L	10	20	50	100
Bitter	Caffeine/water, g/L	0.3	0.6	1.3	2.6
Salty	Sodium chloride/water, g/L	1.0	2.0	5.0	10
Odor					
Alcoholic	3-Methylbutanol/water, mg/L	10	30	80	180
Texture					
Hardness	Cream cheese,[a] American cheese,[a] peanuts, carrot slices[a]				
Fracturability	Corn muffin,[a] Graham cracker, Finn crisp bread, Life Saver				

[a] At ½ in. thickness.

TABLE 9.5
**Scoresheet, Ranking Test
for Intensity**

Rank the salty taste solutions in the coded
cups in ascending order of saltiness.

	Code
Least salty	_____
	_____
	_____
Most salty	_____

TABLE 9.6
Scoresheet, Rating Test for Intensity

Rate the saltiness of each coded solution for intensity/strength of saltiness using the line scale for each.

Code

463	None_____Strong
318	None_____Strong
941	None_____Strong
502	None_____Strong

scoresheets when suitable. If desired, use sequential Triangle tests (Chapter 6, p. 94) to decide acceptance or rejection of candidates. However, as already mentioned, do not rely too much on taste acuity.

3. Ranking/Rating Tests for Intensity

These tests are used to determine candidates' ability to discriminate graded levels of intensity of a given attribute. Use rating on an appropriate scale if this is the method the test panelist will eventually use; otherwise use ranking (Chapter 7, p. 106). Present a series of samples in random order, in which one parameter is present at different levels, which cover the range present in the product(s) of interest. Ask candidates to rank the samples in ascending order (or rate them using the prescribed scale) according to the level of the stated attribute (sweetness, oiliness, stiffness, surface smoothness, etc.); see suggested materials in Table 9.4.

Typical scoresheets are shown in Tables 9.5 and 9.6. The selection sequence may make use of more than one attribute ranking/rating test, especially if the ultimate panel will need to cover several sense modalities, e.g., color, visual surface oiliness, stiffness, and surface smoothness.

4. Interpretation of Results of Screening Tests

Matching tests — Reject candidates scoring less than 75% correct matches. Reject candidates for attribute tests who score less than 60% in choosing the correct descriptor.

Detection/discrimination tests — When using Triangle tests, reject candidates scoring less than 60% on the "easy" tests (6 × threshold) or less than 40% on the "moderately difficult" tests (3 × threshold). When using Duo-trio tests, reject candidates scoring less than 75% on the "easy" tests or less than 60% on the "moderately difficult" tests. Or use the sequential tests procedure, as described in Chapter 6, p. 94.

Ranking/rating tests — Accept candidates ranking samples correctly or inverting only adjacent pairs. In the case of rating, use the same rank order criteria and expect candidates to use a large portion of the prescribed scale when the stimulus covers a wide range of intensity.

B. TRAINING

To ensure development of a professional attitude to sensory analysis on the part of panelists, conduct the training in a controlled professional sensory facility. Instruct subjects how to precondition the sensory modality in question, e.g., not to use perfumed cosmetics and to avoid exposure to foods or fragrances for 30 min before sessions; how to prepare skin or hands for fabric and skinfeel evaluations; and how to notify the panel leader of allergic reactions which affect the test modality. On any day, excuse subjects suffering from colds, headaches, lack of sleep, etc.

From the outset, teach subjects the correct procedures for handling the samples before and during evaluation. Stress the importance of adhering to the prescribed test procedures, reading all instructions and following them scrupulously. Demonstrate ways to eliminate or reduce sensory adaptation, e.g., taking shallow sniffs of fragrances and leaving several tens of seconds between sample evaluations. Stress the importance of disregarding personal preferences and concentrating on the detection of difference.

Begin by presenting samples of the product(s) under study which represent large, easily perceived sensory differences. Concentrate initially on helping panelists to understand the scope of the project and to gain confidence. Repeat the test method using somewhat smaller but still easily perceived sample differences. Allow the panel to learn through repetition until full confidence is achieved.

For attribute difference tests, carefully introduce panelists to the attributes, the terminology used to describe them, and the scale method used to indicate intensity. Present a range of products showing representative intensity differences for each attribute.

Continue to train "on the job" by using the new panelist in regular discrimination tests. Occasionally, introduce training samples to simulate "off-notes" or other key product differences in order to keep the panel on track and attentive.

Be aware of changes in attitude or behavior on the part of one or more panelists who may be confused, losing interest, or distracted by problems at work or outside. The history of sensory testing is full of incredible results, which could have come only from panelists who were "lost" during the test with the sensory analyst failing to anticipate and detect a failure in the "test instrument."

IV. SELECTION AND TRAINING OF PANELISTS FOR DESCRIPTIVE TESTING

A. SELECTION FOR DESCRIPTIVE TESTING

When selecting panelists for descriptive analysis, the panel leader or panel trainer needs to determine each candidate's capabilities in three major areas:

1. For each of the sensory properties under investigation (such as fragrance, flavor, oral texture, or skinfeel) the ability to detect differences in characteristics present and in their intensities
2. The ability to describe those characteristics using (a) verbal descriptors for the characteristics and (b) scaling methods for the differences in intensity
3. The capacity for abstract reasoning, as descriptive analysis depends heavily upon the use of references when characteristics must be quickly recalled and applied to other products

In addition to screening panelists for these descriptive capabilities, panel leaders must prescreen candidates for the following personal criteria:

1. Interest in full participation in the rigors of the training, practice, and ongoing work phases of a descriptive panel
2. Availability to participate in 80% or more of all phases of the panel's work; conflict with work load, travel, or even the candidate's supervisor perhaps eventually causing the panelist to drop off the panel during or after training, thus losing one panelist from an already small number of 10 to 15
3. General good health and no illnesses related to the sensory properties being measured, such as:
   a. Diabetes, hypoglycemia, hypertension, dentures, chronic colds or sinusitis, or food allergies in those candidates for flavor and/or texture analysis of foods, beverages, pharmaceuticals, or other products for internal use
   b. Chronic colds or sinusitis, for aroma analysis of foods, fragrances, beverages, personal care products, pharmaceuticals, or household products
   c. Central nervous system disorders or reduced nerve sensitivity due to the use of drugs affecting the central nervous system, for tactile analysis of personal care skin products, fabrics, or household products

The ability to detect and describe differences, the ability to apply abstract concepts, and the degree of positive attitude and predilection for the tasks of descriptive analysis can all be determined through a series of tests which include

- A set of prescreening questionnaires
- A set of acuity tests
- A set of ranking/rating tests
- A personal interview

The investment in a descriptive panel is large in terms of time and human resources, and it is wise to conduct an exhaustive screening process, rather than training unqualified subjects.

Lists of screening criteria for three descriptive methods (the Flavor Profile, Quantitative Descriptive Analysis, and Texture Profile) can be found in ASTM Special Technical Publication 758 (1981). The following criteria listed are those used to select subjects for training in the Spectrum method of descriptive analysis, as described in Chapter 11. These can be applied to the screening of employees, or for external screening in cases where recruiting from the local community is preferred because of time-consuming panels (20 to 50 h per person per week). The additional prescreening questionnaires are used to select individuals who can verbalize and think in concepts. This reduces the risk of selecting outside panelists who have sensory acuity but cannot acquire the "technical" orientation of panels recruited from inside the company.

1. Prescreening Questionnaires

For a panel of 15, typically 40 to 50 candidates may be prescreened using questionnaires such as those shown in Appendix 9.1. Appendix 9.1A applies to a tactile panel (skinfeel or fabric feel); Appendix 9.1B to a flavor panel; Appendix 9.1C to an oral texture panel; and Appendix 9.1D to a fragrance panel. Appendix 9.1E tests the candidate's potential to learn scaling and can be used with any of the preceding questionnaires in Appendix 9.1. Of the 40 to 50 original candidates, generally 20 to 30 qualify and proceed to the acuity tests.

2. Acuity Tests

To qualify for this stage, candidates should:

- Indicate no medical or pharmaceutical causes of limited perception
- Be available for the training sessions
- Answer 80% of the verbal questions in the prescreening questionnaires in Appendices 9.1A through 9.1D correctly and clearly
- In the questionnaire Appendix 9.1E, assign scalar ratings which are within 10 to 20% of the correct value for all figures

Candidates should demonstrate ability to:

- Detect and describe characteristics present in a qualitative sense
- Detect and describe intensity differences in a quantitative sense

Therefore, although detection tests (e.g., Triangle or Duo-trio tests using variations in formulation or processing of the product to be evaluated) may yield a group of subjects who can detect small product variables, detection alone is not enough for a descriptive panelist. To qualify, subjects must be able to adequately discriminate and describe some key sensory attributes within the modalities used with the product class under test, and also must show ability to use a rating scale correctly to describe differences in intensity.

Detection — The panel trainer presents a series of samples representing key variables within the product class, in the form of Triangle or Duo-trio tests (Zook and Wessmann, 1977). Differences in process time or temperature (roast, bake, etc.), ingredient level (50%, 150% of normal), or packaging can be used as sample pairs to determine acuity in detection. Attempt to present the easier pairs of samples first and follow with pairs of increasing difficulty. Select subjects who achieve 50 to 60% correct replies in Triangle tests, or 70 to 80% in Duo-trio tests, depending on the degree of difficulty of each test.

TABLE 9.7
Scoresheet Containing Two Ranking Tests Used to Screen Candidates for a Texture Panel

Descriptive texture panel screening

1. Place one piece of each product between *molars;* bite through *once;* evaluate for hardness
   Rank the samples from least hard to most hard.

   Least hard   _____

                _____

                _____

                _____

   Most hard   _____

2. Place one piece of each product between molars; bite down once and evaluate for crispness (crunchiness).

   Least crisp _____

                _____

                _____

                _____

   Most crisp   _____

Description — Present a series of products showing distinct attribute characteristics (fragrance/flavor oils, geometrical texture properties [Civille and Szczesniak, 1975]) and ask candidates to describe the sensory impression. Use the fragrance list in Table 9.1 without a list of descriptors from which to choose. The candidate must describe each fragrance using his/her own words. These may include chemical terms (e.g., cinnamic aldehyde), common flavor terms (e.g., cinnamon), or related terms (e.g., like Red Hots candy, Big Red gum, and Dentyne). Candidates should be able to describe 80% of the stimuli using chemical, common, or related terms and should at least attempt to describe the remainder with less specific terms (e.g., sweet, brown spice, hot spice).

3. Ranking/Rating Screening Tests for Descriptive Analysis

Having passed the prescreening tests and acuity tests, the candidate is ready for screening with the actual product class and/or sensory attribute for which the panel is being selected. A good example for a Camembert cheese panel is given by Issanchou et al. (1995). Candidates should rank or rate a number of products on a selection of key attributes, using the technique of the future panel. These tests can be supplemented with a series of samples which demonstrate increasing intensity of certain attributes, such as tastes and odors (see Table 9.4), or oral texture properties (Appendix 11.2, Texture Section D, Scale 5 is suitable, containing hardness standards from cream cheese = 1.0 to hard candy = 14.5; also Scale 10, which contains standards for crispness from Granola Bar at 2.0 to cornflakes at 14.0). A questionnaire such as Table 9.7 is suitable. For certain skinfeel and fabric feel properties, use Appendix 11.2E or 11.2F, or reference samples may need to be selected from among commercial products and laboratory prototypes, representing increasing intensity levels of selected attributes. Choose candidates who can rate all samples in the correct order for 80% of the attributes scaled. Allow for reversal of adjacent samples only, and check that candidates use most of the scale for at least 50% of the attributes tested.

4. Personal Interview

Especially for descriptive panels, a personal interview is necessary in order to determine whether candidates are well suited to the group dynamics and analytical approach. Generally, candidates who have passed the prescreening questionnaire and all of the acuity tests are interviewed individually by the panel trainer or panel leader. The objective of the interview is to confirm the candidate's interest in the training and work phases of the panel including his/her availability with respect to work load, supervisor, and travel, and also communication skills and general personality. Candidates who express little interest in the sensory programs as a whole and in the descriptive panel in particular should be excused. Individuals with very hostile or very timid personalities may also be excluded, as they may detract from the needed positive input of each panelist.

B. TRAINING FOR DESCRIPTIVE TESTING

The important aspect of any training sequence is to provide a structured framework for learning based on demonstrated facts and to allow the students, in this case panelists, to grow in both skills and confidence. Most descriptive panel training programs require between 40 and 120 h of training. The amount of time needed depends on the complexity of the product (wine, beer, and coffee panels require far more time than those evaluating powdered drink mixes or breakfast cereals), on the number of attributes to be covered (a short-version descriptive technique for quality control or storage studies, Chapter 11, p. 175, requires fewer and simpler attributes), and on the requirements for validity and reliability (a more experienced panel will provide greater detail with greater reproducibility).

1. Terminology Development

The panel leader or panel trainer in conjunction with the project team must identify key product variables to be demonstrated to the panel during the initial stages of training. The project team should prepare a prototype or collect from commercially available samples an array of products as a frame of reference, which represents as many as possible of the attribute differences likely to be encountered in the product category. The panel is first introduced to the chemical (olfaction, taste, chemical feeling factors) and physical principles (rheological, geometrical, etc.) which govern or influence the perception of each product attribute. With these concepts and terms as a foundation, the panel then develops procedures for evaluation and terminology with definitions and references for the product class.

Examples of this process are discussed by Szczesniak (1963) for oral texture, Schwartz (1975) and Civille and Dus (1991) for skincare products, McDaniel et al. (1987) for wines, Meilgaard and Muller (1987) for beer, Lyon (1987) for chicken, Johnsen et al. (1988) for peanuts, Johnsen and Civille (1986) for beef, and Johnsen et al. (1987) for catfish. Typically, the first stage of training may require 15 to 20 h as panelists begin to develop an understanding of the broad array of descriptors which fall into the category being studied (appearance, flavor, oral texture, etc.). This first phase is designed to provide them with a firm background in the underlying modality and for them to begin to perceive the different characteristics as they are manifest in different product types.

2. Introduction to Descriptive Scaling

The scaling method of choice may be introduced during the first 10 to 20 h of training. By using a set of products or references which represent 3 to 5 different levels of each attribute, the panel leader reinforces both the sensory characteristic and the scaling method, by demonstrating different levels or intensities across several attributes. Appendix 11.2, p. 195, provides examples of different intensity levels of several sensory attributes for several sensory descriptive categories: flavor (aromatics, tastes, feeling factors), solid and semisolid texture (Muñoz, 1986) (hardness, adhesiveness, springiness, etc.), skinfeel (ASTM, 1997; Civille and Dus, 1991) (wetness, slipperiness, oiliness, etc.), and fabric feel (Civille and Dus, 1990) (slipperiness, grittiness, fuzziness, etc.).

The continued use of intensity reference scales during practice is meant to provide continued reinforcement of both attributes and intensities, so that the panel begins to see the descriptive process as a use of terms and numbers (characteristics and intensities) to define or document any product in the category learned.

3. Initial Practice

The development of a precise lexicon for a given product category is often a three-step process. In the first step a full array of products, prototypes, or examples of product characteristics is presented to the panel as a frame of reference. From this frame of reference the panel generates an original long list of descriptors to which all panelists are invited to contribute. In the second stage, the original list, containing many overlapping terms, is rearranged and reduced into a working list in which the descriptors are comprehensive (they describe the product category completely) and yet discrete (overlapping is minimized). The third and last stage consists of choosing products, prototypes, and external references which can serve to represent good examples of the selected terms.

Once the panel has a grasp on the terminology and a general understanding of the use of each scale, the panel trainer or leader presents a series of samples to be evaluated one at a time, generally two or more of which represent a *very* wide spread in qualitative (attributes) and quantitative (intensity) differences. At this early stage of development, which lasts 15 to 40 h, the panel gains basic skills and confidence. The disparate samples allow the panel to see that the terms and scales are effective as descriptors and discriminators, and help the members to gain confidence both as individuals and as a group.

4. Small Product Differences

With the help of the project/product team, the panel leader collects samples which represent smaller differences within the product class, including variations in production variables and/or bench modifications of the product. The panel is encouraged to refine the procedures for evaluation and terminology with definitions and references to meet the needs of detecting and describing product differences. Care must be taken to reduce variations between supposedly identical samples; panelists in training tend to see variability in results as a reflection of their own lack of skill. Sample consistency contributes to panel confidence. This stage represents 10 to 15 h of panel time.

5. Final Practice

The panel should continue to test and describe several products during the final practice stage of training (15 to 40 h). The earlier samples should be fairly different, and the final products tested should approach the real world testing situations for which the panel will be used.

During all five stages of the training program, panelists should meet after each session and discuss results, resolve problems or controversies, and ask for additional qualitative or quantitative references for review. This type of interaction is essential for developing the common terminology, procedures for evaluation, and scaling techniques which characterize a finely tuned sensory instrument.

V. PANEL PERFORMANCE AND MOTIVATION

Any good measuring tool needs to be checked regularly to determine its ability to perform validly and consistently. In the case of a sensory panel, the individuals need to be monitored, as well as the panel as a whole. Panels are comprised of human subjects, who have other jobs and responsibilities in addition to their participation in the sensory program, and it is necessary to find ways to maintain the panelists' interest and motivation over long periods of product testing.

A. PERFORMANCE

For both difference and descriptive panels, the sensory analyst needs to have a measure of the performance of each panelist and of the panel, in terms of validity and reproducibility. Validity is the correctness of the response. In certain difference tests, such as the Triangle and Duo-trio, and in some directional attribute tests, the analyst knows the correct answer (the odd sample, the coded reference, the sweeter sample) and can assess the number of correct responses over time. The percent of correct responses can be computed for each panelist on a regular monthly or bimonthly basis. Weighted scores can also be calculated, based on the difficulty of each test in which the panelist participated (Aust, 1984). For the panel as a whole, validity can be measured by comparing panel results to other sensory test data, instrumental data, or the known variation in the stimulus, such as increased heat treatment, addition of a chemical, etc.

Reliability, or the ability to reproduce results, can be easily assessed for the individual panelists and for the panel as a whole by replicating the test, using duplicate test samples, or using blind controls.

For descriptive data, which are analyzed statistically by the analysis of variance, the panelists' performance can be assessed across each attribute as part of the data analysis (see ASTM STP 758 [1981], pp. 29–30 or Lea, Næs, and Rødbotten [1997] for a detailed description of this analysis applied to a set of QDA results). It is recognized and accepted in QDA that panelists will use different parts of the scale to express their perceptions of the same sample. It is the relative differences in their ratings and not their absolute values that are considered important. In other descriptive methods, such as Spectrum, panelists are calibrated through the use of references to use the same part of the scale when evaluating the same sample. A descriptive panel of this type is equivalent to an analytical instrument, which requires regular calibration checks. Several approaches, in addition to the ASTM guideline just mentioned, are appropriate for monitoring the individual and combined performance of "calibrated" panelists. Two aspects of performance which require monitoring are the panel's accuracy (bias) and precision (variability) of the panel.

Bias — To assess a panelist's ability to be "on target," the panel leader can determine the panelist's ability to match the accepted intensity of the attributes of a control or reference. The statistical measure of difference from the target or control rating, called bias, is defined as:

$$\text{panelist bias, } d = x - \mu \tag{9.1}$$

where d is the deviation or bias, x is the observed panelist value, and μ is the value for the control or target attribute.

Variability — With several evaluations of a blind control or reference, the panelist's variability about *his/her own mean rating* is calculated using the panelist's standard deviation as follows:

$$\text{panelist SD, } s = \sqrt{\sum_{i=1}^{n} \left(x_i - \bar{x}\right)^2 / (n-1)} \tag{9.2}$$

Good panelists have both low bias and low variability. The bias formula may be modified by removing the sign; this produces the *absolute bias*, calculated as:

$$\text{panelist bias, } |d| = |x - \mu| \tag{9.3}$$

so that large positive and negative deviations do not offset each other. Small values of absolute bias are desirable. The panelists' statistics should be plotted over time to identify those panelists who need retraining or calibration.

When split-plot analysis of variance is used for descriptive data analysis, the judge-by-sample interaction is part of the results. When this interaction is significant, it is necessary to look at plots of the data to determine the source(s). Figure 9.1 shows three plots of judge-by-sample interactions. In each graph, each line represents *one* panelist's average ratings for two samples. In the first plot (A), the judge-by-sample interaction is not significant. All judges tend to rate the samples in the same direction and with the same relative degrees of intensity. Thus the lines are in the same direction and similar in slope. The second plot (B) shows an extreme case of judge-by-sample interaction: several samples are rated quite differently by some of the judges. Thus the lines run in different directions and have different slopes. The third plot (C) shows a few judges whose slopes differ from the rest. In this case, although the judge-by-sample interaction is statistically significant, the problem is less extreme. It is one of slight differences in scale use rather than total reversals, as in plot B. Generally a judge-by-sample interaction indicates the need for more training, more frequent use of reference scales, or review of terminology.

B. Panelist Maintenance, Feedback, Rewards, and Motivation

One of the major sources of motivation for panelists is a sense of doing meaningful work. After a project is completed, panelists should be informed by letter or a posted circular of the project and test objectives, the test results, and the contribution made by the sensory results to the decision taken regarding the product. Immediate feedback after each test also tends to give the individual panelist a sense of "How am I doing?" The fears of some project leaders that panelists might become discouraged in tests with a low probability of success (a Triangle test often has fewer than 50% correct responses) have proven groundless. Panelists do take into account the complexity of the sample, the difficulty of the test, and the probability of success. Panelists do want to know about the test, and can indeed learn from past performance. Discussion of results after a descriptive panel session is highly recommended. The need to constantly refine the terms, procedures, and definitions is best served by regular panel interaction after all the data have been collected.

Feedback to panelists on performance can be provided with data regarding their individual performance over 3 to 5 repeat evaluations of the same product vis-à-vis the panel as a whole. The data in Table 9.8 for a given sample indicates the mean and standard deviation for each panelist (numbers) for each attribute (letters) as well as the panel mean and standard deviation. Panelists

TABLE 9.8
Panel Performance Summary

| Attributes | Panelist | | | | | | | Panel |
	1	2	3	4	5	...	14	$\bar{X}$/SD
A	7.5/02[a]	7.0/2	6.8/2	6.9/1	7.9/2.5		6.2/1.9	6.9/05[b]
B	4.2/1.4	4.8/2	5.5/1.6	5.0/0	4.2/1.2		4.6/1.6	4.8/0.4
C	1.4/1	3/1.3	1.5/1.2	1.0/0.9	1.1/0.8		3/1.3	1.8/0.8
D	9.0/0.5	8.0/0.7	9.0/1.0	6.4/1.2	12/1.1		10/1.3	9.4/1.6
E	4.0/0.7	4.2/0.8	3.5/1	1.9/1.2	4.4/0.9		3.8/2	3.9/1.1

[a] Panelist mean/standard deviation.
[b] Panelist grand mean/grand standard deviation.

The 14 panelists evaluated the same sample in between other samples over a period of 3 weeks. The panel grand mean for Attribute A was 6.9 and the SD over the 14 panelist means was 0.5 or 7.2%, showing satisfactory agreement between panelists for this attribute. Panelist 5 rated the attributes A and E much higher than the panel means and showed a high SD for Attribute A.

can then determine how well the individual means agree with that of the panel as a whole (bias). In addition, the panelist's standard deviation provides an indication of that panelist's reliability (variability) on that attribute. Data for two or three products or samples over 3 to 5 evaluations should be shown to panelists on a regular basis, e.g., every 3 to 4 months. Plots of judge-by-sample interaction, such as those shown in Figure 9.1, may also be shown to panelists to demonstrate both the general agreement among all the panelists and the performance of each panelist relative to the others.

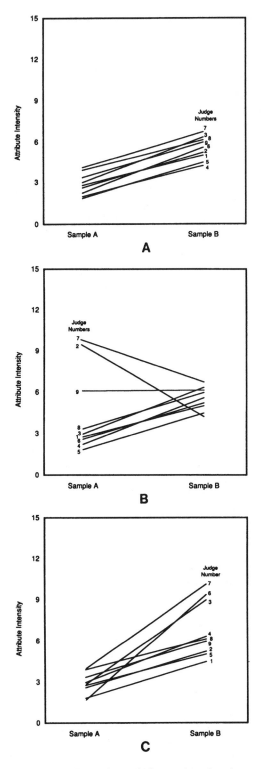

FIGURE 9.1   Judge and sample interaction plots (see text).

In addition to the psychological rewards derived from feedback, panelists also respond positively and are further motivated to participate enthusiastically by a recognition and/or reward system. The presentation of certificates of achievement for:

- High panel attendance
- High panel performance
- Improved performance
- Completion of a training program
- Completion of a special project

stimulates panel performance and communicates to panelists that the evaluation is recognized as worthwhile. Short-term rewards, such as snacks, tokens for company products, and raffle tickets for larger prizes, are often given daily to subjects. Over the longer term, sensory analysts often sponsor parties, outings, luncheons, or dinners for panelists, if possible, with talks by project or company management describing how the results were used. Publicity for panel work in the company newspaper or the local community media serves to recognize the current panel members and stimulates inquiry from potential candidates. The underlying support by management for the full sensory program and for the active participation by panelists is a key factor in recruiting and maintaining an active pool of highly qualified members.

REFERENCES

ASTM, 1981. Committee E-18, *Guidelines for the Selection and Training of Sensory Panel Members,* ASTM Special Technical Publication 758, ASTM, Philadelphia.

ASTM, 1997. *E1490-92 Standard Practice for Descriptive Skinfeel Analysis of Creams and Lotions.* Available from ASTM, 100 Barr Harbor Dr., West Conshohocken, PA 19428, or from www.astm.org.

Aust, L.B., 1984. Computers as an aid in discrimination testing. *Food Technol.* 38(9), 71.

Bressan, L.P. and Behling, R.W., 1977. The selection and training of judges for discrimination testing. *Food Technol.* 31(11), 62.

Civille, G.V. and Dus, C.A., 1990. Development of terminology to describe the handfeel properties of paper and fabrics. *J. Sensory Stud.* 5, 19–32.

Civille, G.V. and Dus, C.A., 1991. Evaluating tactile properties of skincare products: a descriptive analysis technique. *Cosmetics and Toiletries* 106(5), 83.

Civille, G.V. and Szczesniak, A.S., 1975. Guide to training a texture panel. *J. Texture Stud.* 6, 19.

ISO, 1991. *Sensory Analysis — Methodology — Method of investigating sensitivity of taste.* International Organization for Standardization, ISO 3972:1979. Available from American National Standards Institute, 11 West 42nd St., New York, NY 10036, or from ISO, 1 rue Varembé, CH 1211 Génève 20, Switzerland.

ISO, 1993. *Sensory Analysis — General guidance for the selection, training, and monitoring of assessors — Part I: Selected assessors.* International Organization for Standardization, ISO 8586-1:1993. Available from American National Standards Institute, 11 West 42nd St., New York, NY 10036, or from ISO, 1 rue Varembé, CH 1211 Génève 20, Switzerland.

Issanchou, S., Lesschaeve, I., and Köster, E.P., 1995. Screening individual ability to perform descriptive analysis of food products: Basic statements and application to a Camembert cheese descriptive panel. *J. Sensory Stud.* 10, 349–368.

Johnsen, P.B. and Civille, G.V., 1986. A standardized lexicon of meat W.O.F. descriptors. *J. Sensory Stud.* 1(1), 99.

Johnsen, P.B., Civille, G.V., Vercellotti, J.R., 1987. A lexicon of pond-raised catfish flavor descriptors. *J. Sensory Stud.* 2(2), 85.

Johnsen, P.B., Civille, G.V., Vercellotti, J.R., Sanders, T.H., and Dus, C.A., 1988. Development of a lexicon for the description of peanut flavor. *J. Sensory Stud.* 3(1), 9.

Lea, P., Næs, T., and Rødbotten, M., 1997. *Analysis of Variance for Sensory Data.* John Wiley & Sons, Chichester, pp. 34–36.

Lyon, B.G., 1987. Development of chicken flavor descriptive attribute terms by multivariate statistical procedures. *J. Sensory Stud.* 2(1), 55.

McDaniel, M., Henderson, L.A., Watson, B.T., Jr., and Heatherbill, D., 1987. Sensory panel training and screening for descriptive analysis of the aroma of pinot noir wine fermented by several strains of malolactic bacteria. *J. Sensory Stud.* 2(3), 149.

Meilgaard, M.C. and Muller, J.E., 1987. Progress in descriptive analysis of beer and brewing products. *Tech. Q. Master Brew. Assoc. Am.* 24(3), 79–85.

Meilgaard, M.C., Reid, D.S., and Wyborski, K.A., 1982. Reference standards for beer flavor terminology system. *J. Am. Soc. Brewing Chem.* 40, 119–128.

Muñoz, A., 1986. Development and application of texture reference scales. *J. Sensory Stud.* 1(1), 55.

Rainey, B., 1979. *Selection and Training of Panelists for Sensory Panels.* IFT Shortcourse: Sensory Evaluation Methods for the Practicing Food Technologist, St. Louis, MO, Atlanta, GA, Boston, MA, and Portland, OR.

Schwartz, N., 1975. Adaptation of the Sensory Texture Profile methods to skin care products. *J. Texture Stud.* 6, 33.

Zook, K. and Wesmann, C., 1977. The selection and use of judges for descriptive panels. *Food Technol.* 31(11), 56.

Appendix 9.1
Prescreening Questionnaires

Each of the prescreening questionnaires is designed to enable the panel leader or trainer to select from a large group of candidates those individuals who are both verbal with respect to sensory properties to be evaluated and capable of expressing perceived amounts. For each type of panel to be trained (tactile, flavor, oral, texture, or fragrance) use the prescreener for that category *plus* the Scaling Exercises in Appendix 1E.

A. Prescreening Questionnaire for a Tactile Panel (Skinfeel or Fabric Feel)

HISTORY:

Name: _____

Address: _____

Phone (home and business): _____

From what group or organization did you hear about this program?

_____

TIME:

1. Are there any weekdays (M–F) that you will not be available on a regular basis?

   _____

2. How many weeks vacation do you plan to take between June 1 and September 30?

   _____

HEALTH:

1. Do you have any of the following?
   Central nervous system disorder   _____
   Unusually cold or warm hands   _____
   Skin rashes   _____
   Calluses on hands/fingers   _____
   Hypersensitive skin   _____
   Tingling in the fingers   _____

2. Do you take any medications which affect your senses, especially touch?

   _____

GENERAL:

1. Is your sense of touch: (check one)
   Worse than average   _____
   Average   _____
   Better than average   _____
2. Does anyone in your immediate family work for a paper, fiber, or textile company?

   _____

   A marketing research or advertising company? _____

TACTILE/TOUCH QUIZ:

1. What characteristics of feel of a towel make you think it is absorbent?

   _____

2. What is thicker, an oily or greasy film? _____

3. When you rub an oily film on your skin, how do your fingers move?
   slip _____ or drag _____ (check one)

4. What feel properties in a tissue do you associate with its softness?

   _____

5. What specific appearance characteristics of a bath tissue influence your perception of the feel
   of it? _____

   _____

6. Name some things that are sticky. _____

   _____

7. When your skin feels moist, what other words or properties could describe it?

   _____

8. Name some things that are rough. _____

   What makes them rough? _____

9. Briefly, how would you define "fullness"? _____

   _____

10. What do you feel in a fabric or paper product that makes it feel stiff?

    _____

11. What other words would you use to describe a lotion as thin or thick?

    _____

12. What characteristics do you feel when you stroke the surface of a fabric?

    _____

    The back of your hand?_____

B. Prescreening Questionnaire for a Flavor Panel

HISTORY:

Name: _____

Address: _____

Phone (home and business): _____

From what group or organization did you hear about this program?

_____

TIME:

1. Are there any weekdays (M–F) that you will not be available on a regular basis?

   _____

2. How many weeks vacation do you plan to take between June 1 and September 30?

   _____

HEALTH:

1. Do you have any of the following?
   - Dentures   _____
   - Diabetes   _____
   - Oral or gum disease   _____
   - Hypoglycemia   _____
   - Food allergies   _____
   - Hypertension   _____
2. Do you take any medications which affect your senses, especially taste and smell?

   _____

FOOD HABITS:

1. Are you currently on a restricted diet? If yes, explain.

   _____

2. How often do you eat out in a month? _____

3. How often do you eat fast foods out in a month? _____

4. How often in a month do you eat a complete frozen meal? _____

5. What is (are) your favorite food(s)? _____

   _____

6. What is (are) your least favorite food(s)? _____

   _____

7. What foods can you not eat? _____

   _____

8. What foods do you not like to eat? _____

   _____

9. Is your ability to distinguish smell and tastes

	SMELL	TASTE
Better than average	_____	_____
Average	_____	_____
Worse than average	_____	_____

10. Does anyone in your immediate family work for a food company? _____

    _____

11. Does anyone in your immediate family work for an advertising company or a marketing research agency? _____

    _____

FLAVOR QUIZ:

1. If a recipe calls for thyme and there is none available, what would you substitute?

   _____

2. What are some other foods that taste like yogurt? _____

   _____

3. Why is it that people often suggest adding coffee to gravy to enrich it?

   _____

4. How would you describe the difference between flavor and aroma? _____

   _____

5. How would you describe the difference between flavor and texture? _____

   _____

6. What is the best one- or two-word description of grated Italian cheese
   (Parmesan or Romano)? _____

7. Describe some of the noticeable flavors in mayonnaise. _____

   _____

8. Describe some of the noticeable flavors in cola. _____

   _____

9. Describe some of the noticeable flavors in sausage. _____

   _____

10. Describe some of the noticeable flavors in Ritz crackers. _____

   _____

C. Prescreening Questionnaire for an Oral Texture Panel

HISTORY:

Name: _____

Address: _____

Phone (home and business): _____

From what group or organization did you hear about this program?

_____

TIME:

1. Are there any weekdays (M–F) that you will not be available on a regular basis?

   _____

2. How many weeks vacation do you plan to take between June 1 and September 30?

   _____

HEALTH:

1. Do you have any of the following?
   Dentures   _____
   Diabetes   _____
   Oral or gum disease   _____
   Hypoglycemia   _____
   Food allergies   _____
   Hypertension   _____
2. Do you take any medications which affect your senses, especially taste and smell?

   _____

FOOD HABITS:

1. Are you currently on a restricted diet? If yes, explain.

   _____

2. How often do you eat out in a month? _____

3. How often do you eat fast foods out in a month? _____

4. How often in a month do you eat a complete frozen meal? _____

5. What is (are) your favorite food(s)? _____

   _____

6. What is (are) your least favorite food(s)? _____

   _____

7. What foods can you not eat? _____

   _____

8. What foods do you not like to eat? _____

   _____

9. Is your sensitivity to textural characteristics in foods

    Better than average   _____
    Average   _____
    Worse than average   _____

10. Does anyone in your immediate family work for a food company? _____

    _____

11. Does anyone in your immediate family work for an advertising company or a marketing research agency? _____

    _____

TEXTURE QUIZ:

1. How would you describe the difference between flavor and texture? _____

    _____

2. Describe some of the textural properties of foods in general. _____

    _____

3. Describe some of the particles one finds in foods. _____

    _____

4. Describe some of the properties which are apparent when one chews on a food.

    _____

5. Describe the differences between crispy and crunchy. _____

    _____

6. What are some textural properties of potato chips? _____

    _____

7. What are some textural properties of peanut butter? _____

    _____

8. What are some textural properties of oatmeal? _____

    _____

9. What are some textural properties of bread? _____

    _____

10. For what type of products is texture important? _____

    _____

D. Prescreening Questionnaire for a Fragrance Panel

HISTORY:

Name: _____

Address: _____

Phone (home and business): _____

From what group or organization did you hear about this program?
_____

TIME:

1. Are there any weekdays (M–F) that you will not be available on a regular basis?
    _____

2. How many weeks vacation do you plan to take between June 1 and September 30?
    _____

HEALTH:

1. Do you have any of the following?
    Nasal disease   _____
    Hypoglycemia   _____
    Allergies   _____
    Frequent colds or sinus condition _____

2. Do you take any medications which affect your senses, especially smell?
    _____

DAILY LIVING HABITS:

1. a. Do you regularly wear a fragrance or an after-shave/cologne? _____

    b. If yes, what brands? _____

2. a. Do you prefer perfumed or nonperfumed soap, detergents, fabric softeners, etc.?
    _____

    b. Why? _____

3. What are some fragranced products that you like? Types or brands _____
    _____

4. What are some fragranced products that you dislike? Types or brands _____
    _____

5. a. Name some odors that make you feel ill. _____
    _____

    b. In what way do you feel ill from them? _____

6. What odors, smells, or fragrances are most appealing to you? _____
    _____

7. Is your ability to distinguish odors?

   Better than average _____ average _____ worse than average _____

8. Does anyone in your immediate family work for a soap, food, or personal products company or an advertising agency? _____

   If so, which one(s)? _____

9. Members of the trained panel should not use heavy perfumes/colognes on evaluation days, nor should they smoke an hour before the panel meets. Would you be willing to do the above if you are chosen as a panelist? _____

FRAGRANCE QUIZ:

1. If a perfume is "floral" in type, what other words could be used to describe it?

   _____

2. What are some products that have an herbal smell? _____

   _____

3. What are some products that have a sweet smell? _____

   _____

4. What types of odors are associated with clean and fresh? _____

   _____

5. How would you describe the difference between fruity and lemony? _____

   _____

6. Briefly, what words would you use to describe the difference between a feminine fragrance and a masculine fragrance? _____

   _____

7. What are some words which would describe the smell of a hamper full of clothes?

   _____

8. Describe some of the noticeable smells in a bakery. _____

   _____

9. Describe some of the noticeable smells in a liquid dish detergent. _____

   _____

10. Describe some of the noticeable smells in bar soaps. _____

    _____

11. Describe some of the noticeable smells in a basement. _____

    _____

12. Describe some of the noticeable smells in a McDonald's restaurant. _____

    _____

E. SCALING EXERCISES

(To be included with each of the prescreening questionnaires)

INSTRUCTIONS: MARK ON THE LINE AT THE RIGHT TO INDICATE THE PROPORTION OF THE AREA THAT IS SHADED.

EXAMPLES.

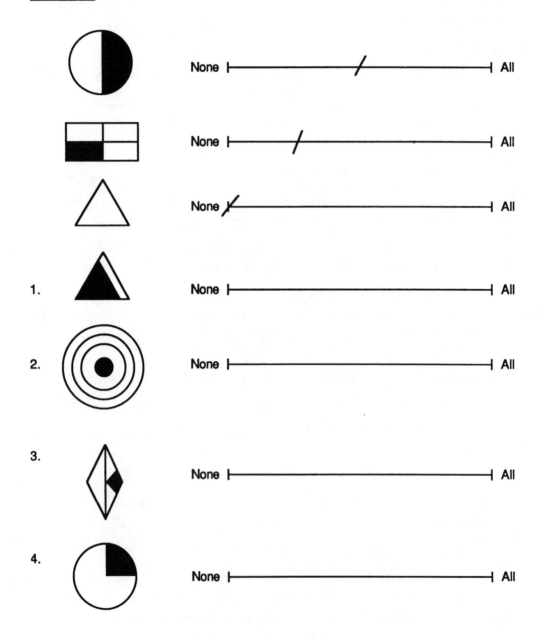

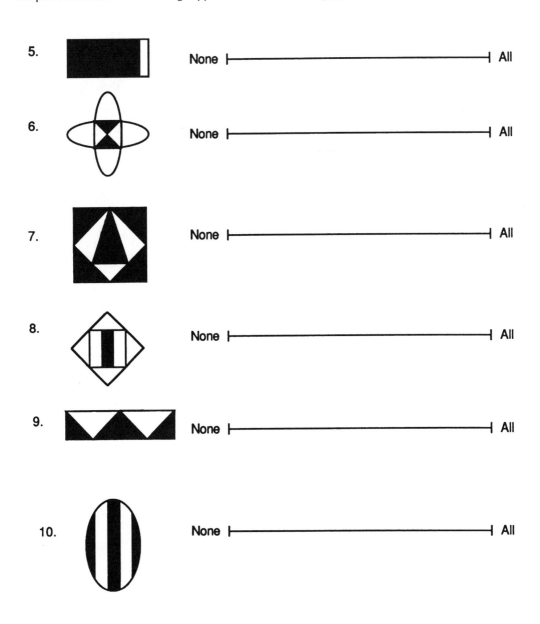

Prescreening questionnaire: scaling exercise. The answers are:

1. 7/8	6. 1/8
2. 1/8	7. 3/4
3. 1/6	8. 1/8
4. 1/4	9. 1/2
5. 7/8	10. 1/2

10 Descriptive Analysis Techniques

CONTENTS

I. DEFINITION

All descriptive analysis methods involve the detection (discrimination) and the description of both the qualitative and quantitative sensory aspects of a product by trained panels of 5 to 100 judges (subjects). Smaller panels of five to ten subjects are used for the typical product on the grocery shelf while the larger panels are used for products of mass production such as beers and soft drinks, where small differences can be very important.

Panelists must be able to detect and describe the perceived sensory attributes of a sample. These qualitative aspects of a product combine to define the product, and include all of the appearance, aroma, flavor, texture or sound properties of a product which differentiate it from others. In addition, panelists must learn to differentiate and rate the quantitative or intensity aspects of a sample and to define to what degree each characteristic or qualitative note is present in that sample. Two products may contain the same qualitative descriptors, but they may differ markedly in the intensity of each, thus resulting in quite different and easily distinctive sensory profiles or pictures of each product. The two samples below have the same qualitative descriptors, but they differ substantially in the amount of each characteristic (quantitatively). The numbers used represent intensity ratings on a 15-cm line scale on which a zero means no detectable amount of the attribute, and 15 cm means a very large amount (Civille, 1979).

The two samples (385 and 408) below are commercially available potato chips.

Characteristic	385	408
Fried potato	7.5	4.8
Raw potato	1.1	3.7
Vegetable oil	3.6	1.1
Salty	6.2	13.5
Sweet	2.2	1.0

Although these two samples of chips have the same attribute descriptors, they differ markedly by virtue of the amount of each flavor note. Sample 385 has distinct fried potato character with underlying oil, sweet, and raw potato notes. Sample 408 is dominated by saltiness with the potato, oil, and sweet notes of lesser impact.

II. FIELD OF APPLICATION

Use descriptive tests to obtain detailed description of the aroma, flavor, and oral texture of foods and beverages, skinfeel of personal care products, handfeel of fabrics and paper products, and the appearance and sound of any product. These sensory pictures are used in research and development (Meilgaard and Muller, 1987) and in manufacturing to:

- Define the sensory properties of a target product for new product development (Szczesniak et al., 1975)
- Define the characteristics/specifications for a control or standard for QA/QC and R&D applications
- Document product attributes before a consumer test to help in the selection of attributes to be included in the consumer questionnaire and to help in explanation of the results of the consumer test
- Track a product's sensory changes over time with respect to understanding shelf life, packaging, etc.
- Map perceived product attributes for the purpose of relating them to instrumental, chemical, or physical properties (Bargmann et al., 1976; Moskowitz, 1979)
- Measure short-term changes in the intensity of specific attributes over time (time-intensity analysis)

III. COMPONENTS OF DESCRIPTIVE ANALYSIS

A. CHARACTERISTICS: THE QUALITATIVE ASPECT

Those perceived sensory parameters which define the product are referred to by various terms such as attributes, characteristics, character notes, descriptive terms, descriptors, or terminology (Johnsen et al., 1988).

These qualitative factors (which are the same as the parameters discussed under classification, Chapter 5, p. 50) include terms which define the sensory profile or picture or thumbprint of the sample. An important aspect is that panelists, unless well trained, may have very different concepts of what a term means. The question of concept formation is reviewed in detail by Lawless and Heymann (1998). The selection of sensory attributes and the corresponding definition of these attributes should be related to the real chemical and physical properties of a product which can be perceived (Civille and Lawless, 1986). Adherence to an understanding of the actual rheology or chemistry of a product make the descriptive data easier to interpret and more useful for decision making. Statistical methods such as ANOVA and multivariate analysis can be used to select the more descriptive terms (Jeltema and Southwick, 1986; ISO, 1994).

The components of a number of different descriptive profiles are given below (a number of examples of each are shown in parentheses). Note that this list is also the key to a more complete list of descriptive terms given in Chapter 11, Appendices 11.1, 11.2, and 11.3. The repeat appearance of certain properties and examples is intentional.

1. Appearance characteristics
   a. Color (hue, chroma, uniformity, depth)
   b. Surface texture (shine, smoothness/roughness)

    c. Size and shape (dimensions and geometry)

    d. Interactions among pieces or particles (stickiness, agglomeration, loose particles)

2. Aroma characteristics

    a. Olfactory sensations (vanilla, fruity, floral, skunky)

    b. Nasal feeling factors (cool, pungent)

3. Flavor characteristics

    a. Olfactory sensations (vanilla, fruity, floral, chocolate, skunky, rancid)

    b. Taste sensations (salty, sweet, sour, bitter)

    c. Oral feeling factors (heat, cool, burn, astringent, metallic)

4. Oral texture characteristics (Szczesniak, 1963; Szczesniak et al., 1963; Brandt et al., 1963)

    a. Mechanical parameters, reaction of the product to stress (hardness, viscosity, deformation/fracturability)

    b. Geometrical parameters, i.e., size, shape and orientation of particles in the product (gritty, grainy, flaky, stringy)

    c. Fat/moisture parameters, i.e., presence, release and adsorption of fat, oil, or water (oily, greasy, juicy, moist, wet)

5. Skinfeel characteristics (Schwartz, 1975; ASTM, 1997a; Civille and Dus, 1991)

    a. Mechanical parameters, reaction of the product to stress (thickness, ease to spread, slipperiness, denseness)

    b. Geometrical parameters, i.e., size, shape, and orientation of particles in product or on skin after use (gritty, foamy, flaky)

    c. Fat/moisture parameters, i.e., presence, release, and absorption of fat, oil, or water (greasy, oily, dry, wet)

    d. Appearance parameters, visual changes during product use (gloss, whitening, peaking)

6. Texture/handfeel of woven and nonwoven fabrics (Civille and Dus, 1990)

    a. Mechanical properties, reaction to stress (stiffness, force to compress or stretch, resilience)

    b. Geometrical properties, i.e., size, shape, and orientation of particles (gritty, bumpy, grainy, ribbed, fuzzy)

    c. Moisture properties, presence and absorption of moisture (dry, wet, oily, absorbent)

Again, the keys to the validity and reliability of descriptive analysis testing are:

- Terms based on a thorough understanding of the technical and physiological principles of flavor or texture or appearance
- Thorough training of all panelists to fully understand the terms in the same way and to apply them in the same way
- Use of references for terminology (see Chapter 11, Appendix 11.2, p. 195) to ensure consistent application of the carrier and descriptive terms to a perception

B. Intensity: the Quantitative Aspect

The intensity or quantitative aspect of a descriptive analysis expresses the degree to which each of the characteristics (terms, qualitative components) is present. This is expressed by the assignment of some *value* along a measurement scale.

    As with the validity and reliability of terminology, the validity and reliability of intensity measurements are highly dependent upon:

- The selection of a scaling technique which is broad enough to encompass the full range of parameter intensities and which has enough discrete points to pick up all the small differences in intensity between samples.

- The thorough training of the panelists to use the scale in a similar way across all samples and across time (see Chapter 9 on panelist training).
- The use of reference scales for intensity of different properties (see Appendix 11.2) to ensure consistent use of scales for different intensities of sensory properties across panelists and repeated evaluations.

Three types of scales are in common use in descriptive analysis (see also Lawless and Heymann, 1998):

1. *Category scales* are limited sets of words or numbers, constructed (as best one can) to maintain equal intervals between categories. A full description can be found in Chapter 5. A category scale from 0 to 9 is perhaps the most used in descriptive analysis, but longer scales are often justified. A good rule of thumb is to evaluate how many steps a panelist can meaningfully employ, and then to adopt a scale twice that length. Sometimes a 100-point scale is justified, e.g., in visual and auditory studies.
2. *Line scales* utilize a line 6 in. or 15 cm long on which the panelist makes a mark; they are described in Chapter 5. Line scales are almost as popular as category scales. Their advantage is that the intensity can be more accurately graded because there are no steps or "favorite numbers"; the chief disadvantage is that it is harder for a panelist to be consistent because a position on a line is not as easily remembered as a number.
3. *Magnitude estimation (ME) scales* are based on free assignment of the first number, after which all subsequent numbers are assigned in proportion (see Chapter 5). ME is used mostly in academic studies in which the focus is on a single attribute that can vary over a wide range of sensory intensities (Moskowitz, 1975, 1978).

Chapter 11, Appendix 11.2 contains sets of reference samples useful for the establishment of scales for various odors and tastes and also for the mechanical, geometrical, and moisture properties of oral texture. All the scales in Appendix 11.2 are based on a 15-cm line scale; however, the same standards can be distributed along a line or scale of any length or numerical value. The scales employ standard aqueous solutions such as sucrose, sodium chloride, citric acid, and caffeine, as well as certain widely available supermarket items which have shown adequate consistency, e.g., Hellmann's Mayonnaise and Welch's Grape Juice.

C. ORDER OF APPEARANCE: THE TIME ASPECT

In addition to accounting for the attributes (qualitative) of a sample, and the intensity of each attribute (quantitative), panels can often detect differences among products in the order in which certain parameters manifest themselves. The order of appearance of physical properties, related to oral, skin, and fabric textures, are generally predetermined by the way the product is handled (the input of forces by the panelist). By controlling the manipulation (one chew, one manual squeeze) the subject induces the manifestation of only a limited number of attributes (hardness, denseness, deformation) at a time (Civille and Liska, 1975).

However, with the chemical senses (aroma and flavor), the chemical composition of the sample and some of its physical properties (temperature, volume, concentration) may alter the order in which certain attributes are detected (IFT, 1981). In some products, such as beverages, the order of appearance of the characteristics is often as indicative of the product profile as the individual aroma and flavor notes and their respective intensities.

Included as part of the treatment of the order of appearance of attributes is aftertaste or afterfeel, which are those attributes that can still be perceived after the product or sample has been used or consumed. A complete picture of a product requires that all characteristics that are perceived after the product use should be individually mentioned and rated for intensity.

Attributes described and rated for aftertaste or afterfeel do not necessarily imply a defect or negative note. For example, the cool aftertaste of a mouthwash or breath mint is a necessary and desirable property. On the other hand, a metallic aftertaste of a cola beverage may indicate a packaging contamination or a problem with a particular sweetener.

When the intensity of one or more (usually not more than three) sensory properties is tracked repeatedly across a designated time span, the technique is called time-intensity analysis. A more detailed description of this technique is given on p. 168 of this chapter.

D. Overall Impression: the Integrated Aspect

In addition to the detection and description of the qualitative, quantitative, and time factors which define the sensory characteristics of a product, panelists are capable of, and management is often interested in, some integrated assessment of the product properties. Ways in which such integration has been attempted include the following four:

Total intensity of aroma or flavor — A measure of the overall impact (intensity) of all the aroma components (perceived volatiles) or a measure of the overall flavor impact, which includes the aromatics, tastes, and feeling factors contributing to the flavor. Such an evaluation can be important in determining the general fragrance or flavor impact which a product delivers to the consumer, who does not normally understand all of the nuances of the contributing odors or tastes which the panel describes. The components of texture are more functionally discrete, and "total texture" is not a property which can be determined.

Balance/blend (amplitude) — A well-trained descriptive panel is often asked to assess the degree to which various flavor or aroma characteristics fit together in the product. Such an evaluation involves a sophisticated understanding, half learned and half intuitive, of the appropriateness of the various attributes, their relative intensity in the complex, and the way(s) in which they harmonize in the complex. Evaluation of balance or blend (or amplitude, as it is called in the Flavor Profile method [Caul, 1957; Cairncross and Sjöstrom, 1950; Keane, 1992]) is difficult even for the highly trained panelist and should not be attempted with naive or less sophisticated subjects. In addition, care must be taken in the use of data on balance or blend. Often a product is not intended to be blended or balanced: a preponderance of spicy aromatics or toasted notes may be essential to the full character of a product. In some products the consumer may not appreciate a balanced composition, despite its well-proportioned notes, as determined by the trained panel. Therefore, it is important to understand the relative importance of blend or balance among consumers for the product in question before measuring and/or using such data.

Overall difference — In certain product situations the key decisions involve determination of the relative differences between samples and some control or standard product. Although the statistical analysis of differences between products on individual attributes is possible with many descriptive techniques, project leaders are often concerned with just how different a sample or prototype is from the standard. The determination of an overall difference (see Difference-from-control test, Chapter 6. VIII, p. 86) allows the project management to make decisions regarding disposition of a sample based on its relative distance from the control; the accompanying descriptive information provides insight into the source and size of the relative attributes of the control and the sample.

Hedonic ratings — It is a temptation to ask the descriptive panel, once the description has been completed, to rate the overall acceptance of the product. In most cases this is a temptation to be resisted, as the panel, through its training process, has been removed from the world of consumers and is no longer representative of any section of the general public. Training tends to change the personal preferences of panelists. As they become more aware of the various attributes of a product, panelists tend to weigh attributes differently from the way a regular consumer would, in terms of the contribution of each attribute to the overall quality, blend, or balance.

IV. COMMONLY USED DESCRIPTIVE TEST METHODS

Over the last 40 years many descriptive analysis methods have been developed, and some have gained and maintained popularity as standard methods (ASTM, 1992, 1996). The fact that these methods are described below is a reflection of their popularity, but it does not constitute a recommendation for use: on the contrary, a sensory analyst who needs to develop a descriptive system for a specific product and project application should study the literature on descriptive methods and should review several methods and combinations of methods before selecting the descriptive analysis system which can provide the most comprehensive, accurate, and reproducible description of each product and the best discrimination between products. See ASTM (loc. cit.) which also contains case studies of four methods, and review the IFT Sensory Evaluation Guide (IFT, 1981), which contains 109 references from different fields.

A. THE FLAVOR PROFILE METHOD

The Flavor Profile method was developed by Arthur D. Little, Inc. in the late 1940s (Keane, 1992). It involves the analysis of perceived aroma and flavor characteristics of a product, their intensities, order of appearance, and aftertaste by a panel of four to six trained judges. An amplitude rating (see previous page) is generally included as part of the profile.

Panelists are selected on the basis of a physiological test for taste discrimination, taste intensity discrimination, and olfactory discrimination and description. A personal interview is conducted to determine interest, availability, and potential for working in a group situation.

For training, panelists are provided with a broad selection of reference samples representing the product range, as well as examples of ingredient and processing variables for the product type. Panelists, with the panel leader's help in providing and maintaining reference samples, develop and define the common terminology to be used by the entire panel. The panel also develops a common frame of reference for the use of the seven-point Flavor Profile intensity scale shown in Chapter 5, p. 54.

The panelists, seated at a round or hexagonal table, individually evaluate one sample at a time for both aroma and flavor, and record the attributes (called "character notes"), their intensities, order of appearance, and aftertaste. Additional samples can be subsequently evaluated in the same session, but samples are not tasted back and forth. The results are reported to the panel leader, who then leads a general discussion of the panel to arrive at a "consensus" profile for each sample. The data are generally reported in tabular form, although a graphic representation is possible.

The Flavor Profile method may be applied when a panel must evaluate many different products, none of which is a major line of a major producer. The main advantage but also a major limitation of the Flavor Profile method is that it only uses five to eight panelists. The lack of consistency and reproducibility which this entails is overcome to some extent by training and by the consensus method. However, the latter has been criticized for one-sidedness. The panel's opinion may become dominated by that of a senior member or a dominant personality, and equal input from other panel members is not obtained. Other points of criticism of the Flavor Profile are that screening methods do not include tests for the ability to discriminate specific aroma or flavor differences which may be of importance in specific product applications, and also that the seven-point scale limits the degree of discrimination between products showing small but important differences.

B. THE TEXTURE PROFILE METHOD

Based somewhat on the principles of the Flavor Profile method, the Texture Profile method was developed by the Product Evaluation and Texture Technology groups at General Foods Corp. to define the textural parameters of foods (Skinner, 1988). Later the method was expanded by Civille and Szczesniak (1973) and Civille and Liska (1975) to include specific attribute descriptors for specific products including semisolid foods, beverages, skinfeel products (Schwartz, 1975; ASTM

1997) and fabric and paper goods (Civille and Dus, 1990). In all cases the terminology is specific for each product type, but it is based on the underlying rheological properties expressed in the first Texture Profile publications (Szczesniak, 1963; Szczesniak et al., 1963; Brandt et al., 1963).

Panelists are selected on the basis of ability to discriminate known textural differences in the specific product application for which the panel is to be trained (solid foods, beverages, semisolids, skin care products, fabrics, paper, etc.). As with most other descriptive analysis techniques, panelists are interviewed to determine interest, availability, and attitude. Panelists, selected for training, are exposed to a wide range of products from the category under investigation, to provide a wide frame of reference. In addition, panelists are introduced to the underlying textural principles involved in the structure of the products under study. This learning experience provides panelists with understanding of the concepts of input mechanical forces and resulting strain on the product. In turn, panelists are able to avoid lengthy discussions about redundant terms and to select the most technically appropriate and descriptive terms for the evaluation of products. Panelists also define all terms and all procedures for evaluation, thus reducing some of the variability encountered in most descriptive testing. The reference scales used in the training of panelists can later serve as references for approximate scale values, thus further reducing panel variability.

Samples are evaluated independently by each panelist using one of the scaling techniques previously discussed. The original Texture Profile method used an expanded 13-point version of the Flavor Profile scale. In the last several years, however, Texture Profile panels have been trained using category, line, and ME scales (see Chapter 11, Appendix 11.2, pp. 180–183, for food texture references for use with a 15-point or 15-cm line scale). Depending on the type of scale used by the panel and on the way the data are to be treated, the panel verdicts may be derived by group consensus, as with the Flavor Profile method, or by statistical analysis of the data. For final reports the data may be displayed in tabular or graphic form.

C. The Quantitative Descriptive Analysis (QDA®) Method

In response to dissatisfaction among sensory analysts with the lack of statistical treatment of data obtained with the Flavor Profile or related methods, the Tragon Corp. developed the QDA® method of descriptive analysis (Stone et al., 1974; Stone and Sidel, 1992). This method relies heavily on statistical analysis to determine the appropriate terms, procedures, and panelists to be used for analysis of a specific product.

Panelists are selected from a large pool of candidates according to their ability to discriminate differences in sensory properties among samples of the specific product type for which they are to be trained. The training of QDA panels requires the use of product and ingredient references, as with other descriptive methods, to stimulate the generation of terminology. The panel leader acts as a facilitator, rather than as an instructor, and refrains from influencing the group. Attention is given to development of consistent terminology, but panelists are free to develop their own approach to scoring, using the 15-cm (6 in.) line scale which the method provides.

QDA panelists evaluate products one at a time in separate booths to reduce distraction and panelist interaction. Panelists enter the data into a computer, or the scoresheets are collected individually from the panelists as they are completed, and the data are entered for computation usually with a digitizer or card reader directly from the scoresheets. Panelists do not discuss data, terminology, or samples after each taste session and must depend on the discretion of the panel leader for any information on their performance relative to other members of the panel and to any known differences between samples.

The results of a QDA test are analyzed statistically, and the report generally contains a graphic representation of the data in the form of a "spider web" with a branch or spoke from a central point for each attribute.

The QDA method was developed in partial collaboration with the Department of Food Science at the University of California at Davis. It represents a large step toward the ideal of this book, the intelligent use of human subjects as measuring instruments, as discussed in Chapter 1. In particular, the use of a graphic scale (visual analog scale), which reduces that part of the bias in scaling resulting from the use of numbers; the statistical treatment of the data; the separation of panelists during evaluation; and the graphic approach to presentation of data have done much to change the way in which sensory scientists and their clients view descriptive methodology. The following are areas which in our view could benefit from a change or further development:

1. The panel, because of lack of formal instruction, may develop erroneous terms. For example, the difference between natural vanilla and pure vanillin should be easily detected and described by a well-trained panel; however, an unguided panel would choose the term "vanilla" to describe the flavor of vanillin. Lack of direction also may allow a senior panelist or stronger personality to dominate the proceedings in all or part of the panel population in the development of terminology.
2. The "free" approach to scaling can lead to inconsistency of results, partly because of particular panelists evaluating a product on a given day and not on another, and partly because of the context effects of one product seen after the other, with no external scale references.
3. The lack of immediate feedback to panelists on a regular basis reduces the opportunity for learning and expansion of terminology for greater capacity to discriminate and describe differences.
4. On a minor point, the practice of connecting "spokes" of the "spider web" can be misleading to some users, who by virtue of their technical training expect the area under a curve to have some meaning. In reality, the sensory dimensions shown in the "web" may be either unrelated to each other, or related in ways which cannot be represented in this manner.

D. THE SPECTRUM™ DESCRIPTIVE ANALYSIS METHOD

This method, designed by Civille, is described in detail in Chapter 11. Its principal characteristic is that the panelist scores the perceived intensities with reference to pre-learned "absolute" intensity scales. The purpose is to make the resulting profiles universally understandable and usable, not only at a later date, but also at any laboratory outside the originating one. The method provides for this purpose an array of standard attribute names ("lexicons"), each with its set of standards that define a scale of intensity, usually from 0 to 15.

E. TIME-INTENSITY DESCRIPTIVE ANALYSIS

For certain products the intensity of perception varies with time over a longer or shorter period, and the time-intensity curve of an attribute may be a key aspect defining the product (Larson-Powers and Pangborn, 1978; Overbosch, 1986; Overbosch et al., 1986; Lee and Pangborn, 1986; ASTM, 1997b). Long-term time-intensity studies measure the reduction of skin dryness periodically over several days of application of a skin lotion. The color intensity of a lipstick can be evaluated periodically over several hours. Shorter term time-intensity studies track certain flavor and/or texture attributes of chewing gum over several minutes. In the shortest term studies, completed within 1 to 3 min, the response can be recorded continuously. Examples include the sweetness of sweeteners (Shamil et al., 1988; Anon., 1988), the bitterness of beer (Pangborn et al., 1983; Schmitt et al., 1984; Leach and Noble, 1986), and the effects of topical analgesics. The response may be recorded using pencil and paper, a scrolling chart recorder (Larson-Powers and Pangborn, 1978), or a computer system (Guinard et al., 1985; Anon., loc. cit.), available commercially in several versions. The panelist should not see the evolving response curve being traced because this may result in bias from preconceived notions of its form.

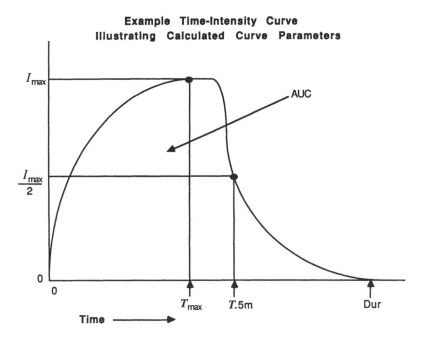

**Example Time-Intensity Curve
Illustrating Calculated Curve Parameters**

FIGURE 10.1 Example of a time-intensity curve illustrating calculated curve parameters. I_{max}, the maximum observed intensity; T_{max}, the time at which the maximum intensity occurs; AUC, the area under the curve; Dur, the intensity duration: the time until the intensity drops back to zero; $T.5m$, the time (after T_{max}) at which intensity has fallen to half of I_{max}.

Current methodology of time-intensity research has been comprehensively reviewed by Lee and Pangborn (1986) and for sweeteners in particular by Booth (1989). Important variables to consider are:

- Protocols for evaluation — type of delivery, amount of product, time to hold in the mouth, type of manipulation, expectoration, or swallow — need to be clearly defined.
- Protocols for coordinating product evaluation (sample holding) and response recording (data entry) need to be worked out in advance to reduce bias from the mode of presentation.
- Panelists may require several training sessions to develop and learn all of the protocols necessary for a well-controlled time-intensity study. Figure 10.1 is an example of the parameters that can be recorded in a time-intensity study; a more detailed example is given by Lee and Pangborn (1986).

Table 10.1 shows an example of responses obtained with three sweeteners.

TABLE 10.1
Time-Intensity Data for Three Sweeteners

Parameter	7.5% Sucrose (conditioning sample)	0.05% Aspartame solution	0.4% Acesulfam-K solution	7.5% Sucrose
Area under the curve, cm^2	121.2	153.7	98.6	154.2
Maximum intensity, I_{max}	7.2	7.6	7.8	7.6
Time of maximum intensity, t_{max}, s	7.4	8.2	4.8	6.2
Duration, s	28.3	33.3	24.7	33.4

F. FREE-CHOICE PROFILING

Free-Choice Profiling is a novel technique developed by Williams and Arnold (1984) at the Agricultural and Food Council (U.K.) as a solution to the problem of consumers using different terms for a given attribute. Free-Choice Profiling allows the panelist to invent and use as many terms as he or she needs to describe the sensory characteristics of a set of samples (Marshall and Kirby, 1988; Guy et al., 1989; Oreskovich et al., 1991). The samples are all from the same category of products, and the panelist develops his or her own scoresheet. The data are analyzed by Generalized Procrustes analysis (Gower, 1975), a multivariate technique which adjusts for the use of different parts of the scale by different panelists and then manipulates the data to combine terms that appear to measure the same characteristic. These combined terms provide a single product profile.

Research comparing Free-Choice Profiling and other descriptive techniques is currently being conducted. The main advantage of the new technique is that it saves much time by not requiring any training of the panelists other than an hour of instruction in the use of the chosen scale. A second advantage is that the panelists, not having been trained, can still be regarded as representing naive consumers. However, questions regarding the ability of the sensory analyst to "interpret" the resulting terms, combined from all panelists, need to be addressed. In order to provide reliable guidance for product researchers, the experimenter/sensory analyst must decide what each combined term actually means. Therefore, the words or terms for each resulting parameter come from the experimenter or sensory analyst rather than from the panelists. The results may be colored more by the perspective of the analyst than by the combined weight of the panelists' verdicts.

REFERENCES

Anon., 1988. Computers tell "how sweet it is." Staff article, *Food Technol.* 42(11), 98.

ASTM, 1992. *Manual on Descriptive Analysis Testing for Sensory Evaluation*, Hootman, R.C., Ed., ASTM Manual 13, 52 pp.

ASTM, 1996. *Sensory Testing Methods*, 2nd ed., Chambers, E., IV and Baker Wolf, M., Eds. ASTM Manual 26, 115 pp.

ASTM, 1997a. *Standard Practice for Descriptive Skinfeel Analysis of Creams and Lotions.* ASTM Standard Practice E1490-92. American Society of Testing and Materials, West Conshohocken, PA.

ASTM, 1997b. *Standard Guide for Time-Intensity Evaluation of Sensory Attributes.* ASTM Standard Guide E1909-97.

Bargmann, R.E., Wu, L., and Powers, J.J., 1976. Search for the Determiners of Food Quality Ratings — Description of Methodology with Application to Blueberries. In: *Correlating Sensory Objective measurements — New Methods for Answering Old Problems*, Powers, J.E. and Moskowitz, H.R., Eds. ASTM Special Technical Publication STP 594, pp. 56–72.

Booth, B., 1989. Time-intensity parameters considered in sweetener research at the NutraSweet Co., presentation to American Society for Testing and Materials (ASTM), Subcommittee E18 on Sensory Evaluation, Kansas City, MO, 1989.

Brandt, M.A., Skinner, E.Z., and Coleman, J.A., 1963. Texture profile method. *J. Food Sci.* 28(4), 404, 1963.

Cairncross, S.E. and Sjöstrom, L.B., 1950. Flavor profiles — a new approach to flavor problems. *Food Technol.* 4, 308.

Caul, J.F., 1957. The profile method of flavor analysis. *Advances in Food Research* 7, 1.

Civille, G.V., 1979. *Descriptive Analysis.* Course notes for IFT Short Course in Sensory Analysis. Institute of Food Technology, Chicago, Chapter 6.

Civille, G.V. and Szczesniak, A.S., 1973. Guide to training a texture profile panel. *J. Texture Stud.* 4, 204.

Civille, G.V. and Liska, I.H., 1975. Modifications and applications to foods of the General Foods sensory texture profile technique. *J. Texture Stud.* 6, 19.

Civille, G.V. and Lawless, H.T., 1986. The importance of language in describing perceptions. *J. Sensory Stud.* 1(3/4), 203.

Civille, G.V. and Dus, C.A., 1990. Development of terminology to describe the handfeel properties of paper and fabrics. *J. Sensory Stud.* 5, 19.

Civille, G.V. and Dus, C.A., 1991. Evaluating tactile properties of skincare products: a descriptive analysis technique. *Cosmetics and Toiletries* 106(5), 83.

Gower, J.C., 1975. Generalized Procrustes Analysis. *Psychometrika* 40, 33.

Guinard, J.-X., Pangborn, R.M., and Shoemaker, C.F., 1985. Computerized procedure for time-intensity sensory measurements. *J. Food Sci.* 50, 543.

Guinard, J.-X., Pangborn, R.M., and Lewis, M.J., 1986. Effect of repeated ingestion on perception of bitterness in beer. *J. Soc. Am. Brew. Chem.* 44, 28.

Guy, C. Piggott, J.R., and Marie, S., 1989. Consumer free-choice profiling of whisky. In: *Distilled Beverage Flavour: Recent Developments*, Piggott, J.R. and Paterson, A., Eds. Ellis Horwood/VCH, Chichester.

IFT, 1981. Sensory Evaluation Guide for Testing Food and Beverage Products, and Guidelines for the Preparation and Review of Papers Reporting Sensory Evaluation Data. *Food Technol.* 35(11), 50.

ISO, 1994. *Sensory-analysis — Identification and selection of descriptors for establishing a sensory profile by a multidimensional approach.* International Organization for Standardization, ISO 11035:1994. Availability, see p. 148.

Jeltema, M.A. and Southwick, E.W., 1986. Evaluation and application of odor profiling. *J. Sensory Stud.* 1(2), 123.

Johnsen, P.B., Civille, G.V., Vercellotti, J.R., Sanders, T.H., and Dus, C.A., 1988. Development of a lexicon for description of peanut flavor. *J. Sensory Stud.* 3(1), 9.

Keane, P., 1992. The Flavor Profile. In: *Descriptive Analysis Testing for Sensory Evaluation*, Hootman, R.C., Ed. ASTM Manual 13, Philadelphia, pp. 5–14.

Larson-Powers, N. and Pangborn, R.M., 1978. Paired comparison and time-intensity measurement of the sensory properties of beverages and gelatins containing sucrose or synthetic sweeteners. *J. Food Sci.* 43, 41.

Lawless, H.T. and Heymann, H., 1998. *Sensory Evaluation of Food. Principles and Practices.* Chapman & Hall, New York, Chapter 10.

Leach, E.J. and Noble, A.C., 1986. Comparison of bitterness of caffeine and quinine by a time-intensity procedure. *Chem. Senses* 11(3), 339.

Lee, W.E., III, and Pangborn, R.M., 1986. Time-intensity: the temporal aspects of sensory perception. *Food Technol.* 40(11), 71.

Marshall, R.J. and Kirby, S.P., 1988. Sensory measurement of food texture by Free-Choice Profiling. *J. Sensory Stud.* 3, 63.

Moskowitz, H.R., 1975. Application of sensory assessment to food evaluation. II. Methods of ratio scaling. *Lebensm.-Wiss. & Technol.* 8(6), 249.

Moskowitz, H.R., 1978. Magnitude estimation: notes on how, what, where and why to use it. *J. Food Qual.* 1, 195.

Moskowitz, H.R., 1979. Correlating sensory and instrumental measures in food texture. *Cereal Foods World* 22, 223.

Oreskovich, D.C., Klein, B.P., and Sutherland, J.W., 1991. Procrustes Analysis and its applications to Free-Choice and other sensory profiling. In: *Sensory Science Theory and Application in Foods*, Lawless, H.T. and Klein, B.P., Eds. Marcel Dekker, New York, pp. 353–393.

Overbosch, P., 1986. A theoretical model for perceived intensity in human taste and smell as a function of time. *Chem. Senses* 11(3), 315–329.

Overbosch, P., van den Enden, J.C., and Keur, B.M., 1986. An improved method for measuring perceived intensity/time relationships in human taste and smell. *Chem. Senses* 11(3), 331–338.

Pangborn, R.M., Lewis, M.J., and Yamashita, J.F., 1983. Comparison of time-intensity with category scaling of bitterness of iso-α-acids in model systems and in beer. *J. Inst. Brew.* 89, 349–355.

Schmitt, D.J., Thompson, L.J., Malek, D.M., and Munroe, J.H., 1984. An improved method for evaluating time-intensity data. *J. Food Sci.* 49, 539.

Schwartz, N., 1975. Method to skin care products. *J. Texture Stud.* 6, 33.

Shamil, S., Birch, G.G., Jackson, A.A.S.F., and Meek, S., 1988. Use of intensity-time studies as an aid to interpreting sweet taste chemoreception. *Chem. Senses* 13(4), 597.

Skinner, E.Z., 1988. The texture profile method. In: *Applied Sensory Analysis of Foods*, Moskowitz, H.R., Ed. CRC Press, Boca Raton, FL, pp. 89–107.

Stone, H. and Sidel, J.L., 1992. *Sensory Evaluation Practices,* 2nd ed. Academic Press, Orlando, FL.

Stone, H., Sidel, J., Oliver, S., Woolsey, A., and Singleton, R.C., 1974. Sensory evaluation by Quantitative Descriptive Analysis. *Food Technol.* 28(11), 24.

Szczesniak, A.S., 1963. Classification of textural characteristics. *J. Food Sci.* 28(4), 397.

Szczesniak, A.S., Brandt, M.A., and Friedman, H.H., 1963. Development of standard rating scales for mechanical parameters of texture and correlation between the objective and the sensory methods of texture evaluation. *J. Food Sci.* 28(4), 397.

Szczesniak, A.S., Loew, B.S., and Skinner, E.Z., Consumer texture profile technique. *J. Food Sci.* 40, 1243, 1975.

Williams, A.A. and Arnold, G.M., 1984. A new approach to sensory analysis of foods and beverages. In: *Progress in Flavour Research, Proc. 4th Weurman Flavour Research Symp.,* Adda, J., Ed. Elsevier, Amsterdam.

11 The Spectrum™ Descriptive Analysis Method

CONTENTS

I. DESIGNING A DESCRIPTIVE PROCEDURE

The name Spectrum covers a procedure designed by Civille and developed over the years in collaboration with a number of companies that were looking for a way to obtain reproducible and repeatable sensory descriptive analysis of their products (Muñoz and Civille, 1992; 1998). The philosophy of Spectrum is pragmatic: it provides the tools with which to design a descriptive procedure for a given product category. The principal tools are the reference lists contained in Appendices 11.1 to 11.3, together with the scaling procedures and methods of panel training described in Chapters 5 and 9. The aim is to choose the most practical system, given the product in question, the overall sensory program, the specific project objective(s) in developing a panel, and the desired level of statistical treatment of the data.

For example, panelists may be selected and trained to evaluate only one product or a variety of products. Products may be described in terms of only appearance, aroma, flavor, texture, or sound characteristics, or panelists may be trained to evaluate all of these attributes. Spectrum is a "custom design" approach to panel development, selection, training, and maintenance.

Courses teaching the basic elements of Spectrum are available and include a detailed manual. Examples of the application are given in Johnsen et al. (1988).

II. TERMINOLOGY

The choice of terms may be broad or narrow according to the panel's objective — only aroma characteristics, or all sensory modalities. However, the method requires that all terminology is developed and described by a panel which has been exposed to the underlying technical principles of each modality to be described. For example, a panel describing color must understand color intensity, hue, and chroma. A panel involved in oral, skinfeel, and/or fabric texture needs to understand what the tactile effects of rheology and mechanical characteristics are and how these in turn are affected by moisture level and particle size. The chemical senses pose an even greater challenge in requiring panelists to demonstrate a valid response to changes in ingredients and processing. Words such as vanilla, cocoa, and distilled orange oil require separate terms and references. If the panel hopes to attain the status of "expert panel" in a given field, it must demonstrate that it can use a concrete list of descriptors based on an understanding of the underlying technical differences among the attributes of a product.

Panelists begin to develop their list of best descriptors by first evaluating a broad array of products (commercial brands, competitors, pilot plant runs, etc.) which define the product category. After some initial experience with the category, each panelist produces a list of terms to describe the set. Additional terms and references may be taken from the literature, e.g., from published flavor lexicons (Johnsen et al., 1988; Civille and Lyon, 1996). The terms are then compiled or organized into a list that is comprehensive yet not overlapping. This process includes using references (see Appendix 11.2) to determine the best choice for a term and to best define that term so that it is understood in the same way by all panelists.

An example of the adaptation of existing underlying terms to a specific product category is the work on noodles by Janto et al., 1998. Several standard terms apply to noodles, but the vast Asian Noodle frame of reference called for additional terms, such as "starch between teeth" and "slipperiness between lips."

III. INTENSITY

Different project objectives may require widely different intensity scales. A key property of a scale is the number of points of discrimination along the scale. If product differences require a large number of points of discrimination to clearly define intensity differences both within and between attributes, the panel leader requires a 15-cm scale, or a category with 30 points or more, or an ME scale.

The Spectrum method is based on extensive use of reference points, which may be chosen according to the guidelines given in Appendix 11.2, p. 195. These are derived from the collective data of several panels over several replicates. Whatever the scale chosen, it must have at least two and preferably three to five reference points distributed across the range. A set of well-chosen reference points greatly reduces panel variability, allowing for a comparison of data across time and products. Such data also allow more precise correlation with stimulus changes (stimulus/response curve) and with instrumental data (sensory/instrumental correlations). The choice of scaling technique may also depend on the available facilities for computer manipulation of data and on the need for sophisticated data analysis.

IV. OTHER OPTIONS

The tools of the Spectrum method include time/intensity tests, the Difference-from-control test, total flavor impact assessment, and others. The basic philosophy, as mentioned, is to train the panel to fully define each and all of a product's attributes, to rate the intensity of each, and to include other relevant characterizing aspects such as changes over time, differences in order of appearance of attributes, and integrated total aroma and/or flavor impact.

The creative and diligent sensory analyst can construct the optimal descriptive technique by selecting from the spectrum of terms, scaling techniques, and other optional components which are available at the start of each panel development.

V. MODIFIED SHORT-VERSION SPECTRUM DESCRIPTIVE PROCEDURES FOR QUALITY ASSURANCE, SHELF-LIFE STUDIES, ETC.

Certain applications of descriptive analysis require evaluation of a few detailed attributes without a full analysis of all the parameters of flavor, texture, and/or appearance. The tracking or monitoring of product changes, necessary in QC/QA sensory work and in shelf-life studies, can provide the required information by logging a small number of selected sensory properties over time. The Modified or Short-Version Descriptive procedure, in any situation, must be based on work done with a fully trained descriptive panel, generally in R&D, which characterizes all of the product's attributes. Once the panel has evaluated a succession of products typical of the full range of sensory properties, e.g., several production samples from all plants and through the practical aging and storage conditions encountered, the sensory analyst and project team can select five to ten key parameters, which together define the range or qualities from "typical" to "off." Future monitoring of just those parameters then permits QA/QC and R&D to identify any changes that require troubleshooting and correction.

Use of the Modified Spectrum Descriptive technique was described by Muñoz et al. (1992) for two applications, a Comprehensive Descriptive procedure and a Difference-from-control procedure. In the Comprehensive Descriptive procedure, a reduced set of characteristics is selected by testing the production variability for most characteristics among consumers and then choosing those characteristics the variability of which *most* affects consumer acceptance. These relationships are used to set sensory specifications that allow the QC sensory program to monitor production. The intensity of the key sensory attributes are measured to determine whether production samples fall in or out of specification, and for what attributes. Such a technique permits detection and definition of any problem areas, which can then be related to processing or raw materials sources. The Comprehensive Descriptive procedure may also be applied to the sensory properties of incoming raw materials and/or in-process batches.

In the second application, the Modified Spectrum Descriptive is coupled with a Difference-from-control test. The modified descriptive panel is trained to recognize the control or standard product along with other samples that the fully trained panel has described as different from the control on the key attributes. The panel is shown the full range of samples and asked to rate them using the normal Difference-from-control scale (see Chapter 6, p. 87). The panel understands that occasionally one of the test samples during normal testing of production will be a blind control and/or one of the original "small difference" or "large difference" demonstration samples. This precaution reduces the likelihood of panelists anticipating too much change in shelf-life studies or too little change in production.

The Difference-from-control test provides an indication of the magnitude of the difference from the standard product. Samples may on occasion show statistical significance for a difference from the control and yet remain acceptable to consumers. The product team can submit to consumer testing three or more products, identified by the panel as showing slight, moderate, and large differences from the control. In place of a "go"/"no go" system based strictly on statistical significance, the company can devise a system of specifications based on known differences that are meaningful to the consumer. The system can be used to track production and storage samples over time in a cost-effective program (see Chapter 12, Example 12.3, p. 251).

REFERENCES

Civille, G.V. and Lyon, B.G., 1996. *Aroma and Flavor Lexicon for Sensory Evaluation. Terms, Definitions, References and Examples.* ASTM Data Series Publication DS 66, West Conshohocken, PA.

Janto, M., Pipatsattayanuwong, S., Kruk, M.W., Hou, G., McDaniel, M.R., 1998. Developing Noodles from US Wheat Varieties for the Far East Market: Sensory Perspective. *Food Quality and Preference* 9(6), 403–412.

Johnsen, P.B., Civille, G.V., Vercellotti, J.R., Sanders, T.H., and Dus, C.A., 1988. Development of a lexicon for the description of peanut flavor. *J. Sensory Stud.* 3(1), 9.

Muñoz, A.M. and Civille, G.V., 1992. The Spectrum descriptive analysis method. In: *ASTM Manual Series MNL 13, Manual on Descriptive Analysis Testing*, Hootman, R.C., Ed. Am. Soc. Testing and Materials, West Conshohocken, PA.

Muñoz, A.M. and Civille, G.V., 1998. Universal, Product and Attribute Scaling and the development of common lexicons in descriptive analysis. *J. Sensory Stud.* 13(1), 57–75.

Muñoz, A.M., Civille, G.V., and Carr, B.T., 1992. *Sensory Evaluation in Quality Control.* Chapman & Hall, New York.

Appendix 11.1
Spectrum Terminology
for Descriptive Analysis

The following lists of terms for appearance, flavor, and texture can be used by panels suitably trained to define the qualitative aspects of a sample.

When required, each of the terms can be quantified using a scale chosen from Chapter 5. Each scale must have at least two, and preferably three to five, reference points chosen, e.g., from Appendix 11.2.

A simple scale can have general anchors:

None - - - - - - - - - - - - - - - - - Strong

or a scale can be anchored using bipolar words (opposites):

Smooth - - - - - - - - - - - - - - -Lumpy
Soft - - - - - - - - - - - - - - - - - - - Hard

Attributes perceived via the chemical senses in general use a unipolar intensity scale (None–Strong), while for appearance and texture attributes, a bipolar scale is best, as shown below.

A. TERMS USED TO DESCRIBE APPEARANCE

1. Color

a. Description   The actual color name or hue, such as red, blue, etc. The description can be expressed in the form of a scale range, if the product covers more than one hue:
[Red -Orange]

b. Intensity   The intensity or strength of the color from light to dark:
[Light - Dark]

c. Brightness   The chroma (or purity) of the color, ranging from dull, muddied to pure, bright color. Fire engine red is a brighter color than burgundy red.
[Dull - Bright]

d. Evenness   The evenness of distribution of the color, not blotchy:
[Uneven/blotchy - Even]

2. Consistency/Texture

a. Thickness   The viscosity of the product:
[Thin -Thick]

b. Roughness   The amount of irregularity, protrusions, grains, or bumps which can be seen on the surface of the product; smoothness is the absence of surface particles:
[Smooth - Rough]
Graininess is caused by small surface particles:
[Smooth - Grainy]
Bumpiness is caused by large particles:
[Smooth -Bumpy]

   c. Particle interaction   The amount of stickiness among particles or the amount of agglomeration of small particles:

      (Stickiness):   [Not sticky - Sticky]

      (Clumpiness):   [Loose particles- Clumps]

3. Size/Shape

   a. Size   The relative size of the pieces or particles in the sample:

                             [Small -Large]

                             [Thin -Thick]

   b. Shape   Description of the predominant shape of particles:

flat, round, spherical, square, etc.

[No scale]

   c. Even distribution   Degree of uniformity of particles within the whole:

[Nonuniform pieces - Uniform pieces]

4. Surface Shine

Amount of light reflected from the product's surface:

[Dull -Shiny]

B. Terms Used to Describe Flavor (General and Baked Goods)

The full list of fragrance and flavor descriptors is too unwieldy to reproduce here; the list of aromatics* alone contains over a thousand words. In the following, aromatics for baked goods are shown as an example.

Flavor is the combined effects of the:

- Aromatics
- Tastes
- Chemical feelings

stimulated by a substance in the mouth. For baked goods it is convenient to subdivide the aromatics into:

- Grainy aromatics
- Grain-related terms
- Dairy terms
- Other processing characteristics
- Sweet aromatics
- Added flavors/aromatics
- Aromatics from shortening
- Other aromatics

Example: Flavor Terminology of Baked Goods

1. Aromatics (of baked goods)

   a. Grainy aromatics   Those aromatics or volatiles which are derived from various grains; the term cereal can be used as an alternative, but it implies finished and/or toasted character and is, therefore, less useful than grainy.

---

* The term *aromatics* is used in this book to cover that portion of the flavor which is perceived by the sense of smell from a substance in the mouth.

Grainy: the general term to describe the aromatics of grains, which cannot be tied to a specific grain by name.

Terms pertaining to a specific grain: corn, wheat, oat, rice, soy, rye. Grain character modified or characterized by a processing note, or lack thereof:

Raw corn	Cooked corn	Toasted corn
Raw wheat	Cooked wheat	Toasted wheat
Raw oat	Cooked oat	Toasted oat
Raw rice	Cooked rice	Toasted rice
Raw soy	Cooked soy	Toasted soy
Raw rye	Cooked rye	Toasted rye

Definitions of processed grain terms:

Raw (name) flour: the aromatics perceived in a particular grain which has not been heat treated.

Cooked (name) flour: the aromatics of a grain which has been gently heated or boiled; Cream of Wheat has cooked wheat flavor; oatmeal has cooked oat flavor.

Baked toasted (name) flour: the aromatics of a grain which has been sufficiently heated to caramelize some of the starches and sugars.

b. Grain-related terms   Green: the aromatic associated with unprocessed vegetation, such as fruits and grains; this term is related to raw, but has the additional character of hexenals, leaves, and grass.

Hay-like/grassy: grainy aromatic with some green character of freshly mowed grass, air-dried grain, or vegetation.

Malty: the aromatics of toasted malt.

c. Dairy terms   Those volatiles related to milk, butter, cheese, and other cultured dairy products. This group includes the following terms:

Dairy: as above.

Milky: more specific than dairy, the flavor of regular or cooked cow's milk.

Buttery: the flavor of high-fat fresh cream or fresh butter; not rancid, butyric, or diacetyl-like.

Cheesy: the flavor of milk products treated with rennet which hydrolyzes the fat, giving it a butyric or isovaleric acid character.

d. Other processing characteristics   Caramelized: a general term used to describe starches and sugars which have been browned; used alone when the starch or sugar (e.g., toasted corn) cannot be named.

Burnt: related to overheating, overtoasting, or scorching the starches or sugars in a product.

e. Added flavors/ aromatics   The following terms relate to specific ingredients which may be added to baked goods to impart specific character notes; in each case, references for the term are needed:

Nutty: peanut, almond, pecan, etc.

Chocolate: milk chocolate, cocoa, chocolate-like.

Spices: cinnamon, clove, nutmeg, etc.

Yeasty: natural yeast (not chemical leavening).

f. Aromatics from shortening   The aromatics associated with oil or fat-based shortening agents used in baked goods:

Buttery: see dairy above.

Oil flavor: the aromatics associated with vegetable oils, not to be confused with an oily film on the mouth surfaces, which is a texture characteristic.

Lard flavor: the aromatics associated with rendered pork fat.

Tallowy: the aromatics associated with rendered beef fat.

g. Other aromatics   The aromatics which are not usually part of the normal product profile and/or do not result from the normal ingredients or processing of the product:

Vitamin: aromatics resulting from the addition of vitamins to the product.

Cardboard flavor: aromatics associated with the odor of cardboard box packaging, which could be contributed by the packaging *or* by other sources, such as staling flours.

Rancid: aromatics associated with oxidized oils, often also described as painty or fishy.

Mercaptan: aromatics associated with the mercaptan class of sulfur compounds. Other terms which panelists may use to describe odors arising from sulfur compounds are skunky, sulfitic, rubbery.

(End of section referring to baked goods only.)

2. **Basic tastes**

a. Sweet   The taste stimulated by sucrose and other sugars, such as fructose, glucose, etc., and by other sweet substances such as saccharin, Aspartame, and Acesulfam K.

b. Sour   The taste stimulated by acids, such as citric, malic, phosphoric, etc.

c. Salty   The taste stimulated by sodium salts, such as sodium chloride and sodium glutamate, and in part by other salts, such as potassium chloride.

d. Bitter   The taste stimulated by substances such as quinine, caffeine, and hop bitters.

3. **Chemical feeling factors**   Those characteristics which are the response of tactile nerves to chemical stimuli.

a. Astringency   The shrinking or puckering of the tongue surface caused by substances such as tannins or alum.

b. Heat   The burning sensation in the mouth caused by certain substances such as capsaicin from red or piperine from black peppers; mild heat or warmth is caused by some brown spices.

c. Cooling   The cool sensation in the mouth or nose produced by substances such as menthol and mints.

C. Terms Used to Describe Semisolid Oral Texture

These terms are those specifically added for semisolid texture. Solid oral texture terms also may be used when applicable to any product or sample. Each set of texture terms includes the procedure for manipulation of the sample.

1. First Compression

Place ¼ tsp. of sample in mouth and compress between tongue and palate.

a. Slipperiness   The amount in which the product slides across the tongue.

[Drag - Slip]

b. Firmness   The force required to compress between tongue and palate.
                        [Soft - Firm]

c. Cohesiveness   The amount the sample deforms rather than shears/cuts.
                        [Shears/short- Deforms/cohesive]

d. Denseness   Compactness of the cross section.
                        [Airy - Dense/compact]

2. *Manipulation*

Compress sample several more times (3 to 8 times).

a. Particle amount   The relative number/amount of particles in the mouth.
                        [None -Many]

b. Particle size   The size of the particle in the mass.
                        [Extremely small- -Large]

3. *Afterfeel*

Swallow or expectorate.

a. Mouthcoating   The amount of film left on the mouth surfaces.
                        [None -Much]

Example: Semisolid Texture Terminology — Oral Texture of Peanut Butter

1. Surface   Hold ¼ tsp. on spoon; feel surface with lips and evaluate for:
                        Oiliness/moistness: amount of oiliness/moistness on surface.
                        [Dry - Oily/moist]
                        Stickiness: amount of product adhering to lips.
                        [Slippery- Sticky]
                        Roughness: amount of particles in surface.
                        [Smooth - Rough]

2. First compression   Place ¼ tsp. of peanut butter in mouth and compress between tongue
                        and palate; evaluate for:
                        Slipperiness: amount in which product slides across tongue.
                        [Drag- Slip]
                        Firmness: force to compress sample.
                        [Soft - Firm]
                        Cohesiveness: amount sample deforms rather than shears/cuts.
                        [Shears/short- Deforms/cohesive]
                        Adhesiveness (palate): amount of force to remove sample from roof
                        of mouth.
                        [No force - High force]
                        Stickiness: amount of product that adheres to oral surfaces.
                        [Not sticky -Very sticky]

3. Breakdown   Manipulate between tongue and palate seven times; evaluate for:
                        Moisture absorption: amount of saliva which mixes with sample.
                        [No mixture - Complete mixture]
                        Semisolid cohesiveness of mass: degree mass holds together.
                        [Loose mass - Cohesive mass]
                        Adhesiveness of mass: degree sample sticks to palate; force to remove
                        from palate.
                        [No force -Large force]

4. Residual   Feel mouth surface and teeth with tongue after product is swallowed
                        or expectorated; evaluate for:

Mouthcoating: amount of particles left on mouth surface.
[None - Extreme]
Oily film: amount of oil film on oral surface.
[None - Extreme]
Adhesiveness to teeth; amount of product left on tooth surfaces.
[None - Extreme]

D. Terms Used to Describe Solid Oral Texture

Each set of texture terms includes the procedure for manipulation of the sample.

1. Surface Texture — Feel surface of sample with lips and tongue.

   a. Geometrical in surface — The overall amount of small and large particles in the surface:
[Smooth - Rough]
Large particles: amount of bumps/lumps in surface:
[Smooth -Bumpy]
Small particles: amount of small grains in surface:
[Smooth - Grainy]

   b. Loose geometrical — Crumbly: amount of loose, grainy particles free of the surface:
[None -Many]

   c. Moistness/dryness — The amount of wetness or oiliness (moistness if both) on surface:
[Dry - Wet/oily/moist]

2. Partial Compression — Compress partially (specify with tongue, incisors, or molars) without breaking, and release.

   a. Springiness (rubberiness) — Degree to which sample returns to original shape after a certain time period:
[No recovery -Very springy]

3. First Bite — Bite through a predetermined size sample with incisors.

   a. Hardness — Force required to bite through:
[Very soft -Very hard]

   b. Cohesiveness — Amount of sample that deforms rather than ruptures:
[Breaks -Deforms]

   c. Fracturability — The force with which the sample breaks:
[Crumbles - Fractures]

   d. Uniformity of bite — Evenness of force throughout bite:
[Uneven, choppy -Very even]

   e. Moisture release — Amount of wetness/juiciness released from sample:
[None - Very juicy]

   f. Geometrical — Amount of particles resulting from bite, or detected in center of sample:
[None -Very grainy (gritty, flaky, etc.)]

4. First Chew — Bite through a predetermined size sample with molars.

   a. Hardness — As above:
[Very soft -Very hard]

   b. Cohesiveness/ fracturability — Both as above:
[Breaks -Deforms]
[Crumbles - Fractures]

    c. Adhesiveness   Force required to remove sample from molars:
                                      [Not sticky -Very sticky]

    d. Denseness   Compactness of cross section:
                                        [Light/airy - Dense]

    e. Crispness   The noise and force with which the sample breaks or fractures:
                                        [Not crisp/soggy - Very crisp]

    f. Geometrical   See definitions in surface texture:
                                        [None -Very grainy (gritty, flaky, etc.)]

    g. Moist/moisture   See definitions in surface texture or first bite texture:
        release   [None - Very juicy]

5. Chew Down   Chew sample with molars for a predetermined number of chews (enough to mix sample with saliva to form a mass):

    a. Moisture absorption Amount of saliva absorbed by product:
                                        [None - All]

    b. Cohesiveness of   Degree to which sample holds together in a mass:
        mass   [Loose mass - Compact mass]

    c. Adhesiveness of   Degree to which mass sticks to the roof of the mouth or teeth:
        mass   [Not sticky -Very sticky]

    d. Flinty/Glassy   The amount of sharp abrasive pieces in the mass:
                                        [None - Very many pieces]

6. Rate of Melt   When applicable: Amount of product melted after a certain number of chews:

                                        [None - All]

    a. Geometrical in mass Roughness/graininess/lumpiness: amount of particles in mass:
                                        [None -Many]

    b. Moistness of mass   Amount of wetness/oiliness/moisture in mass:
                                          [Dry - Moist/oily/wet]

    c. Number of chews   Count number.
        to disintegrate

7. Residual   Swallow or expectorate sample.

    a. Geometrical   (Chalky, particles) amount of particles left in mouth:
                                        [None - Very much]

    b. Oily mouth coating Amount of oil left on mouth surfaces:
                                        [None - Very much]

    c. Sticky mouth   Stickiness/tackiness of coating when tapping tongue on roof of mouth:
        coating   [Not sticky -Very sticky]

    d. Tooth packing   Amount of product left on teeth:
                                        [None - Very much]

Example: Solid Texture Terminology of Oral Texture of Cookies

    1. Surface   Place cookie between lips and evaluate for:
                                        Roughness: degree to which surface is uneven
                                        [Smooth - Rough]
                                        Loose particles: amount of loose particles on surface
                                        [None -Many]
                                        Dryness: absence of oil on the surface
                                        [Oily - Dry]

2. First bite   Place one third of cookie between incisors, bite down, and evaluate for:

Fracturability: force with which sample ruptures

[Crumbly - Brittle]

Hardness: force required to bite through sample

[Soft - Hard]

Particle size: size of crumb pieces

[Small -Large]

3. First chew   Place one third of cookie between molars, bite through, and evaluate for:

Denseness: compactness of cross section

[Airy - Dense]

Uniformity of chew: degree to which chew is even throughout

[Uneven - Even]

4. Chew down   Place one third of cookie between molars, chew 10 to 12 times, and evaluate for:

Moisture absorption: amount of saliva absorbed by sample

[None -Much]

Type of breakdown: thermal, mechanical, salivary

[No scale]

Cohesiveness of mass: degree to which mass holds together

[Loose - Cohesive]

Tooth pack: amount of sample stuck in molars

[None -Much]

Grittiness: amount of small, hard particles between teeth during chew

[None -Many]

5. Residual   Swallow sample and evaluate residue in mouth:

Oily: degree to which mouth feels oily

[Dry -Oily]

Particles: amount of particles left in mouth

[None -Many]

Chalky: degree to which mouth feels chalky

[Not chalky - Very chalky]

E. Terms Used to Describe Skinfeel of Lotions and Creams

1. Appearance

In a Petri dish, dispense the product in a spiral shape. Using a nickel-size circle, fill from edge to center.

a. Integrity of shape   Degree to which product holds its shape.

[Flattens- Retains shape]

b. Integrity of shape after 10 sec   Degree to which product holds its shape after 10 sec.

[Flattens- Retains shape]

c. Gloss   The amount of reflected light from product.

[Dull/flat -Shiny/glossy]

2. Pick Up

Using automatic pipet, deliver 0.1 cc of product to tip of thumb or index finger. Compress product slowly between finger and thumb one time.

   a. Firmness   Force required to fully compress product between thumb and index finger.
[No force - High force]

   b. Stickiness   Force required to separate fingertips.
[Not sticky - Very sticky]

   c. Cohesiveness   Amount sample strings rather than breaks when fingers are separated.
[No strings - High strings]

   d. Amount of peaking Degree to which product makes stiff peaks on fingertips.
[No peaks/flat - Stiff peaks]

3. Rub Out

Using automatic pipet, deliver 0.05 cc of product to center of 2" circle on inner forearm. Gently spread product within the circle using index or middle finger, at a rate of two strokes per second.

After three rubs, evaluate for:

   a. Wetness   Amount of water perceived while rubbing.
[None - High amount]

   b. Spreadability   Ease of moving product over the skin.
[Difficult/drag - Easy/slip]

After 12 rubs, evaluate for:

   c. Thickness   Amount of product felt between fingertip and skin.
[Thin, almost no product - - - - - - - - - - - - Thick, lots of product]

After 15–20 rubs, evaluate for:

   d. Oil   Amount of oil perceived in the product during rub-out.
[None - Extreme]

   e. Wax   Amount of wax perceived in the product during rub-out.
[None - Extreme]

   f. Grease   Amount of grease perceived in the product during rub-out.
[None - Extreme]

Continue rubbing and evaluate for:

   g. Absorbency   The number of rubs at which the product loses wet, moist feel and a resistance to continue is perceived [upper limit = 120 rubs].

4. Afterfeel (Immediate)

   a. Gloss   Amount or degree of light reflected off skin.
[Dull -Shiny]

   b. Sticky   Degree to which fingers adhere to product.
[Not sticky - Very sticky]

   c. Slipperiness   Ease of moving fingers across skin.
[Difficult/drag - Easy/slip]

   d. Amount of residue Amount of product on skin.
[None - Large amount]

   e. Type of residue   Oily, waxy, greasy, powdery, chalky.

F. Terms Used to Describe Handfeel of Fabric or Paper

   1. Force to gather   The amount of force required to collect/gather the sample toward the palm of the hand.
[Low force - High force]

2. Force to compress   The amount of force required to compress the gathered sample into the palm.
[Low force - High force]

3. Stiffness   The degree to which the sample feels pointed, ridged, and cracked; not pliable, round, curved.
[Pliable/round -Stiff]

4. Fullness   The amount of material/paper/fabric/sample felt in the hand during manipulation.
[Low amount of sample/flimsy - - - - - High amount of sample/body]

5. Compression resilience   The force with which the sample presses against cupped hands.
[Creased/folded - Original shape]

6. Depression depth   The amount that the sample depresses when downward force is applied.
[No depression - Full depression]

7. Depression resilience/ springiness   The rate at which the sample returns to its original position after depression is removed.
[Slow - Fast/springy]

8. Tensile stretch   The degree to which the sample stretches from its original shape.
[No stretch - High stretch]

9. Tensile extension   The degree to which the sample returns to original shape, after tensile force is removed. (Note: This is a visual evaluation.)
[No return -Fully returned]

10. Hand friction   The force required to move the hand across the surface.
[Slip/no drag - Drag]

11. Fabric friction   The force required to move the fabric over itself.
[Slip/no drag - Drag]

12. Roughness   The overall presence of gritty, grainy, or lumpy particles in the surface; lack of smoothness.
[Smooth - Rough]

13. Gritty   The amount of small abrasive picky particles in the surface of the sample.
[Smooth/not gritty - Gritty]

14. Lumpy   The amount of bumps, embossing, large fiber bundles in the sample.
[Smooth/not lumpy -Lumpy]

15. Grainy   The amount of small, rounded particles in the sample.
[Smooth/not grainy - Grainy]

16. Fuzziness   The amount of pile, fiber, fuzz on the surface.
[Bald -Fuzzy/nappy]

17. Thickness   The perceived distance between thumb and fingers.
[Thin -Thick]

18. Moistness   The amount of moistness on the surface and in the interior of the paper/fabric. Specify if the sample is oily vs. wet (water) if such a difference is detectable.
[Dry - Wet]

19. Warmth   The difference in thermal character between paper/fabric and hand.
[Cool - Warm]

20. Noise intensity   The loudness of the noise.
[Soft - Loud]

21. Noise pitch   Sound frequency of the noise.
[Low/bass -High/treble]

G. Terms Used to Describe the Feel of Hair (Wet and Dry)

Wet Hair Evaluation Procedure

1. Preparation before Application

Measure length of hair swatch from the end of the card to the end of the hair. Record the measurement. Pull hair swatch taut and measure as above. Record measurement.
    Usually evaluate for:

a.	Sheen	Amount of reflected light.
		[Dull -Shiny]

    Comb through swatch with rattail comb. At third stroke of combing, evaluate for:

b.	Combability (dry) (top half of swatch)	Ease with which comb can be moved down hair shafts without resistance or hair tangling.
		[Difficult - Easy]
c.	Combability (dry) (bottom half of swatch)	Ease with which comb can be moved down hair shafts without resistance or hair tangling.
		[Difficult - Easy]
d.	"Fly away" hair	The tendency of the individual hairs to repel each other during combing after three strokes of combing down hair shafts.
		[None -Much]

2. Application of Lotion

Dip hair swatch into cup of room temperature (72°) tap water. Thoroughly wet hair swatch. Squeeze out excess water. Pipet 0.125 cc of hair lotion onto edge of palm of hand. Using opposite index and middle fingers, rub onto edge of palm 2–3 times to distribute lotion. Pick up hair swatch by the card. Using long, even strokes, from the top to bottom, apply lotion to hair swatch, turning card after each stroke, rubbing ends of swatch with index and middle fingers. Evaluate for:

a.	Ease of distribution	Ease of rubbing product over hair.
		[Difficult - Easy]
b.	Amount of residue	The amount of residue left on the surface of the hands. (Untreated skin = 0)
		[None -Extreme]
c.	Type of residue	Oily, waxy, greasy.

3. Evaluation

Clean hands with water before proceeding. Comb through hair swatch with a rattail comb one time and evaluate for:

a.	Ease of detangling	Ease to comb through hair.
		[Very tangled, hard to comb - - - - - - - - Not tangled, easy to comb]

    At the third stroke of combing evaluate for:

b.	Combability (wet) (top half of swatch)	Ease with which comb can be moved down hair shafts without resistance or hair tangling.
		[Difficult - Easy]
c.	Combability (wet) (bottom half of swatch)	Ease with which comb can be moved down hair shafts without resistance or hair tangling.
		[Difficult - Easy]
d.	Stringiness (visual)	The sticking of individual hairs together in clumps.
		[Unclumped - Clumped]
e.	Wetness (tactile)	The amount of perceived moisture.
		[Dry - Wet]

f.	Coldness (tactile)	Thermal sensation of lack of heat.
		[Hot - Cold]
g.	Slipperiness (tactile)	Lack of drag or resistance as moving along hairs between fingers.
		[Drags - Slips]
h.	Roughness (tactile)	A rough, brittle texture of hair shafts.
		[Smooth - Rough]
i.	Coatedness (tactile)	The amount of residue left on the hair shaft.
		[None, uncoated - Very coated]
j.	Stickiness of hair to skin (tactile)	The tendency of the hair to stick to the fingers.
		[Not sticky - Very sticky]

4. Evaluation after drying

Let hair swatch dry for 30 minutes lying on clean paper towels checking swatch at 5-minute intervals and evaluate earlier if dried. At the third stroke of combing evaluate for:

a.	Combability (dry) (top half of swatch)	Ease with which comb can be moved down hair shafts without resistance or hair tangling.
		[Difficult - Easy]
b.	Combability (dry) (bottom half of swatch)	Ease with which comb can be moved down hair shafts without resistance or hair tangling.
		[Difficult - Easy]
c.	"Fly away" hair	The tendency of the individual hairs to repel each other during combing after three strokes of combing down hair shafts.
		[None -Much]
d.	Stringiness (visual)	The sticking of individual hairs together in clumps.
		[Unclumped - Clumped]
e.	Sheen	Amount of reflected light.
		[Dull -Shiny]
f.	Roughness (tactile)	A rough, brittle texture of hair shafts.
		[Smooth - Rough]
g.	Coatedness (tactile)	The amount of residue left on the hair shaft.
		[None, uncoated - Very coated]

DRY HAIR EVALUATION PROCEDURE

1. Preparation before Application

Measure length of hair swatch from the end of the card to the end of the hair. Record the measurement. Pull hair swatch taut and measure as above. Record measurement. Visually evaluate hair for:

| a. | Sheen | Amount of reflected light. |
| | | [Dull -Shiny] |

Comb through hair with rattail comb. At third stroke of combing, evaluate for:

b.	Combability (dry) (top half of swatch)	Ease with which comb can be moved down hair shafts without resistance or hair tangling.
		[Difficult - Easy]
c.	Combability (dry) (bottom half of swatch)	Ease with which comb can be moved down hair shafts without resistance or hair tangling.
		[Difficult - Easy]
d.	"Fly away" hair	The tendency of the individual hairs to repel each other during combing after three strokes of combing down hair shafts.
		[None -Much]

2. Application of Lotion

Pipet 0.125 cc of hair lotion onto edge of palm of hand. Using opposite index and middle fingers, rub onto edge of palm 2–3 times to distribute lotion. Pick up hair swatch by the card. Using long, even strokes, from the top to bottom, apply lotion to hair swatch, turning card after each stroke, rubbing ends of swatch with index and middle fingers. Evaluate for:

    a. Ease of distribution   Ease of rubbing product over hair.
                                    [Difficult - Easy]
    b. Amount of residue   The amount of residue left on the surface of the hands.
                                      (Untreated skin = 0)
                                      [None -Extreme]
    c. Type of residue   Oily, waxy, greasy.

3. Evaluation

Clean hands with water before proceeding. Comb through hair swatch with a rattail comb. At the third stroke of combing evaluate for:

    a. Combability (wet)   Ease with which comb can be moved down hair shafts without
         (top half of swatch) resistance or hair tangling.
                                      [Difficult - Easy]
    b. Combability (wet)   Ease with which comb can be moved down hair shafts without resis-
         (bottom half of   tance or hair tangling.
         swatch)   [Difficult - Easy]
    c. Stringiness (visual)   The sticking of individual hairs together in clumps.
                                        [Unclumped - Clumped]
    d. Wetness (tactile)   The amount of perceived moisture.
                                          [Dry - Wet]
    e. Coldness (tactile)   Thermal sensation of lack of heat.
                                          [Hot - Cold]
    f. Slipperiness (tactile) Lack of drag or resistance as moving along hairs between fingers.
                                          [Drags - Slips]
    g. Roughness (tactile)   A rough, brittle texture of hair shafts.
                                         [Smooth- Rough]
    h. Coatedness (tactile)   The amount of residue left on the hair shaft.
                                        [None, uncoated - Very coated]
    i. Stickiness of hair   The tendency of the hair to stick to the fingers.
         to skin (tactile)   [Not sticky -Very sticky]

4. Evaluation after drying

Let hair swatch dry for 30 minutes lying on clean paper towels, checking swatch at 5-minute intervals and evaluate earlier if dried. Record drying time. Measure length of hair swatch from the end of the card to the end of the hair. Record the measurement. Pull hair swatch taut and measure as above. Record measurement. Comb through hair swatch with rattail comb. At the third stroke of combing evaluate for:

    a. Combability (dry)   Ease with which comb can be moved down hair shafts without
         (top half of swatch) resistance or hair tangling.
                                      [Difficult - Easy]
    b. Combability (dry)   Ease with which comb can be moved down hair shafts without
                                  bottom half of swatch resistance or hair tangling.
                                      [Difficult - Easy]

   c. "Fly away" hair   The tendency of the individual hairs to repel each other during combing after three strokes of combing down hair shafts.
[None -Much]

   d. Stringiness (visual)   The sticking of individual hairs together in clumps.
[Unclumped - Clumped]

   e. Sheen   Amount of reflected light.
[Dull -Shiny]

   f. Roughness (tactile)   A rough, brittle texture of hair shafts.
[Smooth - Rough]

   g. Coatedness (tactile)   The amount of residue left on the hair shaft.
[None, uncoated - Very coated]

H. Terms Used to Describe the Lather and Skinfeel of Bar Soap

Full Arm Test

1. Preparation for Skinfeel Test

Instruct panelists to refrain from using any type of moisturizing cleansers on evaluation days (these include bar soaps and cleansing creams, lotions, and astringents). Also ask panelists to refrain from applying lotions, creams, or moisturizers to their arms on the day of evaluation. Panelists may, however, rinse their arms with water and pat dry.

Limit panelists to evaluation of no more than two samples per day (1 sample per site, beginning with the left arm). For the second soap sample, repeat the washing procedure on the right arm evaluation site. Wash each site once only.

2. Baseline Evaluation of Site:

Visually evaluate skin for:

   a. Gloss   The amount or degree of light reflected off skin.
[Dull -Shiny]

   b. Visual dryness   The degree to which the skin looks dry (ashy/flaky).
[None -Very dry]

       Stroke cleansed fingers lightly across skin and evaluate for:

   c. Slipperiness   Ease of moving fingers across the skin.
[Drag - Slip]

   d. Amount of residue   The amount of residue left on the surface of the skin.
[None -Extreme]

   e. Type of residue   Indicate the type of residue:
Soap film, oily, waxy, greasy, powder.

   f. Dryness/roughness   The degree to which the skin feels rough.
[Smooth - Rough]

   g. Moistness   The degree to which the skin feels moist.
[Dry -Moist]

   h. Tautness   The degree to which the skin feels taut or tight.
[Loose/Pliable -Very tight]

       Using edge of fingernail, scratch a line through the test site. Visually evaluate for:

   i. Whiteness   The degree to which the scratch appears white.
[None - Very white]

3. *Evaluation of Lather and Skinfeel:*

Application and Washing Procedure. Apply wet soap bar to wet evaluation site. Apply with up-down motion (1 up-down lap = ½ second).

    a. Amount of lather observed during application.

        at 10, 20, 30 laps   [None - Extreme]

At 30 laps continue with

    b. Thickness of lather   Amount of product felt between fingertips and skin.

                                     [Thin -Thick]

    c. Bubble size variation The variation seen within the bubble size.

                                     [Homogeneous - Heterogeneous]

    d. Bubble size   The size of the soap bubbles in the lather (visual).

                                     [Small -Large]

Rinsing procedure. Rinse site by placing arm directly under warm running water. Use free hand to stroke gently with up-down lap over the site. Rinse for 15 laps. (1 lap = 1 second). Also rinse evaluation fingers.

Evaluation before drying.

    a. Rinsability   The degree to which the sample rinses off (visual).

                                     [None - All]

Gently stroke upward on skin site with a clean finger and evaluate for:

    b. Slipperiness   Ease of moving fingers across the skin.

                                     [Drag- Slip]

    c. Amount of residue   The amount of residue left on the surface of the skin.

                                     [None - - - - - - - - - - - - - - - -,- - - - - - - - - - - - - - - - - -Extreme]

    d. Type of residue   Indicate the type of residue: soap film, oily, waxy, greasy, powder.

Evaluation after drying. Dry the site by covering it with a paper towel and patting dry *3 times* along the site. Also thoroughly dry evaluation finger. Tap dry, cleansed finger over treated skin. Evaluate for:

    a. Gloss   Visual: amount of light reflected on the surface of the skin.

                                       [Dull - Shiny/glossy]

    b. Visual dryness   The degree to which the skin looks dry (ashy/flaky).

                                       [None -Very dry]

Gently stroke skin site with clean finger and evaluate for:

    c. Stickiness   The degree to which fingers stick to residual product on the skin.

                                     [Not sticky -Very sticky]

    d. Slipperiness   Ease of moving fingers across the skin.

                                       [Drag- Slip]

    e. Amount of residue   The amount of residue left on the surface of the skin.

                                       [None -Extreme]

    f. Type of residue   Indicate the type of residue: Soap film, oily, waxy, greasy, powder.

    g. Dryness/roughness   The degree to which the skin feels dry/rough.

                                     [Smooth- Dry/rough]

    h. Moistness   The degree to which the skin feels moist, wet.

                                       [Dry -Moist]

    i. Tautness   The degree to which the skin feels taut or tight.

                                       [Loose/pliable- Very taut]

        Using the edge of the fingernail, scratch through test site and evaluate for:

    j. Whiteness   The degree to which the scratch appears white.

                                     [None - Very white]

I. Terms Used to Describe the Skinfeel of Antiperspirants

Roll-On/Solids/Gels

1. Preparation of Skin

Evaluation site (crook of arm) is washed with non-abrasive, non-deodorant soap (such as Neutrogena) more than 1 hour before evaluation. A 6" × 2" rectangle is marked on the crook of the arm so the fold bisects the rectangle.

2. Baseline Evaluation:

Prior to application, instruct panelists to evaluate untreated sites for baseline references.
Visually evaluate skin for:

a. Gloss   The amount or degree of light reflected off skin.
                        [Dull - Shiny]
b. Visual dryness   The degree to which the skin looks dry (ashy/flaky).
                        [None -Very dry]

Stroke cleansed fingers lightly across skin and evaluate for:

c. Slipperiness   Ease of moving fingers across the skin.
                        [Drag - Slip]
d. Amount of residue   The amount of residue left on the surface of the skin.
                        [None - Extreme]
e. Type of residue   Indicate the type of residue:
                        Soap film, oily, waxy, greasy, powder.
f. Dryness/roughness   The degree to which the skin feels rough.
                        [Smooth - Rough]
g. Moistness   The degree to which the skin feels moist.
                        [Dry -Moist]
h. Tautness   The degree to which the skin feels taut or tight.
                        [Loose/pliable -Very tight]

Using edge of fingernail, scratch a line through the test site. Visually evaluate for:

i. Whiteness   The degree to which the scratch appears white.
                        [None - Very white]

3. Application of Antiperspirant

Roll-on gels: Pipet 0.05 cc of product at 2 spots along the 2" bottom and top of the 2" × 6" rectangle evaluation site. Spread the product on the site using 12 rubs (6 laps) with a vinyl-covered finger.

Solids/Gels: Apply the product by stroking up the arm once through the 2" × 6" rectangle (force to apply), then back down and up the arm three times (ease to spread), using a consistent pressure to get the product on the arm. A tare weight is taken of each application and recorded.

4. Immediate Evaluation

Immediately after application, evaluate for:

a. Coolness   The degree to which the sample feels "cool" on the skin (somesthetic).
                        [Not at all cool - Very cool]
b. Gloss   The amount of reflected light from the skin.
                        [Not at all shiny - Very shiny]
c. Whitening   The degree to which the skin turns white.
                        [None - Very white]

d. Amount of residue   The amount of product visually perceived on the skin (visual).

[None - Large amount]

e. Tautness   The degree to which the skin feels taut or tight.

[Loose/pliable - Very tight]

Fold arm to make contact. Hold 5 seconds. Unfold arm and evaluate for:

f. Stickiness (fold)   Degree to which arm sticks to itself.

[Not at all - Very sticky]

Stroke finger lightly across skin on one section of rectangle and evaluate for:

g. Wetness   The amount of water perceived on the skin.

[None - High amount]

h. Slipperiness   Ease of moving fingers across the skin.

[Drag - Slip]

i. Amount of residue   The amount of residue perceived on skin (tactile).

Evaluate by stroking finger across site.

[None - Extreme]

j. Oil   The amount of oil perceived on skin.

[None - Extreme]

k. Wax   The amount of wax perceived on skin.

[None - Extreme]

l. Grease   The amount of grease perceived on skin.

[None - Extreme]

m. Powder/chalk/grit   The amount of powder, chalk and/or grit perceived on skin.

[None - Extreme]

n. Silicone   The amount of silicone perceived on skin.

[None - Occluded]

5. *After 5, 10, 15, and 30 minutes, evaluate for:*

a. Occlusion   The degree to which the sample occludes or blocks the air passage to the skin.

[None - Occluded]

b. Whitening   The degree to which the skin turns white.

[None - Large amount]

c. Amount of residue   The amount of product visually perceived on skin (visual).

[None - Large amount]

d. Tautness   The degree to which the skin feels taut or tight.

[Loose/pliable - Very tight]

Fold arm to make contact. Hold 5 seconds. Unfold arm and evaluate for:

e. Stickiness   The degree to which arm sticks to itself.

[Not at all sticky - Very sticky]

Stroke fingers lightly across skin on one section of rectangle and evaluate for:

f. Wetness   The amount of water perceived on the skin.

[None - High amount]

g. Slipperiness   Ease of moving fingers across the skin.

[Drag - Slip]

h. Amount of residue   The amount of residue perceived on skin (tactile).

[None - Extreme]

i. Oil   The amount of oil perceived on skin.

[None - Extreme]

j. Wax   The amount of wax perceived on skin.

[None - Extreme]

k. Grease — The amount of grease perceived on skin.
[None - Extreme]

l. Powder/Chalk/Grit — The amount of powder, chalk, and/or grit perceived on skin.
[None - Extreme]

m. Silicone — The amount of silicone perceived on skin.
[None - Extreme]

6. After 30 minutes, evaluate as follows:

Place a swatch of black fabric over test site. Fold arm so fingertips touch the shoulder. Pull fabric from crook.

a. Rub-off whitening — The amount of residue on the dark fabric.
[None - Large amount]

Appendix 11.2
Spectrum Intensity Scales
for Descriptive Analysis

The scales below (all of which run from 0 to 15) contain intensity values for aromatics (A), and for tastes and chemical feeling factors (B), which were derived from repeated tests with trained panels at Hill Top Research, Inc., Cincinnati, Ohio, and also for various texture characteristics (C), which were obtained from repeated tests at Hill Top Research or which were developed at Bestfoods Technical Center, Somerset, New Jersey.

New panels can be oriented to the use of the 0 to 15 scale by presentation of the concentrations of caffeine, citric acid, NaCl, and sucrose which are listed under Section B. If a panel is developing a descriptive system for an orange drink product, the panel leader can present three "orange" references:

1. Fresh squeezed orange juice labeled "Orange Complex 7.5"
2. Reconstituted Minute Maid concentrate labeled "Orange Complex 6.5 *and* Orange Peel 3.0"
3. Tang labeled "Orange Complex 9.5 *and* Orange Peel 9.5"

At each taste test of any given product, labeled reference samples related to its aromatic complex can be presented, so as to standardize the panel's scores and keep panel members from drifting.

A. Intensity Scale Values (0 to 15) for Some Common Aromatics

Term	Reference	Scale value
Astringency	Grape juice (Welch's)	6.5
	Tea bags/1 h soak	6.5
Baked wheat	Sugar cookies (Kroger)	4
	Brown Edge cookies (Nabisco)	5
Baked white wheat	Ritz crackers (Nabisco)	6.5
Caramelized sugar	Brown Edge cookies (Nabisco)	3
	Sugar cookies (Kroger)	4
	Social Tea (Nabisco)	4
	Bordeaux cookies (Pepperidge Farm)	7
Celery	V-8 vegetable juice (Campbell)	5
Cheese	American cheese (Kraft)	5
Cinnamon	Big Red gum (Wrigley)	12
Cooked apple	Applesauce (Mott)	5
Cooked milk	Butterscotch pudding (Royal)	4
Cooked orange	Frozen orange concentrate (Minute Maid) — reconstituted	5.0
Cooked wheat	Pasta (De Cecco) — cooked	5.5
Egg	Mayonnaise (Hellmann's)	5
Egg flavor	Hard-boiled egg	13.5

A. Intensity Scale Values (0 to 15) for Some Common Aromatics (continued)

Term	Reference	Scale value
Grain Complex	Cream of Wheat (Nabisco)	4.5
	Spaghetti (De Cecco) — cooked	4.5
	Ritz cracker (Nabisco)	6
	Whole wheat spaghetti (De Cecco) — cooked	6.5
	Triscuit (Nabisco)	8
	Wheatina cereal	9
Grape	Kool-Aid	5.0
	Grape juice (Welch's)	10
Grapefruit	Bottled grapefruit juice (Kraft)	8.0
Lemon	Brown Edge cookies (Nabisco)	3
	Lemonade (Country Time)	5
Milky Complex	American cheese (Kraft)	3
	Powdered milk (Carnation)	4
	Whole milk	5
Mint	Doublemint gum (Wrigley)	11
Oil	Potato chips (Pringles)	1
	Potato chips (Frito-Lay)	2
	Heated oil (Crisco)	4.0
Orange Complex	Orange drink (Hi-C)	3
	Frozen orange concentrate (Minute Maid) — reconstituted	7.0
	Fresh-squeezed orange juice	8.0
	Orange concentrate (Tang)	9.5
Orange Peel	Soda (Orange Crush)	2
	Frozen orange concentrate (Minute Maid) — reconstituted	3
	Orange concentrate (Tang)	9.5
Peanut, med. roasted	(Planters)	7
Potato	Potato chips (Pringles)	4.5
Roastedness	Coffee (Maxwell House)	7
	Espresso coffee (Medaglia D'Oro)	14
Soda (Baking)	Saltines (Nabisco)	2
Spice Complex	Spice cake (Sara Lee)	7.5
Tuna	Canned light tuna (Bumble Bee)	11
Vanillin	Sugar cookies (Kroger)	7

B. Intensity Scales Values (0 to 15) for the Four Basic Tastes

	Sweet	Salt	Sour	Bitter
American Cheese (Kraft)		7	5	
Applesauce, natural (Mott)	5		4	
Applesauce, regular (Mott)	8.5		2.5	
Big Red gum (Wrigley)	11.5			
Bordeaux cookies (Pepperidge Farm)	12.5			
Basic Taste Blends				
5% Sucrose/0.1% Citric Acid	6		7	

B. Intensity Scales Values (0 to 15) for the Four Basic Tastes (continued)

	Sweet	Salt	Sour	Bitter
5% Sucrose/0.55% NaCl	7	9		
0.1% Citric Acid/0.55% NaCl		11	6	
5% Sucrose/0.1% Citric Acid/0.3% NaCl	5	5	3.5	
5% Sucrose/0.1% Citric Acid/0.55% NaCl	4	11	6	
Caffeine, solution in water				
0.05%				2
0.08%				5
0.15%				10
0.20%				15
Celery seed				9
Chocolate bar (Hershey)	10		5	4
Citric acid, solution in water				
0.05%			2	
0.08%			5	
0.15%			10	
0.20%			15	
Coca-Cola Classic	9			
Endive, raw				7
Fruit punch (Hawaiian)	10		3	
Grape juice (Welch's)	6		7	2
Grape Kool-Aid	10		1	
Grapefruit juice, bottled (Kraft)	3.5		13	2
Kosher dill pickle (Vlasic)		12	10	
Lemon juice (ReaLemon)			15	
Lemonade (Country Time)	7		5.5	
Mayonnaise (Hellmann's)		8	3	
NaCl, solution in water				
0.2%		2.5		
0.35%		5		
0.5%		8.5		
0.7%		15		
Orange (fresh-squeezed juice)	6		7.5	
Soda (Orange Crush)	10.5		2	
Frozen orange concentrate (Minute Maid) — reconstituted	5.5		5	
Potato chips (Frito-Lay)		9.5		
Potato chips (Pringles)		8.5		
Ritz cracker (Nabisco)	4	8		
Soda cracker (Premium)		5		
Spaghetti sauce (Ragu)	8	12		
Sucrose, solution in water				
2.0%	2			
5.0%	5			
10.0%	10			
16.0%	15			
Sweet pickle (Vlasic)	8.5		8	
Orange concentrate (Tang)	9.5		4.5	
Tea bags/1 h soak				8
Triscuit (Nabisco)		9.5		
V-8 vegetable juice (Campbell)		8		
Wheatina cereal		6		2.5

C. Intensity Scale Values (0 to 15) for Semisolid Oral Texture Attributes

Scale value	Reference	Brand/type/mfg.	Sample size

1. Slipperiness

2.0	Baby food — beef	Gerber	1 oz
3.5	Baby food — peas	Gerber	1 oz
7.5	Vanilla yogurt	Whitney's	1 oz
11.0	Sour cream	Breakstone	1 oz
13.0	Miracle Whip	Kraft Foods	1 oz

2. Firmness

3.0	Aerosol whipped cream	Redi-Whip	1 oz
5.0	Miracle Whip	Kraft	1 oz
8.0	Cheese Whiz	Kraft	1 oz
11.0	Peanut butter	CPC Best Foods	1 oz
14.0	Cream cheese	Kraft/Philadelphia Light	1 oz

3. Cohesiveness

1.0	Instant gelatin dessert Jello	Kraft-General Foods	½ in. cube
5.0	Instant vanilla pudding Jello	Kraft-General Foods	1 oz
8.0	Baby food — bananas	Gerber or Beechnut	1 oz
11.0	Tapioca pudding	Canned	1 oz

4. Denseness

1.0	Aerosol whipped cream	Reddi-Whip	1 oz
2.5	Marshmallow Fluff	Fluff	1 oz
5.0	Nougat center	3 Musketeers Bar/M&M/Mars	½ in. cube
13.0	Cream cheese	Kraft/Philadelphia Light	½ in. cube

5. Particle Amount

0	Miracle Whip	Kraft-General Foods	1 oz
5.0	Sour cream &	Breakstone	1 oz
	instant Cream of Wheat	Nabisco	1 oz
10.0	Mayonnaise & corn flour	Hellmann's & Argo	1 oz

6. Particle Size

0.5	Lean cream	Sealtest	1 oz
3.0	Cornstarch	Argo	1 oz
10.0	Sour cream &	Breakstone	1 oz
	instant Cream of Wheat	Nabisco	1 oz
15.0	Baby rice cereal	Gerber	1 oz

7. Mouth Coating

3.0	Cooked cornstarch	Argo	1 oz
8.0	Pureed potato		1 oz
12.0	Tooth powder	Brand available	1 oz

D. Intensity Scale Values (0 to 15) for Solid Oral Texture Attributes

Scale value	Reference	Brand/type/manufacturer	Sample size
		1. Standard Roughness Scale[a]	
0.0	Gelatin dessert	Jello	2 tbsp.
5.0	Orange peel	Peel from fresh orange	½ in. piece
8.0	Potato chips	Pringles	5 pieces
12.0	Hard granola bar	Quaker Oats	½ bar
15.0	Rye wafer	Finn Crisp	½ in. square

Technique: Hold sample in mouth; feel the surface to be evaluated with the lips and tongue.
Definition: The amount of particles *in* the surface.
[Smooth - Rough]

2. Standard Wetness Scale

Scale value	Reference	Brand/type/manufacturer	Sample size
0.0	Unsalted Premium cracker	Nabisco	1 cracker
3.0	Carrots	Uncooked, fresh, unpeeled	½ in. slice
7.5	Apples	Red Delicious, uncooked, fresh, unpeeled	½ in. slice
10.0	Ham	Oscar Mayer	½ in. piece
15.0	Water	Filtered, room temp.	½ tbsp

Technique: Hold the sample in mouth; feel surface with lips and tongue.
Definition: The amount of moisture, due to an aqueous system, on the surface.
[Dry - Wet]

3. Standard Stickiness to Lips Scale

Scale value	Reference	Brand/type/manufacturer	Sample size
0.0	Cherry tomato	Uncooked, fresh, unpeeled	½ in. slice
4.0	Nougat	Three Musketeers/	½ in. cake
		M&M-Mars	Remove chocolate first
7.5	Breadstick	Stella D'oro	½ stick
10.0	Pretzel rod	Bachmans	1 piece
15.0	Rice Krispies	Kellogg's	1 tsp

Technique: Hold sample near mouth; compress sample lightly between lips and release.
Definition: The degree to which the surface of the sample adheres to the lips.
[None -Very]

4. Standard Springiness Scale

Scale value	Reference	Brand/type/manufacturer	Sample size
0.0	Cream cheese	Kraft/Philadelphia Light	½ in. cube
5.0	Frankfurter	Cooked 10 min/Hebrew National	½ in. slice
9.5	Marshmallow	Miniature marshmallow/Kraft	3 pieces
15.0	Gelatin dessert	Jello, Knox (see *Note*)	½ in. cube

Technique: Place sample between molars; compress partially without breaking the sample structure; release.
Definition: (1) The degree to which sample returns to original shape or
(2) The rate with which sample returns to original shape.
[Not springy - Very springy]

Note: One package Jello and one package Knox gelatin are dissolved in 1½ cups hot water and refrigerated for 24 h.

[a] The roughness scale measures the amount of irregular particles *in* the surface. These may be small (chalky, powdery), medium (grainy), or large (bumpy).

D. Intensity Scale Values (0 to 15) for Solid Oral Texture Attributes (continued)

Scale value	Reference	Brand/type/manufacturer	Sample size
		5. Standard Hardness Scale	
1.0	Cream cheese	Kraft/Philadelphia Light	½ in. cube
2.5	Egg white	Hard cooked	½ in. cube
4.5	Cheese	Yellow American pasteurized process/Land O'Lakes	½ in. cube
6.0	Olives	Goya Foods/giant size, stuffed	1 olive pimento removed
7.0	Frankfurter	Large, cooked 5 min/Hebrew National	½ in. slice
9.5	Peanuts	Cocktail type in vacuum tin/Planters	1 nut, whole
11.0	Carrots	Uncooked, fresh, unpeeled	½ in. slice
11.0	Almonds	Shelled/Planters	1 nut
14.5	Hard candy	LifeSavers	3 pieces, one color

Technique: For solids, place food between the molars and bite down evenly, evaluating *the force required to compress the food.* For semisolids, measure hardness by compressing the food against palate with tongue. When possible, the same height for hardness standards is ½ in.

Definition: The force to attain a given deformation, such as:
- force to compress between molars, as above
- force to compress between tongue and palate
- force to bite through with incisors

[Soft- -Hard]

6. Standard Cohesiveness Scale

Scale value	Reference	Brand/type/manufacturer	Sample size
1.0	Corn muffin	Pepperidge Farm	½ in. cube
5.0	Cheese	Yellow American pasteurized process/Land O'Lakes	½ in. cube
8.0	Pretzel	Soft pretzel	½ in. piece
10.0	Dried fruit	Sun dried seedless raisins/Sun-Maid	1 tsp
12.5	Candy chews	Starburst/M&M/Mars	1 piece
15.0	Chewing gum	Freedent	1 stick

Technique: Place sample between molars; compress fully (can be done with incisors).

Definitions: The degree to which sample deforms rather than crumbles, cracks, or breaks.

[Rupturing - Deforming]

7. Standard Fracturability Scale

Scale value	Reference	Brand/type/manufacturer	Sample size
1.0	Corn muffin	Thomas'	½ in. cube
2.5	Egg Jumbos	Stella D'oro	½ in. cube
4.2	Graham crackers	Nabisco	½ in. cube
6.7	Melba toast	Plain, rectangular/Devonsheer, Melba Co.	½ in. square
8.0	Ginger snaps	Nabisco	½ in. square
10.0	Rye wafers	Finn Crisp/Shaffer, Clark & Co.	½ in. square
13.0	Peanut brittle	Kraft	½ in. square candy part
14.5	Hard candy	LifeSavers	1 piece

Technique: Place food between molars and bite down evenly until the food crumbles, cracks, or shatters.

Definition: The force with which the sample breaks.

[Crumbly -Brittle]

D. Intensity Scale Values (0 to 15) for Solid Oral Texture Attributes (continued)

Scale value	Reference	Brand/type/manufacturer	Sample size

8. Standard Viscosity Scale

Scale value	Reference	Brand/type/manufacturer	Sample size
1.0	Water	Bottled Mountain Spring	1 tsp
2.2	Light cream	Sealtest Foods	1 tsp
3.0	Heavy cream	Sealtest Foods	1 tsp
3.9	Evaporated milk	Carnation Co.	1 tsp
6.8	Maple syrup	Vermont Maid, R.J. Reynolds	1 tsp
9.2	Chocolate syrup	Hershey Chocolate	1 tsp
11.7	Mixture: ½ cup condensed milk + 1 tsp heavy cream	Magnolia/Borden Foods	1 tsp
14.0	Condensed milk	Magnolia/Borden Foods	1 tsp

Technique: (1) Place 1 tsp of product close to lips; draw air in gently to induce flow of liquid; measure the force required.
(2) Once product is in mouth, allow to flow across tongue by moving tongue slowly to roof of mouth; measure rate of flow (the force here is gravity).

Definition: The rate of flow per unit force:
• the force to draw between lips from spoon
• the rate of flow across tongue
[Not viscous - Viscous]

9. Standard Denseness Scale

Scale value	Reference	Brand/type/manufacturer	Sample size
0.5	Cool Whip	Birds Eye/General Foods	2 tbsp
2.5	Marshmallow Fluff	Fluff-Durkee-Mower	2 tbsp
4.0	Nougat	Three Musketeers/M&M/Mars	½ in. cube Remove chocolate first
6.0	Malted milk balls	Whopper, Leaf Confectionery	5 pieces
9.0	Frankfurter	Cooked 5 min, Oscar Mayer	5½ in. slices
13.0	Fruit jellies	Chuckles/Hershey	3 pieces

Technique: Place sample between molars and compress.
Definition: The compactness of the cross section.
[Airy - Dense]

10. Standard Crispness Scale

Scale value	Reference	Brand/type/manufacturer	Sample size
2.0	Granola Bar	Quaker Low Fat Chewy Chunk	⅓ bar
5.0	Club Cracker	Keeblers Partner Club Cracker	½ cracker
6.5	Graham Cracker	Honey Maid	1" sq.
7.0	Oat Cereal	Cheerios	1 oz.
9.5	Bran Flakes	Kellogg's Bran Flakes Cereal	1 oz.
11.0	Cheese Crackers Goldfish	Pepperidge Farm Cheddar Cheese Crackers	1 oz.
14.0	Corn Flakes	Kellogg's Corn Flakes Cereal	1 oz.
17.0	Melba Toast	Devonsheer Melba Toast	½ cracker

Technique: Place sample between molar teeth and bite down evenly until the food breaks, crumbles, cracks or shatters.
Definition: The force and noise with which a product breaks or fractures (rather than deforms) when chewed with the molar teeth (first and second chew).
[Not crisp/soggy - Very crisp]

D. Intensity Scale Values (0 to 15) for Solid Oral Texture Attributes (continued)

Scale value	Reference	Brand/type/manufacturer	Sample size

11. Standard Juiciness Scale

Scale value	Reference	Brand/type/manufacturer	Sample size
1.0	Banana	Banana	½" slice
2.0	Carrot	Raw Carrot	½" slice
4.0	Mushroom	Raw Mushroom	½" slice
7.0	Snap Bean	Raw Snap Bean	5 pieces
8.0	Cucumber	Raw Cucumber	½" slice
10.0	Apple	Red Delicious Apple	½" wedge
12.0	Honeydew Melon	Honey Dew Melon	½" cubes
15.0	Orange	Florida Juice Orange	½" wedge
15.0	Watermelon	Watermelon	½" cube (no seeds)

Technique: Chew sample with the molar teeth for up to 5 chews.
Definition: The amount of juice/moisture perceived in the mouth.
              [None- -Very]

12. Standard Flinty/Glassy Scale

Scale value	Reference	Brand/type/manufacturer	Sample size
2.0	Bugles Corn Snacks	Bugles Corn Snacks	1 oz.
4.0	Triples	Triples Cereal	1 oz.
8.0	Frosted Flakes	Kellogg's Frosted Flakes	1 oz.
12.5	Hard Candy	Candy Canes, Ribbon candy	1 piece

Technique: Chew sample 3 times and using the tongue measure the degree of pointiness of pieces and amount of pointy
              shards present.
Definition: The degree to which the sample breaks into pointy shards and the amount present after 3 chews.
              [None- Very/many]

13. Standard Crispness Scale

Scale value	Reference	Brand/type/manufacturer	Sample size
2.0	Granola Bar	Quaker Low Fat Chewy Chunk	⅓ bar
5.0	Club Cracker	Keeblers Partner Club Cracker	½ cracker
6.5	Graham Cracker	Honey Maid	1" square
7.0	Oat Cereal	Cheerios	1 oz.
9.5	Bran Flakes	Kellogg's Bran Flakes Cereal	1 oz.
11.0	Cheese Crackers	Pepperidge Farm Cheddar Cheese Goldfish	1 oz.
14.0	Corn Flakes	Kellogg's Corn Flakes Cereal	1 oz.
17.0	Melba Toast	Devonsheer Melba Toast	½ cracker

Technique: Place sample between molar teeth and bite down evenly until the food crumbles, cracks or shatters.
Definition: The force with which a product breaks or fractures (rather than deforms) when chewed with the molar teeth
              (first chew and second chew).
              [Deforms - Breaks}

14. Standard Moisture Absorption Scale

Scale value	Reference	Brand/type/manufacturer	Sample size
0.0	Licorice	Shoestring	1 piece
4.0	Licorice	Twizzlers/Red Licorice/Hershey	1 piece
7.5	Popcorn	Bagged popcorn/Bachman	2 tbsp
10.0	Potato chips	Wise	2 tbsp
13.0	Cake	Pound Cake, frozen type/Sara Lee	1 slice
15.0	Saltines	Unsalted top Premium cracker/Nabisco	1 cracker

Technique: Chew sample with molars for up to 15 to 20 chews.
Definition: The amount of saliva absorbed by sample during chew down.
              [No absorption -Large amount of absorption]

D. Intensity Scale Values (0 to 15) for Solid Oral Texture Attributes (continued)

Scale value	Reference	Brand/type/manufacturer	Sample size
		15. Standard Cohesiveness of Mass Scale	
0.0	Licorice	Shoestring	1 piece
2.0	Carrots	Uncooked, fresh, unpeeled	½" slice
4.0	Mushroom	Uncooked, fresh	½" slice
7.5	Frankfurter	Cooked 5 min/Hebrew National	½" slice
9.0	Cheese	Yellow American pasteurized process/Land O'Lakes	½" cube
13.0	Soft brownie	Archway Cookies	½" cube
15.0	Dough	Pillsbury/Country Biscuit Dough	1 tbsp

Technique: Chew sample with molars for up to 15 chews.
Definition: The degree to which chewed sample (at 10 to 15 chews) holds together in a mass.
[Loose mass - Tight mass]

		16. Standard Tooth Packing Scale	
0.0	Mini-clams	Geisha/Nozaki America	3 pieces
1.0	Carrots	Uncooked, fresh, unpeeled	½" slice
3.0	Mushrooms	Uncooked, fresh, unpeeled	½" slice
7.5	Graham cracker	Nabisco	½" square
9.0	Cheese	Yellow American pasteurized process/Land O'Lakes	½" cube
11.0	Cheese Snacks	Wise-Borden Cheese Doodles	5 pieces
15.0	Candy	Jujubes	3 pieces

Technique: After sample is swallowed, feel the tooth surfaces with tongue.
Definition: The degree to which product sticks on the surface of teeth.
[None stuck - Very much stuck]

E. Intensity Scale Values (0 to 10) for Skinfeel Texture Attributes

Scale value	Product	Manufacturer
	1. Integrity of Shape (Immediate)	
0.7	Baby Oil	Johnson & Johnson
4.0	Therapeutic Keri Lotion	Westwood Pharmaceut.
7.0	Vaseline Intensive Care	Chesebrough-Pond's
9.2	Lanacane	Combe Inc.
	2. Integrity of Shape (After 10 sec)	
0.3	Baby Oil	Johnson & Johnson
3.0	Therapeutic Keri Lotion	Westwood Pharmaceut.
6.5	Vaseline Intensive Care	Chesebrough-Pond's
9.2	Lanacane	Combe Inc.
	3. Gloss	
0.5	Gillette Foamy Reg. Shave Cream	Gillette Co.
3.6	Fixodent	Richardson Vicks
6.8	Neutrogena Hand Cream	Neutrogena
8.0	Vaseline Intensive Care	Chesebrough-Pond's
9.8	Baby Oil	Johnson & Johnson
	4. Firmness	
0	Baby Oil	Johnson & Johnson
1.3	Oil of Olay	Olay Company, Inc.
2.7	Vaseline Intensive Care	Chesebrough-Pond's
5.5	Ponds Cold Cream	Chesebrough-Pond's
8.4	Petrolatum	generic
9.8	Lanolin Wax	Amerchol
	5. Stickiness	
0.1	Baby Oil	Johnson & Johnson
1.2	Oil of Olay	Olay Company, Inc.
2.6	Vaseline Intensive Care	Chesebrough-Pond's
4.3	Jergens Aloe & Lanolin	Jergens Skin Care Laboratories
8.4	Petrolatum	generic
9.9	Lanolin Wax	Amerchol
	6. Cohesiveness	
0.2	Noxema Skin Care	Noxell
0.5	Vaseline Intensive Care	Chesebrough-Pond's
5.0	Jergens	Jergens Skin Care Laboratories
7.9	Zinc Oxide	generic
9.2	Petrolatum	generic
	7. Peaking	
0	Baby Oil	Johnson & Johnson
2.2	Vaseline Intensive Care	Chesebrough-Pond's
4.6	Curel	S.C. Johnson & Son
7.7	Zinc Oxide	generic
9.6	Petrolatum	generic

E. Intensity Scale Values (0 to 10) for Skinfeel Texture Attributes (continued)

Scale value	Product	Manufacturer
	8. Wetness	
0	Talc	Whitaker, Clark & Daniels, Inc.
2.2	Petrolatum	generic
3.5	Baby Oil	Johnson & Johnson
6.0	Vaseline Intensive Care	Chesebrough-Pond's
8.8	Aloe Vera Gel	Nature's Family
9.9	Water	—
	9. Spreadability	
0.2	AAA Lanolin	Amerchol
2.9	Petrolatum	generic
6.9	Vaseline Intensive Care	Chesebrough-Pond's
9.7	Baby Oil	Johnson & Johnson
	10. Thickness	
0.5	Isopropyl alcohol	generic
3.0	Petrolatum	generic
6.5	Vaseline Intensive Care	Chesebrough-Pond's
8.7	Neutrogena Hand Cream	Neutrogena
	11. Amount of Residue	
0	Untreated skin	—
1.5	Vaseline Intensive Care	Chesebrough-Pond's
4.1	Therapeutic Keri Lotion	Westwood Pharmaceut.
8.5	Petrolatum	generic

F. Intensity Scale Values (0 to 15) for Fabricfeel Attributes

Scale value	Fabric type	Testfabrics ID#[a]
	1. Stiffness	
1.3	Polyester/cotton 50/50 single knit tubular	7421
4.7	Mercerized cotton print cloth	400M
8.5	Mercerized combed cotton poplin	407
14.0	Cotton organdy	447
	2. Force to Gather	
1.5	Polyester cotton 50/50 single knit tubular	7421
3.5	Cotton cloth greige	400R
7.5	Cotton terry cloth	420
14.5	#10 Cotton duck greige	426
	3. Force to Compress	
1.5	Polyester/cotton 50/50 single knit tubular	7421
3.4	Cotton cloth greige	400R
9.3	Cotton terry cloth	420
14.5	#10 Cotton duck greige	426
	4. Depression Depth	
0.7	Cotton print cloth	400
1.8	S.N. cotton duck	464
6.4	Texturized polyester interlock knit fabric	730
12.4	Texturized polyester double knit twill	719
15.0	Cotton terry cloth	420
	5. Springiness	
0.7	Cotton print cloth	400
1.8	S.N. cotton duck	464
6.2	Texturized polyester interlock knit fabric	730
10.0	Cotton terry cloth	420
12.6	Texturized polyester double knit twill	719
	6. Fullness/Body	
1.6	Combed cotton batiste	735
4.9	Bleached mercerized cotton	409
7.8	Cotton single knit	473
13.3	Cotton fleece	484
	7. Tensile Stretch	
0.5	#8 Cotton duck greige	474
2.6	Spun viscose challis	266W
10.6	Texturized polyester double knit twill	719
15.0	Texturized polyester interlock knit fabric	730
	8. Compression Resilience: Intensity	
0.9	Polyester/cotton 50/50 single knit fabric	7421
3.8	Cotton cloth greige	400R
9.5	Acetate satin bright ward, delustered filling	105B
14.0	#10 Cotton duck greige	426

F. Intensity Scale Values (0 to 15) for Fabricfeel Attributes (continued)

Scale value	Fabric type	Testfabrics ID#[a]
	9. Compression Resilience: Rate	
1.0	Polyester/cotton 50/50 single knit tubular	7421
7.0	Filament nylon 6.6 semidull taffeta	306A
14.0	Dacron	738
	10. Thickness	
1.3	Filament nylon 6.6 semidull taffeta	306A
3.3	Cotton print cloth	400
6.5	Blended, mercerized cotton sheeting	409
13.0	#10 Cotton duck greige	426
	11. Fabric to Fabric Friction	
1.7	Filament nylon 6.6 semidull taffeta	306A
5.0	Dacron	738
10.0	Acetate satin bright ward, delustered filling	105B
15.0	Cotton fleece	484
	12. Fuzzy	
0.7	Dacron	738
3.6	Cotton crinkle gauze	472
7.0	Cotton T-shirt, tubular	437W
13.6	Cotton fleece	484
	13. Hand Friction	
1.4	Filament nylon 6.6 semidull taffeta	306A
3.5	Cotton Egyptian shirting	490
7.2	Cotton print cloth	400
10.0	Cotton flannel	425
14.2	Cotton terry cloth	420
	14. Noise intensity	
1.6	Cotton flannel	425
2.7	Cotton crinkle gauze	472
6.3	Cotton organdy	447
14.5	Dacron 56 taffeta	738
	15. Noise Pitch	
1.5	Cotton flannel	427
2.5	Cotton crinkle gauze	472
7.2	Cotton organdy	447
14.5	Dacron 56 taffeta	738
	16. Gritty	
1.5	Filament arnel tricot	116
6.0	Cotton cloth, greige	400R
10.0	Cotton print cloth	400
11.5	Cotton organdy	447

F. Intensity Scale Values (0 to 15) for Fabricfeel Attributes (continued)

Scale value	Fabric type	Testfabrics ID#[a]
	17. Grainy	
2.1	Mercerized combed cotton poplin	407
4.9	Carded cotton sateen bleached	428
9.5	Cotton tablecloth fabric	455
13.6	#8 Cotton duck greige	474

[a] Testfabrics identification numbers are the product numbers of Testfabrics Inc., P.O. Box 26, West Pittston, PA 18643

Appendix 11.3
Spectrum Descriptive Analysis
Product Lexicons

A. WHITE BREAD FLAVOR

1. Aromatics
Grain Complex
   Raw white wheat (dough)
   Cooked white wheat
   Toasted
   Cornstarch
   Whole grain
Yeasty/fermented
Dairy Complex
   Milk, cooked milk
   Buttery, brown butter
Eggy
Sweet Aromatic Complex:
   Caramelized/honey/malty/fruity
Mineral: inorganic, stones, cement, metallic
Baking Soda
Vegetable Oil
Other Aromatics: Mushroom, carrot, earthy, fermented, acetic, plastic, cardboard, chemical leavening

2. Basic Tastes
Salty
Sweet
Sour
Bitter

3. Chemical Feeling Factors
Metallic
Astringent/drying
Phosphate
Baking soda feel

B. WHITE BREAD TEXTURE

1. Surface
Crumb texture
   Roughness
   Loose particles
   Moistness
Crust Texture
   Roughness
   Loose particles
   Moistness

2. First Chew
Crumb denseness
Crumb cohesiveness
Crumb firmness
Crust hardness
Crust denseness
Crust cohesiveness

3. Partial Compression
Crumb springiness

4. Chewdown
Moisture absorption
Moistness of mass
Adhesive to palate
Cohesiveness of mass
Lumpy
Grainy

5. Residual
Loose particles
Toothstick
Toothpack
Tacky film

C. TOOTHPASTE FLAVOR

1. Before Expectoration
Aromatics
Mint Complex
  Peppermint/Menthol
  Spearmint
  Wintergreen
Base/Chalky
Bicarbonate
Anise
Fruity
Brown Spice
Soapy

2. After Rinsing
Aromatics
Minty
Fruity
Brown Spice
Anise

3. Basic Tastes
Sweet
Bitter
Salty

4. Chemical Feeling Factors
Burn
Bicarbonate feel
Cool
Astringent
Metallic

D. TOOTHPASTE TEXTURE

1. Brush on Front Teeth 10×
Firmness
Sticky
Number of brushes to foam
Ease to disperse

2. Expectorate
Chalky
Slickness of teeth

3. 20 Brushes (Back teeth)
Grittiness between teeth
Amount of foam
Slipperiness of foam
Denseness of foam

4. Rinse
Slickness of teeth

E. POTATO CHIP FLAVOR

1. Aromatics
Potato Complex
  Raw Potato/Green
  Cooked Potato
  Browned
  Dehydrated
Earthy/Potato Skins
Sweet Potato
Oil Complex
  Heated Vegetable Oil
  Overheated/Abused Oil
Sweet Caramelized
Cardboard
Painty
Spice

2. Basic Tastes
Salty
Sweet
Sour
Bitter

3. Chemical Feeling Factors
Tongue Burn
Astringent

F. POTATO CHIP TEXTURE

1. Surface
Oiliness
Roughness, macro
Roughness, micro
Loose Crumbs

2. First Bite/First Chew
Hardness
Crispness
Denseness
Particles after 4–5 chews

3. Chewdown
Moisture absorption
Chews to bolus
Persistence of crisp
Abrasiveness of mass
Moistness of mass
Cohesiveness of mass

4. Residual
Toothpack
Chalky mouth
Oily film

G. MAYONNAISE FLAVOR

1. Aromatics
Vinegar (type)
Cooked egg/eggy
DairyMilky/cheesey/butter
Mustard (type)
Onion/garlic
Lemon/citrus
Pepper (black/white)
Lemon juice
Fruity (grape/apple)
Brown spice (clove)
Paprika
Vegetable oil (aromatic)

2. Basic Tastes
Salty
Sweet
Sour
Bitter

3. Chemical Feeling Factors
Astringent
Tongue burn/heat
Prickly/pungent

Other Aromatics: Cardboard (stale oil), starch, paper, nutty/woody, sulfur, painty (rancid oil), caramelized, fishy

H. MAYONNAISE TEXTURE

1. Surface Compression
Slipperiness

2. First Compression
Firmness
Cohesiveness
Stickiness to palate

3. Manipulation
Cohesiveness of mass
Lumpy mass
Adhesive mass
Rate of breakdown

4. Residual
Oily film
Sticky/tacky film
Chalky film

I. CORN CHIP FLAVOR

1. Aromatics
Corn Complex
  Raw Corn
  Cooked Corn
  Toasted/Browned Corn
  Masa/Fermented
Caramelized
Oil Complex
  Heated Oil
  Heated Corn Oil
  Hydrogenated
Other grain (type)
Burnt
Earthy/Green Husks

2. Basic Tastes
Salty
Sweet
Sour
Bitter

3. Chemical Feeling Factors
Astringent
Burn

J. CORN CHIP TEXTURE

1. Surface
Roughness, Macro
Roughness, Micro
Manual Oiliness
Oiliness on Lips
Loose Particles

2. First Bite/First Chew
Hardness
Crispness/Crunchiness
Denseness
Amount of Particles

3. Chewdown
Moisture Absorption
Chews to Bolus
Moistness of Mass
Persistence of Crunch/Crisp
Cohesiveness of Mass
Graininess of Mass

4. Residual
Toothpack
Grainy Particles
Chalky Mouthfeel
Oily/Greasy Mouthfeel

K. CHEESE FLAVOR

1. Aromatics
Dairy Complex
  Cooked Milk/Caramelized
  Butterfat
  Butyric/Soured
  NFDM
  Cultured/Diacetyl
Smoky
Nutty/Woody
Fruity
Degraded Protein/Casein/Animal
Plastic/Vinyl

2. Basic Tastes
Sweet
Sour
Salty
Bitter

3. Chemical Feeling Factors
Astringent
Bite/Sharp
Burn

L. CHEESE TEXTURE

1. Surface

Rough Macro-bumpy

Rough Micro-grainy/gritty or chalky

Wetness

Oily/Fatty

Loose Particles

2. First Bite/First Chew

Firmness

Hardness

Denseness

Cohesiveness

Toothstick

Number of Pieces

3. Partial Compression

Springiness

4. Chewdown

Mixes with Saliva

Rate of Melt

Cohesiveness of Mass

Moistness of Mass

Adhesiveness of Mass

Lumpiness of Mass

Grainy mass

Toothstick

5. Residual

Toothstick

Mouthcoat

Oily Film

Chalky Film

Tacky

Dairy Film

Particles Left

Sticky film

M. CARAMEL/CONFECTIONS FLAVOR

1. Aromatics

Caramelized Sugar

Dairy Complex

  Baked Butter

  Cooked Milk

Sweet Aromatics

  Vanilla

  Vanillin

Diacetyl

Scorched

Yeasty (dough)

Other (cellophane)

2. Basic Tastes

Sweet

Sour

Salty

3. Chemical Feeling Factors

Tongue Burn

N. CARAMEL TEXTURE

1. Surface

Lipstick

Moistness

Roughness

2. First Bite/First Chew

Hardness

Denseness

Cohesiveness

Toothstick

3. Chewdown

# of Chews to Bolus

Mixes with Saliva

Cohesiveness of Mass

Moistness of Mass

Roughness of Mass

Toothpull

Adhesiveness to Palate

# of Chews to Swallow

4. Residual

Oily/Greasy Film

Tacky Film

Toothstick

O. CHOCOLATE CHIP COOKIE FLAVOR

1. Aromatics
White Wheat Complex
  Raw White Wheat
  Cooked White Wheat
  Toasted/Browned
Chocolate/Cocoa Complex
  Chocolate
  Cocoa
Dairy Complex
  NFDM
  Baked Butter
  Cooked Milk
Sweet Aromatics Complex
  Brown Sugar/Molasses
  Vanilla, Vanillin
  Caramelized
  Coconut
Nutty
Fruity
Baked egg
Shortening (Heated Oil, Hydrogenated Vegetable Fat)
Baking soda
Cardboard

2. Basic Tastes
Sweet
Salty
Bitter

3. Chemical Feeling Factors
Burn

P. CHOCOLATE CHIP COOKIE TEXTURE

1. Surface
Roughness, Micro
Roughness, Macro
Loose Crumbs/Particles
Oiliness
Surface Moisture

2. First Bite/First Chew
Firmness/Hardness
Crispness
Denseness
Cohesiveness
Crumbly

3. Chewdown
Chews to Bolus
Moisture Absorption
Cohesiveness of Mass
Moistness of Mass
Awareness of Chips
Roughness of Mass
Persistence of Crisp

4. Residual
Toothpack
Toothstick
Oily/Greasy Film
Grainy Particles
Loose Particles
Mouthcoating

Q. SPAGHETTI SAUCE FLAVOR

1. Aromatics
Tomato Complex
  Raw
  Cooked
Tomato Character
  Seedy/Skin
  Fruity
  Fermented/Soured
  Viney
  Skunky
Caramelized
Vegetable Complex
  Bell Pepper, Mushroom, Other
Onion/Garlic
Green Herb Complex
  Oregano, Basil, Thyme
Black Pepper
Cheese/Italian
Other
  Fish, Meat, Metallic

2. Basic Tastes
Salty
Sweet
Sour
Bitter

3. Chemical Feeling Factors
Astringent
Heat
Bite

R. SPAGHETTI SAUCE TEXTURE

1. Surface
Wetness
Oiliness
Particulate

2. First Compression
Viscosity/Thickness
Cohesiveness
Pulpy Matrix/Base
  Amount
  Size
Amount Large Particles
Amount of Small Particles

3. Manipulation
Amount of Particles/Chunks
  Largest Size
  Smallest Size
Chew Particles
  Hardness
  Crispness
  Fibrousness (Vegetables and Herbs)
Manipulate 5 times
  Mixes with Saliva
  Amount of Particles

4. Residual
Oily Mouthcoat
Loose Particles

Appendix 11.4
Spectrum Descriptive Analysis
Examples of Full Product Descriptions

A. WHITE BREAD

	Standard	Premium
1. APPEARANCE	Golden Brown	Golden Brown
Color of Crust	10	12
Evenness Color of Crust	12	12
Color of Crumb	Yellow	Yellow
Chroma of Crumb	10	9
Cell Size	7	11
Cell Uniformity	12	8
Uniformity of Shape	12	9
Thickness	10	7
Distinctiveness of Cap	2	7
2. FLAVOR		
AROMATICS		
Grain Complex		
Raw	5.5	7
Cooked	2	0
Browned	1	2.5
Bran	0	0
Dairy/Buttery	0	3.5
Soured (Milky, Cheese, Grain)	2.5	0
Caramelized	0	3
Yeasty/Fermented	2	4
Plastic	1	0
Chemical Leavening	4	0
Baking Soda	0	0
BASIC TASTES		
Sweet	2.5	5
Salty	8	7
Sour	3	2
Bitter	1.5	0
FEELING FACTORS		
Metallic	1.5	0
Astringent	3	1.5
Baking Soda Feel	0	0
3. TEXTURE		
SURFACE		
Roughness of Crumb	6	5
Initial Moistness	6.5	9

A. WHITE BREAD (continued)

	Standard	Premium
FIRST CHEW		
Crust Firmness	5	3.5
Crust Cohesiveness	7	2
Firmness of Crumb	3	3.5
Denseness of Crumb	3	8
Cohesiveness of Crumb	10	6.5
Uniformity of Chew	6.5	12
CHEWDOWN (10 chews)		
Moisture Absorption	12	14
Cohesiveness of Mass	10	11
Moistness of Mass	8	12
Roughness of Mass	6	4
Lumpy	5	1.5
Grainy	1	3
Adhesiveness to Palate	6	4
Stickiness to Teeth	4	2
RESIDUAL		
Loose Particles	3	1
Tacky Film	2	0
Little Fibers Between Teeth	0	0

B. TOOTHPASTE

	Standard Mint Paste	Mint Gel
1. APPEARANCE		
Extruded	5	6
Cohesive	9	20
Shape	9	8
Gloss	6.5	15
Particulate	0	0
Opacity	15	2
Color	3.5	9
Chroma	10	12
2. FLAVOR		
2.1 First Foam		
Mint Complex	11	6
Peppermint/Menthol	0	6
Spearmint	0	0
Wintergreen	11	0
Brown Spice Complex	3.5	0
Cinnamon	1	0
Clove	2	0
Anise	0	3.5
Floral	0	2
Base/Chalky	3.5	3
Soapy	1.5	2.5
Sweet	9	9
Salty	2	0
Bitter	3	5
Sour	0	0

B. TOOTHPASTE (continued)

	Standard Mint Paste	Mint Gel
2.2 Expectorate		
Minty	7	1.5
Brown Spice	1	0
Floral	0	2
Burn	2	4
Cool	9	14
Astringency	4	7
Base	1.5	3
2.3 Rinse		
Brown Spice	1.5	0
Fruity	0	0
Minty	3.5	1.5
Base	1.5	2
Salty	0	0
Sweet	4	4
Burn	1.5	2.5
Cool	8	11
Bitter	1.5	4
Soapy	0	1
2.4 Five Minutes		
Fruity	0	0
Minty	3	1
Soapy	1.5	1
Cool	7	6
Bitter	2	5
Brown Spice	0	0
Anise	0	3
3. TEXTURE		
3.1 Brush on front teeth 10×		
Firmness	4.5	6
Sticky	8	9
3.2 First Foam		
Amount of Foam	8	7
Slipperiness of foam	7	4
Denseness of Foam	11	9.5
3.3 Expectorate		
Chalky	4.5	7
Slickness of teeth	5	3.5

C. PEANUT BUTTER

	Local Brand	National Brand
1. APPEARANCE		
Color Intensity	7.0	7.5
Chroma	5.4	6.0
Gloss	5.2	5.1
Visible Particles	2.5	2.0

C. PEANUT BUTTER (continued)

	Local Brand	National Brand
2. FLAVOR		
Roasted Peanut	3.0	6.1
Raw/Beany	2.3	1.3
Over Roasted	0.6	3.0
Sweet Aromatic	3.1	4.5
Woody/Hull/Skins	4.4	1.6
Fermented Fruit	0	0
Phenol	0	0
Cardboard	0.4	0
Burnt	0	0
Musty	0.3	0
Green	0.1	0
Painty	0.1	0
Soy	1.0	0
Salt	11.9	9.1
Sweet	9.2	7.4
Sour	1.9	1.1
Bitter	3.1	1.6
Astringent	2.5	2.0
3. TEXTURE		
Surface Roughness	2.5	1.3
Firmness	7.0	5.7
Cohesiveness	6.9	7.0
Denseness	15	15
Adhesive	11.4	9.8
Mixes with Saliva	8.4	9.9
Adhesiveness of Mass	4.9	2.6
Cohesiveness of Mass	5.4	4.1
Roughness of Mass	1.8	1.0
4. RESIDUAL		
Loose Particles	0.1	0
Oily Film	1.6	1.5
Chalky Film	1.7	1.1

D. MAYONNAISE

	National Brand Mayonnaise	National Brand Dressing
1. APPEARANCE		
Color	Cream/Yellow	White
Color Intensity	2	1
Chroma	12	10
Shine	10	12.5
Lumpiness	9	4
Bubbles	5	2

D. MAYONNAISE (continued)

	National Brand Mayonnaise	National Brand Dressing
2. FLAVOR		
2.1 Aromatics		
Eggy	6.8	1.5
Mustard	4.5	3.5
Vinegar	4.5	9
Lemon	3.5	1
Oil	1.5	0
Starchy	0	1.5
Onion	1.5	0
Clove	0	4.8
2.2 Basic Tastes		
Salty	8	7
Sour	3	8
Sweet	3	8
2.3 Chemical Feeling Factors		
Burn	2	3
Pungent	2	3
Astringent	3.5	6
3. TEXTURE		
Adhesiveness to Lips	6	10
Firmness	8.5	9
Denseness	11	12.5
Cohesiveness	6	10
Cohesiveness of Mass	7	8.5
Adhesiveness of Mass	7	5
Mixes with Saliva	11.5	8
Oily Film	4	1.5
Tackiness	0	0
Chalkiness	0	1

E. MARINARA SAUCE

	Shelf-Stable (Jar)	Fresh-Refrigerated
1. APPEARANCE		
Color	Red/orange	Red/orange
Color Intensity	11	13
Chroma	12	8
Shine	7.5	7.5
Total Particles		
Micro Particles	10	8
Macro Particles	5	12

E. MARINARA SAUCE (continued)

	Shelf-Stable (Jar)	Fresh-Refrigerated
2. FLAVOR		
2.1 Aromatics		
Tomato Complex	8	7
Raw	1.5	5
Cooked	6.8	3
Tomato Character	8	7
Seedy/Skin	1	2.5
Fruity	6	3
Fermented/Soured	0	0
Viney	2.5	2
Skunky	1	0
Caramelized	4	2
Vegetable Complex		
Bell Pepper, Mushroom, Other	2	4
Onion/Garlic	5	6.5
Green Herbs Complex		
Oregano, Basil, Thyme	5	7.8
Black Pepper	1.5	4
Cheese/Italian	3.5	1
2.2 Basic Tastes		
Sweet	7	5.5
Sour	2.5	2
Salty	9	7
2.3 Chemical Feeling Factors		
Astringent	4	4.5
Heat	1.5	4
3. TEXTURE		
3.1 First Compression		
Cohesiveness	3	1
Pulpy Matrix/Base	5.5	9.5
3.2 Manipulation		
Amount of Particles/Chunks	4	10
Largest Size	3	8
Smallest Size	1	2.5
3.3 Chew Particles		
Hardness	3	5.5
Crispness	2	6
Fibrousness (vegetables & herbs)	4	5
3.4 Manipulate 5 Times		
Mixes with Saliva	11	12
4. RESIDUAL		
Oily Mouthcoat	2	4
Loose Particles	1	4

Appendix 11.5
Spectrum Descriptive Analysis
Training Exercises

A. BASIC TASTE COMBINATIONS EXERCISE

1. Scope

This exercise serves as a basic panel calibration tool. A product's flavor often includes a combination of two or three taste modalities, and the blends of salt, sweet, and sour provide the panel with an opportunity to develop the skill of rating taste intensities without the distraction of aromatics.

2. Test Design

Trainees begin by familiarizing themselves with the Reference Set, consisting of 6 cups with single component solutions. The cups carry labels such as Sweet 5, Salt 10, etc., where 5 = weak, 10 = medium, and 15 = very strong. The Reference Set remains available for the duration of the exercise.

The Evaluation Set consists of equal proportion blends of two or three of the reference solutions. The panel leader can prepare some or all of the blends in the Evaluation Set. The panel leader hands out one blend at a time, and the trainees record their impressions using the score sheet below.

At the end of the exercise, the sheet marked Average Results is made available. The panel leader should expect the panel means to fall within one point of these averages.

3. Materials

Assume 15 participants and 10 ml serving size: Prepare 1 liter of each reference solution which requires 150 g white sugar, 8.5 g salt, and 3 g citric acid. Serving items needed are:

    300 plain plastic serving cups, 2-oz size
    15 individual serving trays
    15 large opaque plastic cups with lid (spit cups), e.g., 16-oz size
    15 water rinse cups, 6-oz size
    6 water serving pitchers
    1 packet napkins
    60 tasting spoons (white plastic) if anyone requires those

4. Reference Set

Label	Content
Salt — 5	0.3% NaCl
Salt — 10	0.55% NaCl
Sweet — 5	5% Sucrose
Sweet — 10	10% Sucrose
Sour — 5	0.1% Citric Acid
Sour — 15	0.2% Citric Acid

Prepare solutions using water free of off flavors. Solutions may be prepared 24–36 hours prior to use. Refrigerate prepared samples. On day of evaluation, allow to warm to 70°F and serve 10 ml per participant.

5. *Evaluation Set*

Contents	Code
5% Sucrose/0.1% Citric Acid	232
5% Sucrose/0.2% Citric Acid	715
10% Sucrose/0.1% Citric Acid	115
5% Sucrose/0.3% NaCl	874
5% Sucrose/0.55% NaCl	903
10% Sucrose/0.3% NaCl	266
0.1% Citric Acid/0.3% NaCl	379
0.2% Citric Acid/0.3% NaCl	438
0.1% Citric Acid/0.55% NaCl	541
5% Sucrose/0.1% Citric Acid/0.3% NaCl	627
10% Sucrose/0.2% Citric Acid/0.55% NaCl	043
10% Sucrose/0.1% Citric Acid/0.3% NaCl	210
5% Sucrose/0.2% Citric Acid/0.3% NaCl	614
5% Sucrose/0.1% Citric Acid/0.55% NaCl	337

Prepare solutions by mixing equal quantities of the appropriate reference solutions. Solutions may be prepared 24–36 hours prior to use. Refrigerate prepared samples. On day of evaluation, allow to warm to 70°F and serve 10 ml per participant.

BASIC TASTE COMBINATIONS EXERCISE: COMPOSITION OF EVALUATION SET

CODE	% SUCROSE	% CITRIC ACID	% NaCl
232	5	.10	
715	5	.20	
115	10	.10	
874	5		.3
903	5		.55
266	10		.3
379		.10	.3
438		.20	.3
541		.10	.55
627	5	.10	.3
043	10	.20	.55
210	10	.10	.3
614	5	.20	.3
337	5	.10	.55

BASIC TASTE COMBINATIONS EXERCISE: SCORESHEET

PARTICIPANT NO. _____   DATE _____

CODE	SWEET	SOUR	SALTY
232	_____	_____	_____
715	_____	_____	_____
115	_____	_____	_____
874	_____	_____	_____
903	_____	_____	_____
266	_____	_____	_____
379	_____	_____	_____
438	_____	_____	_____
541	_____	_____	_____
627	_____	_____	_____
043	_____	_____	_____
210	_____	_____	_____
614	_____	_____	_____
337	_____	_____	_____

BASIC TASTE COMBINATIONS EXERCISE: AVERAGE RESULTS

SAMPLE	SWEET	SOUR	SALTY
232	6	7	
715	4	8.5	
115	9.5	4	
874	6		6
903	7		9
266	11		7
379		9	9
438		10	6.5
541		6	11
627	5	3.5	5
043	8	8	9
210	9	4	6
614	3	9	8
337	4	6	11

B. Cookie Variation Exercise

1. Scope

This exercise teaches the Spectrum lexicon (list of terms) for baked cookies by exposing the trainees to a set of samples of increasing complexity, adding one ingredient at a time. Many products that are combinations of ingredients can be handled in this manner, by constructing the flavor complex one or two terms at a time.

2. Test Design

Trainees begin by evaluating cookie #1, baked from flour and water. They are asked to suggest terms to describe this sample. Together, the panel leader and the trainees discuss the terms, for example cooked wheat/pasta-like/cream of wheat/breadcrumb, and doughy/raw/raw wheat/raw flour. They then select a single descriptor to represent each set of linked terms, for example cooked wheat and raw wheat. Trainees record the results on the scoresheet marked "vocabulary construction."

The panel leader hands out cookie #2, baked from flour, water and butter, and trainees suggest terms for the added aromatics. Again, the group selects a single descriptor to cover each sequence of linked (overlapping) terms.

Once the lexicon is developed, it can be validated by comparing any two of the reference samples and determining whether the lexicon works to discriminate and describe the samples appropriately.

The scoresheet marked "possible full vocabulary" can then be used to describe any pair of the samples, using a scale of 0 = absent, 5 = weak, 10 = medium, and 15 = very strong for the intensity of each attribute.

3. Reference Set

1. Flour, Water
2. Flour, Water, Butter
3. Flour, Water, Margarine
4. Flour, Water, Shortening
5. Flour, Water, Shortening, Salt
6. Flour, Water, Shortening, Baking Soda
7. Flour, Water, Sugar
8. Flour, Water, Brown Sugar
9. Flour, Water, Butter, Sugar
10. Flour, Water Margarine, Sugar
11. Flour, Water, Shortening, Sugar
12. Flour, Water, Sugar, Egg, Margarine
13. Flour, Water, Sugar, Egg, Margarine, Vanilla Extract
14. Flour, Water, Sugar, Egg, Margarine, Almond Extract

4. Cookie Recipes

Prepare each recipe as shown in the table on the next page. Spread dough into 9 × 13 oblong non-stick baking pan. Precut into squares before baking. Bake at 350° to 375° for 35 minutes (or more, until slightly browned). ***Ovens may vary for temperature and time.***

Each cookie recipe should be the same color for serving. All recipe amounts are based on 15 participants. Adjust if needed. Store in labeled airtight containers. Samples may be stored for 24–36 hrs.

1
2½ cups flour
1 cup water
	2
	2½ cups flour
	¼ cup water
	½ cup + 2 tablespoons
	  butter
		3
		2½ cups flour
		¼ cup water
		½ cup + 2 tablespoons
		  margarine
			4
			2½ cups flour
			¼ cup water
			½ cup + 2 tablespoons
			  shortening

5
2½ cups flour
¼ cup water
½ cup + 2 tablespoons
  shortening
1 teaspoon salt
	6
	2½ cups flour
	¼ cup water
	½ cup + 2 tablespoons
	  shortening
	⅓ teaspoon baking soda
		7
		2½ cups flour
		¾ cup water
		1 cup white granulated
		  sugar
			8
			2½ cups flour
			¾ cup water
			1 cup brown sugar

9
2½ cups flour
¼ cup water
½ cup + 2 tablespoons
  butter
1 cup white granulated
  sugar
	10
	2½ cups flour
	¼ cup water
	½ cup + 2 tablespoons
	  margarine
	1 cup white granulated
	  sugar
		11
		2½ cups flour
		¼ cup water
		½ cup + 2 tablespoons
		  shortening
		1 cup white granulated
		  sugar
			12
			2½ cups flour
			¼ cup water
			1 cup white granulated
			  sugar
			1 egg
			½ cup + 2 tablespoons
			  margarine

13
2½ cups flour
¼ cup water
1 cup white granulated
  sugar
1 egg
½ cup + 2 tablespoons
  margarine
1 teaspoon pure vanilla
  extract
	14
	2½ cups flour
	¼ cup water
	1 cup granulated white
	  sugar
	1 egg
	½ cup + 2 tablespoons
	  margarine
	1 teaspoon almond extract		

5. Materials at Each Participant's Station

Styrofoam cup with lid (spit cup)
Paper water rinse cup
Rinse water
Napkin
Cupcake paper cups coded: 1–14
Rinse water serving pitchers
Tasting spoons

6. Groceries and Paper Products

Purchase the total amount to serve the appropriate amount of each sample to each participant.

All purpose flour
Butter
Margarine   Cupcake paper cups (16 per participant)
Shortening   Individual serving trays (1 per participant)
White granulated sugar   Styrofoam (opaque) cups with lids (spit cups)
Light brown sugar   Water rinse cups
Eggs   Napkins
Baking soda   Water serving pitchers
Salt
Pure vanilla extract
Almond extract

COOKIE VARIATION EXERCISE — VOCABULARY CONSTRUCTION

1. Flour, Water   _____

2. Flour, Water, Butter   _____

3. Flour, Water, Margarine   _____

4. Flour, Water, Shortening   _____

5. Flour, Water, Shortening, Salt   _____

6. Flour, Water, Shortening,
   Baking Soda   _____

7. Flour, Water, Sugar   _____

8. Flour, Water, Brown Sugar   _____

9. Flour, Water, Butter, Sugar   _____

10. Flour, Water, Margarine, Sugar   _____

11. Flour, Water, Shortening, Sugar   _____

12. Flour, Water, Sugar, Egg,
    Margarine   _____

13. Flour, Water, Sugar, Egg,
    Margarine, Vanilla Extract   _____

14. Flour, Water, Sugar, Egg,
    Margarine, Almond Extract   _____

COOKIE VARIATION EXERCISE — EXAMPLE OF RESULTS

1. Flour, Water	raw wheat/dough/raw flour
	cooked wheat/pasta/cream of wheat/breadcrumb
2. Flour, Water, Butter	as #1 plus: butter/baked butter/browned/butter
	toasted wheat
3. Flour, Water, Margarine	as #1 plus: heated vegetable oil; toasted wheat
4. Flour, Water, Shortening	as #1 plus: heated vegetable fat/Crisco
	toasted wheat/pie crust
5. Flour, Water, Shortening, Salt	as #4 plus: salty
6. Flour, Water, Shortening, Baking Soda	as #5 plus: baked soda aromatic, salty
	baking soda feeling factor
7. Flour, Water, Sugar	as #1 plus caramelized, sweet
	toasted wheat
8. Flour, Water, Brown Sugar	as #7 plus molasses
9. Flour, Water, Butter, Sugar	as #2 plus sweet, caramelized
10. Flour, Water, Margarine, Sugar	as #3 plus sweet, caramelized
11. Flour, Water, Shortening, Sugar	as #4 plus sweet, caramelized
12. Flour, Water, Sugar, Egg, Margarine	as #11 plus baked eggy
13. Flour, Water, Sugar, Egg, Margarine, Vanilla	as #12 plus: vanilla/vanillin/cake
14. Flour, Water, Sugar, Egg, Margarine, Almond Extract	as #12 plus cherry/almond

COOKIE VARIATION EXERCISE — POSSIBLE FULL VOCABULARY

CHARACTERISTICS	#379	#811
White Wheat Complex		
Raw		
Cooked		
Toasted		
Eggy		
Shortening Complex		
Butter, Baked		
Heated Vegetable Oil		
Sweet Aromatics		
Caramelized		
Vanilla/Vanillin		
Almond/Cherry		
Molasses		
Other Aromatics (Baking Soda, etc.)		
Sweet		
Salty		
Baking Soda Feel		

CONTENTS

I. PURPOSE AND APPLICATIONS

The primary purpose of affective tests is to assess the personal response (preference and/or acceptance) by current or potential customers of a product, a product idea, or specific product characteristics.

Affective tests are used mainly by producers of consumer goods, but also by providers of services such as hospitals and banks, and even the Armed Forces, where many tests were first developed (see Chapter 1, p. 1). Each year, consumer tests are used more and more. They have proven highly effective as a principal tool in designing products or services that will sell in larger quantity and/or attract a higher price. The companies that prosper are seen to excel in consumer testing know-how and consequently in knowledge about their consumers.

This chapter gives rough guidelines for the design of consumer tests and in-house affective tests. More detailed discussions are given by Amerine et al., 1965; Schaefer, 1979; Civille et al., 1987; Gatchalian, 1981; Lawless and Heymann, 1998; Moskowitz, 1983; Resurreccion, 1998; Stone and Sidel, 1993; and Wu and Gelinas, 1989, 1992. A question that divides these authors is the use of in-house panels for acceptance testing. Our opinion is that this depends on the product: Baron Rothschild does not rely on consumer tests for his wines, but Nabisco and Kraft Foods need them. For the average company's products, the amount of testing generated by intended and unavoidable variations in process and raw materials far exceeds the capacity of all the consumer panels in the world, so one has no choice but to use in-house panels for most jobs and then calibrate against consumer tests as often as possible.

Most people today have participated in some form of consumer tests. Typically, a test involves 100 to 500 target consumers divided over three or four cities, for example, males legal age to 34 who made a purchase of imported beer within the last 2 weeks. Potential respondents are screened by phone or in a shopping mall. Those selected and willing are given a variety of beers together with a scorecard requesting their preference or liking ratings and the reasons, along with past buying habits and various demographic questions such as age, income, employment, ethnic background, etc. Results are calculated in the form of preference scores overall and for various subgroups.

Study designs need to be carefully tailored to the expected consumer group. The globalization of products often requires different study designs for different audiences. As this is written, a task group of ASTM E18 is developing guidelines for consumer research across countries and cultures.

The most effective tests for preference or acceptance are based on carefully designed test protocols run among carefully selected subjects with representative products. The choice of test protocol and subjects is based on the project objective. Nowhere in sensory evaluation is the definition of the project objective more critical than with consumer tests which often cost from $10,000 to $100,000 or more. In-house affective tests are also expensive; the combined cost in salaries and overhead can run $400 to $2000 for a 20-min test involving 20 to 40 people or more.

From a project perspective, the reasons for conducting consumer tests usually fall into one of the following categories:

- Product maintenance
- Product improvement/optimization
- Development of new products
- Assessment of market potential
- Product category review
- Support for advertising claims

A. PRODUCT MAINTENANCE

In a typical food or cosmetics company, a large proportion of the product work done by R&D and marketing deals with the maintenance of current products and their market shares and sales volumes. Research and Development projects may involve cost reduction, substitution of ingredients, process and formulation changes, and packaging modifications, in each case *without* affecting the product characteristics and overall acceptance. Sensory evaluation tests used in such cases are often difference tests for similarity and/or descriptive tests. However, when a match is not possible, it is necessary to take one or more "near misses" out to the consumer, in order to determine if these prototypes will at least achieve parity (in acceptance or preference) with the current product and, perhaps, with the competition.

Product maintenance is a key issue in quality control/quality assurance and shelf-life/storage projects. Initially it is necessary to establish the "affective status" of the standard or control product with consumers. Once this is done, internal tests can be used to measure the magnitude and type of change over time, condition, production site, raw material sources, etc. with the aid of QC or storage testing. The sensory differences detected by internal tests, large and small, may then be evaluated again by consumer testing in order to determine how large a difference is sufficient to reduce (or increase) the acceptance rating or percent preference vis-à-vis the control or standard.

B. PRODUCT IMPROVEMENT/OPTIMIZATION

Because of the intense competition among consumer products, companies constantly seek to improve and optimize products, so that they deliver what the consumer is looking for and thus fare better than the competition. A product improvement project generally seeks to "fix" or upgrade one or two key product attributes, which consumers have indicated need some improvement. A product optimization project typically attempts to manipulate a few ingredient or process variables so as to improve the desired attributes and hence the overall consumer acceptance. Both types of projects require the use of a good descriptive panel: (1) to verify the initial consumer needs and (2) to document the characteristics of the successful prototype. Examples of projects to improve product attributes are:

- Increasing a key aroma and/or flavor attribute, such as lemon, peanut, coffee, chocolate, etc.
- Increasing an important texture attribute, such as crisp, moist, etc., or reducing negative properties such as soggy, chalky, etc.
- Decreasing a perceived off note (e.g., crumbly dry texture, stale flavor or aroma, artificial rather than natural fruit flavor).
- Improving perceived performance characteristics, such as longer lasting fragrance, brighter shine, more moisturized skin, etc.

In *product improvement,* prototypes are made, tested by a descriptive or attribute panel to verify that the desired attribute differences are perceptible, and then tested with consumers to determine the degree of perceived product improvement and its effect on overall acceptance or preference scores.

For *product optimization* (Carr, B.T., in Wu and Gelinas, 1989; Gacula, 1993; Institute of Food Technologists, 1979; Moskowitz, 1983; Resurreccion, 1998; Sidel and Stone, 1979) ingredients or process variables are manipulated; the key sensory attributes affected are identified by descriptive analysis, and consumer tests are conducted to determine if consumers perceive the change in attributes and if such modifications improve the overall ratings.

The study of attribute changes together with consumer scores enables the company to identify and understand those attributes and/or ingredients or process variables that "drive" overall acceptance in the market.

C. DEVELOPMENT OF NEW PRODUCTS

During the typical new product development cycle, affective tests are needed at several critical junctures, e.g., focus groups to evaluate a concept or a prototype; feasibility studies in which the test product is presented to consumers, allowing them to see and touch it; central location tests during product development to confirm that the product characteristics do confer the expected advantage over the competition; controlled comparisons with the competition during test marketing; renewed comparisons during the reduction-to-practice stage to confirm that the desired characteristics survive into large-scale production; and finally central location and home use tests during the growth phase to determine the degree of success enjoyed by the competition as it tries to catch up.

Depending on test results at each stage, and the ability of R&D to reformulate or scale up at each step, the new product development cycle can take from a few months to a few years. This process requires the use of several types of affective tests, designed to measure, e.g., responses to the first concepts, chosen concepts vs. protoypes, different prototypes, and competition vs. proto-types. At any given time during the development process, the test objective may resemble those of a product maintenance project, e.g., a pilot plant scale-up, or an optimization project, as described above.

D. ASSESSMENT OF MARKET POTENTIAL

Typically, the assessment of market potential is a function of the Marketing Department, which in turn will consult Sensory Evaluation about aspects of the questionnaire design (such as key attributes which describe differences among products), the method of testing, and data previously collected by Sensory Evaluation. Questions about intent to purchase; purchase price; current purchase habits; consumer food habits (Barker, 1982; Meiselman, 1984); and the effects of packaging, advertising, and convenience are critical for the acceptance of branded products. The sensory analyst's primary function is to guide research and development. Whether the sensory analyst should also include market-oriented questions in consumer testing is a function of the structure of the individual company, including the ability of the marketing group to provide such data, and the ability of the sensory analyst to assume responsibility for assessing market conditions.

E. CATEGORY REVIEW

When a company wishes to study a product category for the purpose of understanding the position of its brand within the competitive set or for the purpose of identifying areas within a product category where opportunities may exist, a category review is recommended (Lawless and Heymann, 1998, p. 605). Descriptive analysis of the broadest array of products and/or prototypes that *defines* or covers the category yields a category map. Using mutivariate analysis techniques, the relative position of *both* the products and the attributes can be displayed in graph form (see Chapter 14, p. 310). This permits researchers to learn: (1) how products and attributes cluster within the product/attribute space; (2) where the opportunities may be in that space for new products; and (3) which attributes best define which products. A detailed example of a category appraisal is that of frankfurters by Muñoz et al. (1996) in which consumer data and descriptive panel data are related statistically.

Additional testing of several of the same products with consumers can permit projection of other vectors into the space. These other vectors may represent consumers' overall liking and/or consumers' integrated terms, such as creamy, rich, fresh, or soft.

F. SUPPORT FOR ADVERTISING CLAIMS

Product and service claims made in print, or on radio, TV, or the Internet, require valid data to support the claims. Sensory claims of parity ("tastes as good as the leading brand") or superiority ("cleans windows better than the leading brand") need to be based on consumer research and/or panel testing using customers, products, and test designs that provide credible evidence of the claim. A good, detailed guide is that of ASTM (1998); see also Gacula (1993), Chapter 9.

II. THE SUBJECTS/CONSUMERS IN AFFECTIVE TESTS

A. SAMPLING AND DEMOGRAPHICS

Whenever a sensory test is conducted, a group of subjects is selected as a *sample* of some larger population, about which the sensory analyst hopes to draw some conclusion. In the case of discrimination tests (difference tests and descriptive tests), the sensory analyst samples individuals with average or above-average abilities to detect differences. It is assumed that if these individuals cannot "see" a difference, the larger human population will be unable to see it. In the case of affective tests, however, it is not sufficient to merely select or sample from the vast human population. Consumer goods and services try to meet the needs of target populations, select markets, or carefully chosen segments of the population. Such criteria require that the sensory analyst first determine *the population for whom the product (or service) is intended;* e.g., for a sweetened breakfast cereal, the target population may be children between the ages of 4 and 12; for a sushi and yogurt blend, the select market may be Southern California; and for a high-priced jewelry item, clothing, or an automobile, the segment of the general population may be young, 25 to 35, upwardly mobile professionals, both married and unmarried.

Consumer researchers need to balance the need to identify and use a sample of consumers who represent the target population against the cost of having a very precise demographic model. With widely used products such as regular cereals, soft drinks, beer, cookies, and facial tissues, research guidance consumer tests may require selection only of users or potential users of the product brand or category. The cost of stricter demographic criteria described as follows may be justified for the later stages of consumer research guidance or for marketing research tests. Among the demographics to be considered in selecting sample subjects are:

User group — Based on the rate of consumption of a product by different groups within the population, brand managers often classify users as light, moderate, or heavy users. These terms are highly dependent on the product type and its normal consumption (see Table 12.1). For specialty products or new products with low incidence in the population, the cost of consumer testing radically increases, because many people must be contacted before the appropriate sample of users can be found.

Age — The ages of 4 to 12 are the ones to choose toys, sweets, and cereals; teenagers at 12 to 19 buy clothes, magazines, snacks, soft drinks, and entertainment. Young adults at 20 to 35 receive the most attention in consumer tests: (1) because of population numbers; (2) because of

TABLE 12.1
Typical Frequency of Use of Various Consumer Products

User classification	Product			
	Coffee	Peanut butter, air freshener	Macaroni and cheese	Rug deodorizer
Light	Up to 1 cup/day	1–4 ×/month	Once/2 months	1 ×/year
Moderate	2–5 cups/day	1–6 ×/week	1–4 ×/month	2–4 ×/year
Heavy	5 cups/day	1 × or more/day	Over 2 ×/week	1 ×/month or more

higher consumption made possible by the absence of family costs; and (3) because lifelong habits and loyalties are formed in this age range. Above 35 we buy houses and raise families; above 65 we use health care; and in consumables we tend to look for value for money with an eagle eye. If a product, such as a soft drink, has a broad age appeal, the subjects should be selected by age *in proportion* to their representation in the user population.

Sex — Although women still buy more consumer goods and clothes, and men buy more automobiles, alcohol, and entertainment, the differences in purchasing habits between the sexes continue to diminish. Researchers should use very current figures on users by sex for products such as convenience foods, snacks, personal care products, and wine.

Income — Meaningful groups for most items marketed to the general population per family and year are:

- Under $20,000
- $20,000 to $40,000
- $40,000 to $70,000
- Over $80,000

Different groups may be relevant at times, e.g., $200,000, $300,000, etc. for yachts over 50 ft.

Geographic location — Because of the regional difference in preference for many products, e.g., across the U.S., it is often necessary to test products in more than one location, and to avoid testing (or to use proportional testing) of products for the general population in areas with distinct local preferences, e.g., New York, the Deep South, Southern California.

Nationality, region, race, religion, education, employment — These and other factors, such as marital status, number and ages of children in family, pet ownership, size of domicile, etc. may be important for sampling of some products or services. The product researcher, brand manager, or sensory analyst must carefully consider all the parameters which define the target population before choosing the demographics of the sample for a given test.

Examples of step-by-step questionnaires used by marketing researchers to screen prospective respondents may be found in Meilgaard (1992) and Resurreccion (1998).

B. Source of Test Subjects: Employees, Local Residents, the General Population

The need to sample properly from the consuming population excludes, in principle, the use of employees and residents local to the company offices, technical center, or plants. However, because of high cost and long turnaround time of consumer tests, companies see a real advantage in using the employees or local population for at least part of their affective testing.

In situations where the project objective is product maintenance (see p. 233) employees and local residents do not represent a great risk as the test group. In a project oriented to maintaining "sensory integrity" of a current product, employees or local residents familiar with the characteristics of the product can render evaluations which are a good measure of the reaction of regular users. In this case, the employee or local resident judges the relative difference in acceptability or preference of a test sample vis-à-vis the well-known standard or control.

Employee acceptance tests can be a valuable resource when used correctly and when limited to maintenance situations. Because of their familiarity with the product and with testing, employees can handle more samples at a time and can give better discrimination, faster replies, and cheaper service. Employee acceptance tests can be carried out at work in a laboratory, in the style of a central location test, or the employees may take the product home.

However, for new product development, product optimization, or product improvement, employees or local residents should not be used to represent the consumer. The following are some examples of biases which may result from conducting affective tests with employees:

1. Employees tend to find reasons to prefer the products which they and their fellow employees helped to make, or if morale is bad, find reasons to reject such products. It is therefore imperative that products be disguised. If this is not possible, a consumer panel must be used.
2. Employees may be unable to weight desirable characteristics against undesirable ones in the same way a consumer would. For example, employees may know that a recent change was made in the process to produce a paler color, and this will make them prefer the paler product and give too little weight to other characteristics. Again, in such a case the color must be disguised, or if this is not possible, outside testing must be used.
3. Where a company makes separate products for different markets, outside tests will be distributed to the target population, but this cannot be done with employees. The way out may be to tell the employee that the product is destined for X market, but sometimes this cannot be done without violating the requirement that the test be blind. If so, again outside testing must be used.

In summary, the test organizer must plan the test imaginatively and must be aware of every conceivable source of bias. In addition, validity of response must be assured by frequent comparisons with real consumer tests on the same samples. In this way, the organizer and the employee panel members slowly build up knowledge of what the market requires, and this in turn makes it easier to gauge the pitfalls and avoid them.

III. CHOICE OF TEST LOCATION

The test location or test site has numerous effects on the results, not only because of the geographic location, but also because the place in which the test is conducted defines several other aspects of the way the product is sampled and perceived. It is possible to get different results from different test sites with a given set of samples and consumers. These differences occur as a result of differences in:

1. The length of time the products are used/tested
2. Controlled preparation vs. normal-use preparation of the product
3. The perception of the product alone in a central location vs. in conjunction with other foods or personal care items in the home
4. The influence of family members on each other in the home
5. The length and complexity of the questionnaire

For a more detailed discussion, see Resurreccion (1998).

A. LABORATORY TESTS

The advantages of laboratory tests are:

1. Product preparation and presentation can be carefully controlled.
2. Employees can be contacted on short notice to participate.
3. Color and other visual aspects which may not be fully under control in a prototype can be masked so that subjects can concentrate on the flavor or texture differences under investigation.

The disadvantages of laboratory tests are:

1. The location suggests that the test products originate in the company or specific plant, which may influence biases and expectations because of previous experience.
2. The lack of normal consumption (e.g., sip test rather than consumption of a full portion) may influence the detection or evaluation of positive or negative attributes.
3. Product tolerances in preparation or use may be different from those of home use (e.g., the product may lose integrity under some types of home use).

B. CENTRAL LOCATION TESTS

Central location tests are usually conducted in an area where many potential purchasers congregate or can be assembled. The organizer sets up a booth or rents a room at a fair, shopping mall, church, or test agency. A product used by schoolchildren may be tested in the school playground; a product for analytical chemists, at a professional convention. Respondents are intercepted and screened in the open, and those selected for testing are led to a closed-off area. Subjects can also be prescreened by phone and invited to a test site. Typically, 50 to 300 responses are collected per location. Products are prepared out of sight and served on uniform plates (cups, glasses) labeled with three-digit codes. The potential for distraction may be high, so instructions and questions should be clear and concise; examples of scoresheets are given in Appendix 12.1. In a variant of the procedure, products are dispensed openly from original packaging, and respondents are shown storyboards with examples of advertising and descriptions of how products will be positioned in the market.

The advantages of central location tests are:

1. Respondents evaluate the product under conditions controlled by the organizer; any misunderstandings can be cleared up and a truer response obtained.
2. The products are tested by the end users themselves which assures the validity of the results.
3. Conditions are favorable for a high percentage return of responses from a large sample population.
4. Several products may be tested by one consumer during a test session, thus allowing for a considerable amount of information for the cost per consumer.

The main disadvantages of central location tests are:

1. The product is being tested under conditions which are quite artificial compared to normal use at home or at parties, restaurants, etc. in terms of preparation, amount consumed, and length and time of use.
2. The number of questions that can be asked may be quite limited. This in turn limits the information obtainable from the data with regard to the preferences of different age groups, socioeconomic groups, etc.

C. Home Use Tests

In most cases, home use tests (or home placement tests) represent the ultimate in consumer tests. The product is tested under its normal conditions of use. The participants are selected to represent the target population. The entire family's opinion is obtained, and the influence of one family member on another is taken into account. In addition to the product itself the home use test provides a check on the package to be used and the product preparation instructions, if applicable. Typical panel sizes are 75 to 300 per city in 3 or 4 cities. Generally two products are compared. The first is used for 4 to 7 days and the scoresheet filled in, after which the second is supplied and rated. The two products should not be provided together because of the opportunities for using the wrong clues as the basis for evaluation, or assigning responses to the wrong scoresheet. Examples of scoresheets are given in Appendix 12.1.

The advantages of home use tests are (Moskowitz, 1983; Resurreccion, 1998):

1. The product is prepared and consumed under natural conditions of use.
2. Information regarding preference between products will be based on stabilized (from repeated use) rather than first impressions alone as in a mall intercept test.
3. Cumulative effect on the respondent from repeated use can provide information about the potentials for repeat sale.
4. Statistical sampling plans can be fully utilized.
5. Because more time is available for the completion of the scoresheet, more information can be collected regarding the consumer's attitudes toward various characteristics of the product, including sensory attributes, packaging, price, etc.

The disadvantages of the home use tests are:

1. A home use test is time consuming, taking from 1 to 4 weeks to complete.
2. It uses a much smaller set of respondents than a central location test; to reach many residences would be unnecessarily lengthy and expensive.
3. The possibility of nonresponse is greater; unless frequently reminded, respondents forget their tasks; haphazard responses may be given as the test draws to a close.
4. A maximum of three samples can be compared; any larger number will upset the natural use situation which was the reason for choosing a home use test in the first place. Thus multisample tests, such as optimization and category review, do not lend themselves to home use tests.
5. The tolerance of the product for mistakes in preparation is tested. The resulting variability in preparation along with variability from the time of use, and from other foods or products used with the test product, combine to produce a large variability across a relatively small sample of subjects.

IV. AFFECTIVE METHODS: QUALITATIVE

A. Applications

Qualitative affective tests are those (e.g., interviews and focus groups) which measure subjective responses of a sample of consumers to the sensory properties of products by having those consumers talk about their feelings in an interview or small group setting. Qualitative methods are used in the following situations:

- To uncover and understand consumer needs that are unexpressed (example: Why do people buy 4-wheel-drive cars to drive on asphalt?). Researchers that include anthropologists and ethnographers conduct open-ended interviews. This type of study, often called "the fuzzy front end," can help marketers identify trends in consumer behavior and product use.

- To assess consumers' initial responses to a product concept and/or a product prototype. When product researchers need to determine if a concept has some general acceptance or, conversely, some obvious problems, a qualitative test can allow consumers to discuss freely the concept and/or a few early prototypes. The results, a summary and a tape of such discussions, permit the researcher to understand better the consumers' initial reactions to the concept or prototypes. Project direction can be adjusted at this point, in response to the information obtained.
- To learn consumer terminology to describe the sensory attributes of a concept, prototype or commercial product, or product category. In the design of a consumer questionnaire and advertising it is critical to use consumer-oriented terms rather than those derived from marketing or product development. Qualitative tests permit consumers to discuss product attributes openly *in their own words*.
- To learn about consumer behavior regarding use of a particular product. When product researchers wish to determine how consumers use certain products (package directions) or how consumers respond to the use process (dental floss, feminine protection), qualitative tests probe the reasons and practices of consumer behavior.

In the qualitative methods discussed below, a highly trained interviewer/moderator is required. Because of the high level of interaction between the interviewer/moderator and the consumers, the interviewer must learn group dynamics skills, probing techniques, techniques for appearing neutral, and summarizing and reporting skills.

B. TYPES OF QUALITATIVE AFFECTIVE TESTS

1. Focus Groups

A small group of 10 to 12 consumers, selected on the basis of specific criteria (product usage, consumer demographics, etc.) meet for 1 to 2 hours with the focus group moderator. The moderator presents the subject of interest and facilitates the discussion using group dynamics techniques to uncover as much specific information from as many participants as possible directed toward the focus of the session.

Typically, two or three such sessions, all directed toward the same project focus, are held in order to determine any overall trend of responses to the concept and/or prototypes. Note is also made of unique responses apart from the overall trend. A summary of these responses plus tapes, audio or visual, are provided to the client researcher. Purists will say that $3 \times 12 = 36$ verdicts are too few to be representative of any consumer trend, but in practice if a trend emerges that makes sense, modifications are made based on this. The modifications may then be tested in subsequent groups.

The literature on marketing is a rich source of details on focus groups, e.g., Casey and Krueger (1994); Krueger (1988); Resurreccion (1998).

2. Focus Panels

In this variant of the focus group, the interviewer utilizes the same group of consumers two or three more times. The objective is to make some initial contact with the group, have some discussion on the topic, send the group home to use the product, and then have the group return to discuss its experiences.

3. One-on-One Interviews

Qualitative affective tests in which consumers are individually interviewed in a one-on-one setting are appropriate in situations in which the researcher needs to understand and probe a great deal from each consumer *or* in which the topic is too sensitive for a focus group.

The interviewer conducts successive interviews with up to 50 consumers, using a similar format with each, but probing in response to each consumer's answers.

One unique variant of this method is to have a person use or prepare a product at a central interviewing site or in the consumer's home. Notes or a video are taken regarding the process, which is then discussed with the consumer for more information. Interviews with consumers regarding how they use a detergent or prepare a packaged dinner have yielded information about consumer behavior which was very different from what the company expected or what consumers *said* they did.

One-on-one interviews or observations of consumers can give researchers insights into unarticulated or underlying consumer needs, and this in turn can lead to innovative products or services that meet such needs.

V. AFFECTIVE METHODS: QUANTITATIVE

A. APPLICATIONS

Quantitative affective tests are those which determine the responses of a large group (50 to several hundred) of consumers to a set of questions regarding preference, liking, sensory attributes, etc. Quantitative affective methods are applied in the following situations:

- To determine *overall preference or liking* for a product or products by a sample of consumers who represent the population for whom the product is intended. Decisions about whether to use acceptance and/or preference questions are discussed under each test method below.
- To determine *preference or liking for broad aspects* of product sensory properties (aroma, flavor, appearance, texture). Studying broad facets of product character can provide insight regarding the factors affecting overall preference or liking.
- To measure *consumer responses to specific sensory attributes* of a product. Use of intensity, hedonic, or "just right" scales can generate data which can then be related to the hedonic ratings discussed previously and to descriptive analysis data.

B. TYPES OF QUANTITATIVE AFFECTIVE TESTS

Affective tests can be classified into two main categories on the basis of the primary task of the test:

Task	Test and type	Questions
Choice	Preference tests	Which sample do you prefer?
		Which sample do you like better?
Rating	Acceptance tests	How much do you like the product?
		How acceptable is the product?

In addition to these questions, which can be asked in several ways using various questionnaire forms (see as follows), the test design often asks secondary questions about the reasons for the expressed preference or acceptance (see pp. 245–247 on attribute diagnostics).

1. Preference Tests

The choice of preference or acceptance for a given affective test should be based again on the project objective. If the project is specifically designed to pit one product *directly* against another in situations such as *product improvement* or *parity with competition,* then a preference test is indicated. The preference test forces a choice of one item over another or others. What it does not do is indicate whether any of the products are liked or disliked. Therefore, the researcher must have prior knowledge of the "affective status" of the current product or competitive product, against which he or she is testing.

Preference tests can be classified as follows:

Test type	No. of samples	Preference
Paired preference	2	A choice of one sample over another (A-B)
Rank preference	3 or more	A relative order of preference of samples (A-B-C-D)
Multiple paired preference (all pairs)	3 or more	A series of paired samples with all samples paired with all others (A-B, A-C, A-D, B-C, B-D,C-D)
Multiple paired preference (selected pairs)	3 or more	A series of paired samples with one or two select samples (e.g., control) paired with two or more others (not paired with each other) (A-C, A-D, A-E, B-C, B-D, B-E)

See Chapter 7, pp. 99–106 for a discussion of principles, procedures, and analysis of paired and multipaired tests.

a. Example 12.1: Paired Preference — Improved Peanut Butter

Problem/situation — In response to consumer requests for a product "with better flavor with more peanutty character," a product improvement project has yielded a prototype which was rated significantly more peanutty in an attribute difference test (such as discussed in Chapter 7, pp. 99–121). Marketing wishes to confirm that the prototype is indeed preferred to the current product, which is enjoying large volume sales.

Test objective — To determine whether the prototype is preferred over the current product.

Test design — This test is one-sided as the prototype was developed to be more peanutty in response to consumer requests. A group of 100 subjects, prescreened as users of peanut butter, are selected and invited to a central location site where they receive the two samples in simultaneous presentation, half in the order A-B, the other half B-A. All samples are coded with three-digit random numbers. Subjects are encouraged to make a choice (see discussion of forced choice, Chapter 7.II.B, p. 100). The scoresheet is shown in Figure 12.1. The null hypothesis is H_0: the preference for the higher-peanut flavor prototype $\leq 50\%$. The alternative hypothesis is H_a: the preference for the prototype $> 50\%$.

Screen samples — Samples used are those already subjected to the attribute difference test described earlier, in which a higher level of peanut flavor was confirmed.

Conduct test — The method described in Chapter 7.II.D, p. 101, was used; 62 subjects preferred the prototype. It is concluded from Table T8 that a significant preference exists for the prototype over the current product.

Interpret results — The new product can be marketed in place of the current with a label stating: More Peanut Flavor.

2. Acceptance Tests

When a product researcher needs to determine the "affective status" of a product, i.e., how well it is liked by consumers, an acceptance test is the correct choice. The product is compared to a well-liked company product or that of a competitor, and a hedonic scale, such as those shown in Figure 12.2, is used to indicate degrees of unacceptable to acceptable, or dislike to like. The two lower scales, "KIDS" and "Snoopy," are commonly used with children of grade school age.

From relative acceptance scores one can infer preference; the sample with the higher score is preferred. The best (more discriminating, more actionable) results are obtained with scales that are balanced, i.e., have an equal number of positive and negative categories and have steps of equal size. The scales shown in Figure 12.3 are not as widely used because they are unbalanced, unevenly spaced, or both. The six-point excellent scale in Figure 12.3, for example, is heavily loaded with

PEANUT BUTTER

INSTRUCTIONS
1. Taste the product on the left first, and the product on the right second.

Now that you've tasted both products, which one do you prefer? Please choose one:

☐ 463 ☐ 189

2. Please comment on the reasons for your choice: ____________________

____

Name ____________________ Date ________

FIGURE 12.1 Scoresheet for Paired Preference test. Example 12.1: improved peanut butter.

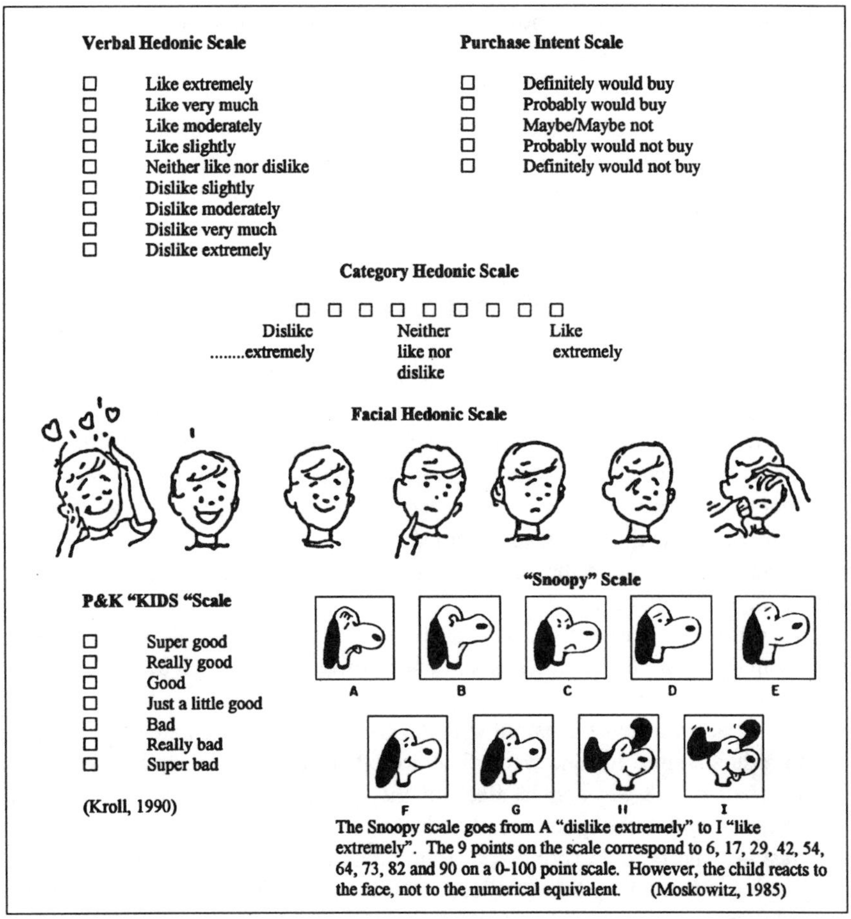

FIGURE 12.2 Scales used in acceptance tests. The last two scales are used with children.

Nine-point wonderful	Nine-point quartermaster (unbal.)	Six-point wonderful (unbal.)
Think it's wonderful	Like extremely	Wonderful, think it's great
Like it very much	Like strongly	I like it very much
Like it quite a bit	Like very well	I like it somewhat
Like it slightly	Like fairly well	So-so, it's just fair
Dislike it slightly	Like moderately	I don't particularly like it
Dislike it quite a bit	Like slightly	I don't like it at all
Dislike it very much	Dislike slightly	
Think it's terrible	Dislike moderately	
	Dislike intensely	

Seven-point excellent	Five-point excellent	Six-point (unbal.)
Excellent	Excellent	Excellent
Very good	Good	Extremely good
Good	Fair	Very good
Fair	Poor	Good
Poor	Terrible	Fair
Very poor		Poor
Terrible		

FIGURE 12.3 Examples of hedonic scales that are unclear in balance or spacing.

positive (Good to Excellent) categories, and the space between Poor and Fair is clearly larger than that between Extremely Good and Excellent. The difference between the latter may be unclear to many people. Acceptance tests are in fact very similar to attribute difference tests (see Chapter 7, pp. 99–121) except that the attribute here is *acceptance* or *liking*. Different types of scales such as category (as shown in Figures 12.2 and 12.3), line scales, or ME scales can be used to measure the degree of liking for a product.

a. Example 12.2: Acceptance of Two Prototypes Relative to a Competitive Product— High Fiber Breakfast Cereal

Problem/situation — A major cereal manufacturer has decided to enter the high fiber cereal market and has prepared two prototypes. Another major cereal producer already has a brand on the market that continues to grow in market share and leads among the high fiber brands. The researcher needs to obtain acceptability ratings for his two prototypes compared to the leading brand.

Project objective — To determine whether one or the other prototype enjoys sufficient acceptance to be test marketed against the leading brand.

Test objective — To measure the acceptability of the two prototypes and the market leader among users of high fiber cereals.

Screen the samples — During a product review, several researchers, the brand marketing staff, and the sensory analyst taste the prototypes and competitive cereal which are to be submitted to a home placement test.

Test design — Each prototype is paired with the competitor in a separate sequential evaluation, in which each product is used for one week. The prototypes and the competitive product are each evaluated first in half of the test homes. Each of the 150 qualified subjects is asked to rate the products on the nine-point verbally anchored hedonic scale shown in Figure 12.2.

Conduct test — One product (prototype or competition) is placed in the home of each pre-screened subject for one week. After the questionnaire is filled in and the first product removed, the second product is given to the subject to use for the second week. The second questionnaire and remaining samples are collected at the end of the second week.

Analyze results — Separate paired *t*-tests (see Chapter 11) are conducted for each prototype vs. the competition. The mean acceptability scores of the samples were

	Prototype	**Competition**	**Difference**
Prototype 1	6.6	7.0	–0.4
Prototype 2	7.0	6.9	+0.1

The average difference between Prototype 1 and the competition was significantly different from zero, that is, the average acceptability of Prototype 1 is significantly less than the competition. There was no significant difference between Prototype 2 and the competition.

Interpret results — The project manager concludes that Prototype 2 did as well as the competition, and the group recommends it as the company entry into the high fiber cereal field.

C. Assessment of Individual Attributes (Attribute Diagnostics)

As part of a consumer test, researchers often endeavor to determine the *reasons* for any preference or rejection by asking additional questions about the sensory attributes (appearance, aroma/fragrance, sound, flavor, texture/feel). Such questions can be classified into the following groups:

1. Affective responses to attributes:
   Preference — Which sample do you prefer for fragrance?
   Hedonic — How do you like the texture of this product?
   [Dislike extremely - Like extremely]
2. Intensity response to attribute:
   Strength — How strong/intense is the crispness of this cracker?
   [None - Very strong]
3. Appropriateness of intensity:
   Just right — Rate the sweetness of this cereal:
   [Not at all sweet enough - - - - - - - - - - - - - - - - Much too sweet]

Figure 12.4 shows examples of attribute questions; others are discussed in Section VI.A, pp. 247–249. In the first example, a preference questionnaire with two samples, respondents are asked, for each attribute, which sample they prefer. In the second example, an "attribute diagnostics" questionnaire with a single sample, respondents rate each attribute on a scale from "like extremely" to "dislike extremely." Such questionnaires are considered less effective in determining the importance of each attribute, because subjects often rate the attributes similar to the overall response, and the result is a series of attributes which have a "halo" of the general response. In addition, if one attribute does receive a negative rating, the researcher has no way of determining the direction of the dislike. If a product texture is disliked, is it "too hard" or "too soft"? — "too thick" or "too thin"?

The "just right" scales shown in the third and fourth examples (see also Vickers, 1988) allow the researcher to assess the intensity of an attribute relative to some mental criterion of the subjects. "Just right" scales cannot be analyzed by calculating the mean response, as the scale might be unbalanced or unevenly spaced, depending on the relative intensities and appropriateness of each attribute in the mind of the consumer. The following procedure is recommended:

1. Calculate the percentage of subjects who respond in each category of the attribute. Example:

	Example of Results for Attribute "Just Right" Scales				
% Response	5	15	40	25	15
Category	Much too little	Somewhat too little	Just right	Somewhat too much	Much too much

Attribute Diagnostics: Example of Attribute-by-Preference Questions

1.	Which sample did you prefer overall?	467____	813____
2.	Which did you prefer for color?	467____	813____
3.	Which did you prefer for cola impact?	467____	813____
4.	Which did you prefer for citrus flavor? \	467____	813____
5.	Which did you prefer for spicy flavor?	467____	813____
6.	Which did you prefer for sweetness?	467____	813____

Attribute Diagnostics Questionnaire with a Single Sample Using Hedonic Rating of Each Attribute

☐ Like extremely
☐ Like very much
☐ Like moderately
☐ Like slightly
☐ Neither like nor dislike
☐ Dislike slightly
☐ Dislike moderately
☐ Dislike very much
☐ Dislike extremely

Using the above scale rate the following:
[Scale could be repeated after each question]
How do you feel *overall* about this beverage?_____
How do you feel about the color?_____
How do you feel about the cola impact?_____
How do you feel about the citrus flavor?_____
How do you feel about the spice flavor?_____
How do you feel about the sweetness?_____
How do you feel about the body?_____

"Just Right" Scales for Attributes (Stew)

Please indicate your opinion about the following characteristics:

Gravy color	☐	☐	☐	☐	☐
	Too light		Just right		Too dark
Amount of vegetables	☐	☐	☐	☐	☐
	Too few		Just right		Too many
Amount of beef flavor	☐	☐	☐	☐	☐
	Too low		Just right		Too high
Amount of saltiness	☐	☐	☐	☐	☐
	Too low		Just right		Too high
Spiciness	☐	☐	☐	☐	☐
	Too low		Just right		Too high
Thickness of gravy	☐	☐	☐	☐	☐
	Too thin		Just right		Too thick

Attribute Diagnostics: Implied "Just Right" Scales

1. Color   ☐ ☐ ☐ ☐ ☐ ☐ ☐
   much too light   much too dark

2. Cola Flavor   ☐ ☐ ☐ ☐ ☐ ☐ ☐
   much too weak   much too strong

3. Citrus Flavor ☐ ☐ ☐ ☐ ☐ ☐ ☐
   much too weak   much too strong

4. Sweetness   ☐ ☐ ☐ ☐ ☐ ☐ ☐
   not at all sweet enough   much too sweet

5. Thickness   ☐ ☐ ☐ ☐ ☐ ☐ ☐
   much too thin   much too thick

6. Carbonation ☐ ☐ ☐ ☐ ☐ ☐ ☐
   not at all carbonated enough   much too carbonated

FIGURE 12.4   Examples of scales used in attribute diagnostics tests.

Attribute Diagnostics: Simple Intensity Scales

Please indicate the intensity of the following attributes of the sample of pasta:

<u>Appearance</u>

1. Color Intensity   □ □ □ □ □ □ □ □ □
                    light   dark

2. Surface Smoothness   □ □ □ □ □ □ □ □ □
                    rough   smooth

3. Broken Pieces   □ □ □ □ □ □ □ □ □
                    none   many

<u>Flavor</u>

4. Cooked Paste   □ □ □ □ □ □ □ □ □
      Flavor/Taste   none   strong

5. Saltiness   □ □ □ □ □ □ □ □ □
                    none   strong

6. Eggy Flavor/Taste   □ □ □ □ □ □ □ □ □
                    none   strong

7. Fresh Flavor/Taste   □ □ □ □ □ □ □ □ □
                    none   strong

<u>Texture</u>

8. Initial Stickiness   □ □ □ □ □ □ □ □ □
                  not sticky   very sticky

9. Firmness   □ □ □ □ □ □ □ □ □
                 very soft   very firm

10. Springiness   □ □ □ □ □ □ □ □ □
              very mushy   very springy

11. Starchy   □ □ □ □ □ □ □ □ □
                  none   very starchy

FIGURE 12.4   (continued)

2. Using a χ^2-test (see Chapter 11), compare the distribution of responses to that obtained by a successful brand.

A similar approach is to use an intensity scale (without midpoint) for each attribute (the fifth example). In order to assess how appropriate each of these attributes is, the intensity values must be related to overall acceptance or to acceptance for that attribute. The studies done by General Foods on the Consumer Texture Profile method (Szczesniak et al., 1975) related consumer intensity ratings to their own ratings for an ideal; it showed high correlations between acceptance ratings and the degree to which various products approached the consumer's ideal.

VI. DESIGN OF QUANTITATIVE AFFECTIVE TESTS

A. QUESTIONNAIRE DESIGN

In designing questionnaires for affective testing the following guidelines are recommended:

1. Keep the length of the questionnaire in proportion to the amount of time the subject *expects* to be in the test situation. Subjects can be contracted to spend hours testing several products with extensive questionnaires. At the other extreme, a few questions may be enough information for some projects. Design the questionnaire to ask the minimum number of questions to achieve the project objective; then set up the test so that the respondents expect to be available for the appropriate time span.

2. Keep the questions clear and somewhat similar in style. Use the same type of scale — whether preference, hedonic, "just right," or intensity scale — within the same section of the questionnaire. Intensity and hedonic questions may be asked in the same questionnaire (see examples in Appendix 12.1), but should be clearly distinguished. The questions and their responses should follow the same general pattern in each section of the questionnaire. Have the scales go in the same direction, e.g., [Too little---------Too much] for each attribute, so that the subject does not have to stop and decode each question.

3. Direct the questions to address the primary differences between/among the products in the test. Attribute questions should relate to the attributes which are detectable in the products and which differentiate among them, as determined by previously conducted descriptive tests. Subjects will not give clear answers to questions about attributes they cannot perceive or differences they cannot detect.

4. Use only questions which are actionable. Do not ask questions to provide data for which there is no appropriate action. If one asks subjects to rate the attractiveness of a package and the answer comes back that the package is somewhat unattractive, does the researcher know what to "fix" or change to alter that rating?

5. Always provide spaces on a scoresheet for open-ended questions. For example, ask why a subject responded the way he/she did to a preference or acceptance question, immediately following that question.

6. Place the overall question for preference or acceptance in the place on the scoresheet which will elicit the most considered response. In many cases the overall acceptance is of primary importance, and analysts rightly tend to place it first on the scoresheet. However, in cases where a consumer is asked several specific questions about appearance and/or aroma before the actual consumption of the product, it is necessary to wait until those attributes are evaluated and rated before addressing the total acceptance or preference question. Appendix 12.1 provides two examples of acceptance questionnaires.

B. Protocol Design

Sensory tests are difficult enough to control in a laboratory setting (see Chapter 3.II, pp. 24–32). Outside the laboratory, in a central location or home use setting, the need for controls of test design, of product handling, and of subject/consumer selection is even greater. In developing and designing outside affective tests the following guidelines are recommended:

Test facility — In a central location test, the facility and test administrators must adhere to strict protocols regarding the size, flexibility, location and environmental controls at each test site. The test should be conducted in locations which provide high access to the target population and subjects should be able to reach the test site easily.

Based on the design of the study, consideration should be given to the ability of each facility to provide adequate space, privacy for each consumer/subject, proper environmental controls (lighting, noise control, odor control, etc.), space for product handling and preparation, and a sufficient number of administrators and interviewers.

Test administrators — The administrators are required to be both trained and experienced in the specific type of test design developed by the sensory analyst. In addition to familiarity with the test design, test administrators must be given a detailed set of instructions for the handling of questionnaires, subjects and samples for a specific study.

Test subjects — Each test site requires careful selection of subjects based on demographic criteria which define the population of interest (see Section II, pp. 235–237). Once selected, subjects are made aware of the location, duration of the test, type and number of products to be tested, and type of payment. Consumers do not respond well to surprises regarding exactly what is expected of them.

Screen samples — Prior to any affective test, samples must be screened to determine:

- Exact sample source to be tested (bench, pilot plant, production, code date)
- The storage condition under which samples are to be held and shipped
- Packaging requirements for storage and shipping
- Shipping method (air, truck, refrigerated, etc.)
- The descriptive analysis, i.e., documentation of the product sensory characteristics, for use in questionnaire design and in final data interpretation for the study.

Sample handling — As part of the test protocol, which is sent to the test site, detailed and specific instructions regarding storage, handling, preparation and presentation of samples are imperative for proper test execution.

Appendix 12.2 provides worksheets for the development of a protocol for an affective test, and an example of a completed protocol.

VII. USING OTHER SENSORY METHODS TO SUPPLEMENT AFFECTIVE TESTING

A. RELATING AFFECTIVE AND DESCRIPTIVE DATA

Product development professionals handling both the R&D and marketing aspects of a product cycle recognize that the consumer's response in terms of overall acceptance and purchase intent is the bottom line in the decision to go or not go with a product or concept (Beausire et al., 1988).

Despite the recognition of the need for affective data, the product development team is generally unsure about what the consumer means when asked about actual sensory perceptions. When a consumer rates a product as too dry or not chocolatey enough, is he really responding to perceived moistness/dryness or perceived chocolate flavor, or is he responding to words that are associated in his mind with goodness or badness in the product? Too many researchers are taking the consumer's response at face value (as the researcher uses the sensory terms) and these researchers end up fixing attributes that may not be broken.

One key to decoding consumer diagnostics and consumer acceptance is to measure the perceived sensory properties of a product using a more objective sensory tool (Shepherd et al., 1988). The trained descriptive or expert panel provides a thumbprint or spectrum of product sensory properties. This sensory documentation constitutes a list of real attribute characteristics or differences among products which can be used both to design relevant questionnaires and to interpret the resulting consumer data after the test is completed. By relating consumer data with panel data and when possible with ingredient and processing variables, or with instrumental or chemical analyses, the researcher can discover the relationships between product attributes and the ultimate bottom line, consumer acceptance.

When data are available for several samples (15 to 30) which span a range of intensities for several attributes (see Candy Bar example in Appendices 12.1 and 12.2), it is possible to study relationships in the data, using the statistical methods described in Chapter 14, pp. 306–324. Figure 12.5 shows four examples. Graph A shows how consumer overall acceptance varies with the intensity of a descriptive panel attribute (e.g., color intensity); this allows the researcher to understand the effect of different intensities of a characteristic and to identify acceptable limits. In Graph B the abscissa depicts the intensity of an undesirable attribute, e.g., an off-flavor, and the ordinate is consumer acceptance of flavor; the steep slope indicates a strong effect on liking for one facet of the product. From the type of relationship in Graph C the researcher can learn how consumers use certain words relative to the more technically precise descriptive terms; we note

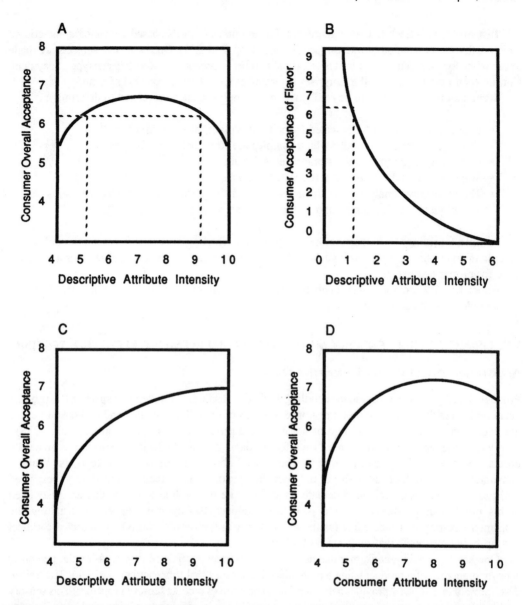

FIGURE 12.5 Examples of data relationships extracted from a consumer study. (A) (top left) Consumer overall acceptance vs. descriptive attribute intensity (color intensity); (B) (top right) Consumer acceptance for flavor vs. descriptive attribute intensity (flavor off-note); (C) (bottom left) Consumer intensity crispness vs. descriptive attribute intensity (crispness); (D) (bottom right) Consumer overall acceptance vs. consumer attribute intensity (sweetness).

that the descriptive panel's rating for crispness correlates well with the consumer's rating, but the latter rises less steeply. Finally, Graph D relates two consumer ratings, showing the range of intensities of an attribute which the consumer finds acceptable. Such a relationship is tantamount to a "just right" assessment.

The data relationships in Figure 12.5 are univariate. Consumer data often show interaction between several variables (products, subjects, and one or more attributes). This type of data requires multivariate statistical methods such as Principal Component Analysis (PCA) or Partial Least Squares (PLS) (see Muñoz et al., 1996 and Chapter 14.)

B. Using Affective Data to Define Shelf-Life or Quality Limits

In Chapter 11, pp. 175–176, we described a "modified" or short-version descriptive procedure where the principal use is to define QA/QC or shelf-life limits. In a typical case, the first step is to send the fresh product out for an acceptability test in a typical user group. This initial questionnaire may contain additional questions asking the consumer to rate a few important attributes.

The product is also rated for acceptability and key attributes by the modified panel, and this evaluation is repeated at regular intervals during the shelf storage period, each time comparing the stored product with a control, which may be the same product stored under conditions that inhibit perceptible deterioration (e.g., deep freeze storage under nitrogen) or if this is not possible, fresh product of current production.

When a significant difference is found by the modified panel, in overall difference from the control and/or in some major attribute(s), the samples are sent again to the user group to determine if the statistically significant difference is meaningful to the consumer. This is repeated as the difference grows with time of shelf storage. Once the size of a panel difference can be related to what reduces consumer acceptance or preference, the internal panel can be used in future to monitor regular production in shelf-life studies, with assurance that the results are predictive of consumer reaction.

1. Example 12.3: Shelf Life of Sesame Cracker

Problem/situation — A company wishes to define the shelf life of a new sesame cracker in terms of the "sell by" date which will be printed on packages on the day of production.

Project objective — To determine at what point during shelf storage the product will be considered — "off," "stale," or "not fresh" by the consumer.

Test objective — (1) Using a research panel trained for the purpose of determining the key attributes of the product at various points during shelf storage and (2) submitting the product to consumer acceptance tests: (a) initially; (b) when the research panel first establishes a difference; and (c) at intervals thereafter, until the consumers establish a difference.

Test design — Samples of a single batch of the sesame crackers were held for 2, 4, 6, 8, and 12 weeks under four different sets of conditions: "control" = near freezing in airtight containers; "ambient" = 70°F/50% RH; "humid" = 85°F/70% RH; and "'hot" = 100°F/30% RH.

Subjects — Panelists (25) from the R&D lab are selected for ability to recognize the aromatics of stale sesame crackers, i.e., the cardboard aromatic of the stale base cracker and the painty aromatic of oxidized oil from the seeds. Consumers (250) must be users of snack crackers and are chosen demographically to represent the target population.

Sensory methods — The research panel used the questionnaire in Figure 12.6 and was trained to score the test samples on the seven line scales, which represent key attributes of appearance, flavor, and texture related to the shelf life of crackers and sesame seeds. Research panelists also received a sample marked "control" with instructions to use the last line of the form as a Difference-from-control test (see Chapter 6.VIII, p. 86). The panelists were informed that these samples were part of a shelf-life study and that occasional test samples would consist of freshly prepared "control product" (such information reduces the tendency of panelists in shelf-life testing to anticipate more and more degradation in products).

The consumers on each occasion received two successive coded samples (the test product and the control, in random order), each with the scoresheet in Figure 12.7, which they filled in on the spot and returned to the interviewer.

Analyze results — The initial acceptance test, in which the 250 consumers received two fresh samples, provided a baseline rating of 7.2 for both, and the accompanying attribute ratings indicated that the crackers were perceived fresh and crisp.

EVALUATION OF SESAME CRACKER

INSTRUCTIONS

1. Evaluate the cracker for appearance, flavor and texture by placing a mark on each line below.

Appearance
Surface color
├──┤
light   dark

Flavor
Toasted wheat
├──┤
none   strong

Sesame seed
├──┤
none   strong

Cardboard
├──┤
none   strong

Painty
├──┤
none   strong

Texture
Hardness
├──┤
soft   hard

Crispness
├──┤
soggy   crisp

2. Compare the cracker with the control and indicate the amount of difference between them by placing a mark on the line below:

├──┤
no difference   very different

Comments _____
_____
_____

Name _____ Date _____

FIGURE 12.6 Research panel scoresheet showing attribute rating and difference rating for Example 12.3: shelf life of sesame cracker.

```
┌─────────────────────────────────────────────────────────────────────┐
│                        SESAME CRACKER                                 │
├─────────────────────────────────────────────────────────────────────┤
```

INSTRUCTIONS

1. Overall evaluation. Place a mark in the box which you feel best describes how you like the product:

☐   ☐   ☐   ☐   ☐   ☐   ☐   ☐   ☐

| Like extremely | Like very much | Like moderately | Like slightly | Neither like nor dislike | Dislike slightly | Dislike moderately | Dislike very much | Dislike extremely |

2. Indicate by placing a mark how you feel the product rates in each category below:

<u>Appearance</u>

Color   ☐ ☐ ☐ ☐ ☐ ☐
             light   dark

<u>Flavor</u>

Salty   ☐ ☐ ☐ ☐ ☐ ☐
        not at all salty   very salty

Sesame flavor   ☐ ☐ ☐ ☐ ☐ ☐
        no sesame flavor   strong flavor

Fresh toasted flavor   ☐ ☐ ☐ ☐ ☐ ☐
        stale, not fresh   very fresh

<u>Texture</u>

Crispness   ☐ ☐ ☐ ☐ ☐ ☐
        soggy   crisp

Aftertaste   ☐ ☐ ☐ ☐ ☐ ☐
        unpleasant   pleasant

Comments _____

_____

_____

Name _____ Date _____

FIGURE 12.7 Consumer scoresheet for Example 12.3: shelf life of sesame cracker.

The same two identical samples were rated 3.2 (out of 15) on the difference-from-control scale by the research panel. The 2- and 4-week samples showed no significant differences. At the 6-week point, the "humid" sample received a difference-from-control rating of 5.9, which was significantly different from 3.2. In addition, the "humid" sample was rated 4.2 in cardboard flavor (against 0 for the fresh control) and 5.1 in crispness (against 8.3 for the fresh control), both significant differences by ANOVA.

The 6-week "humid" samples were then tested by the consumers and were rated 6.7 on acceptance, against 7.1 for the control $p < 0.05$). The rating for "fresh toasted flavor" also showed a significant drop.

The product researcher decided to conduct consumer tests with the other two test samples ("ambient" and "hot") as soon as the difference-from-control ratings by the research panel exceeded

5.0. Subsequent tests showed that consumers were only sensitive to differences which were rated 5.5 or above by the research panel. All further shelf-life testing on sesame crackers used the 5.5 difference-from-control rating as the critical point above which differences were not only statistically significant, but potentially meaningful to the consumer.

REFERENCES

Amerine, M.A., Pangborn, R.M., and Roessler, E.G., 1965. *Principles of Sensory Evaluation of Food.* Academic Press, New York, Chapter 9.

ASTM E-18, 1998. *Standard Guide for Sensory Claim Substantiation,* E1958–98. American Society for Testing and Materials, West Conshohocken, PA.

Barker, L., 1982. *The Psychobiology of Human Food Selection.* AVI Publishing, Westport, CT.

Beausire, R.L.W., Norback, J.P., and Maurer, A.J., 1988. Development of an acceptability constraint for a linear programming model in food formulation. *J. Sensory Stud,* 3(2):137.

Casey, M.A. and Krueger, R.A., 1994. Focus group interviewing. In: *Measurement of Food Preferences,* MacFie, H.J.H. and Thomson, D.M.H., Eds. Blackie Academic & Professional, London, pp. 77–96.

Civille, G.V., Muñoz, A., and Chambers, E., IV, 1987. Consumer testing considerations. In: *Consumer Testing. Course Notes.* Sensory Spectrum, Chatham, NJ.

Gacula, M.C., Jr., 1993. *Design and Analysis of Sensory Optimization.* Food & Nutrition Press, Westport, CT, 301 pp.

Gatchalian, M.M., 1981. *Sensory Evaluation Methods with Statistical Evaluation.* College of Home Economics, University of the Philippines, Diliman, Quezon City, p. 230.

Institute of Food Technologists, 1979. *Sensory Evaluation Short Course.* IFT, Chicago.

Kroll, B.J., 1990. Evaluation rating scales for sensory testing with children. *Food Technology* 44(11), 78–86.

Krueger, R.A., 1988. *Focus Groups. A Practical Guide for Applied Research.* Sage Publications, Newbury Park, CA, 197 pp.

Lawless, H.T. and Heymann, H., 1998. *Sensory Evaluation of Food. Principles and Practices.* Chapman & Hall, New York, Chapters 13, 14, and 15, pp. 430–547, and Chapter 18, pp. 605–606.

Meilgaard, M.C., 1992. Basics of consumer testing with beer in North America. *Proc. Ann. Meet. Inst. Brew., Australia & New Zealand Sect.,* Melbourne, 37–47. See also *The New Brewer* 9(6), 20–25.

Meiselman, H.L., 1984. Consumer studies of food habits. In: *Sensory Analysis of Foods,* Piggott, J.R., Ed. Elsevier Applied Science, London, Chapter 8.

Moskowitz, H.R., 1983. *Product Testing and Sensory Evaluation of Foods. Marketing and R&D Approaches.* Food & Nutrition Press, Westport, CT.

Moskowitz, H.R., 1985. Product testing with children. In: *New Directions for Product Testing and Sensory Analysis,* Food & Nutrition Press, Westport, CT, 147–164.

Muñoz, A.M., Chambers, E., IV, and Hummer, S., 1996. A multifaceted category research study: how to understand a product category and its consumer responses. *J. Sensory Stud.* 11, 261–294.

Resurreccion, A.V.A., 1998. *Consumer Sensory Testing for Product Development.* Aspen Publishers, Gaithersburg, MD, 254 pp.

Schaefer, E.E., Ed., 1979. *ASTM Manual on Consumer Sensory Evaluation.* American Society for Testing and Materials, Special Technical Publication 682. ASTM, Philadelphia, PA.

Shepherd, R., Griffiths, N.M., and Smith, K., 1988. The relationship between consumer preference and trained panel responses. *J. Sensory Stud.* 3(1), 19.

Sidel, J.L. and Stone, H., 1979. In: *Sensory Evaluation Methods for the Practicing Food Technologist,* Johnson, M.R., Ed. Institute of Food Technologists, Chicago, p. 10-1.

Stone, H. and Sidel, J.L., 1993. *Sensory Evaluation Practices,* 2nd ed. Academic Press, San Diego, CA.

Szczesniak, A.S., Skinner, E.Z., and Loew, B.J., 1975. Consumer textile profile method. *J. Food Sci.* 40, 1253.

Vickers, Z., 1988. Sensory specific satiety in lemonade using a just right scale for sweetness. *J. Sensory Stud.* 3(1), 1.

Wu, L.S. and Gelinas, A.D., Eds., 1989 and 1992. *Product Testing with Consumers for Research Guidance,* Vols. 1 and 2, ASTM Standard Technical Publications STP 1035 and 1155, ASTM, Philadelphia, 90 and 95 pp.

Appendix 12.1
Questionnaires for Consumer Studies

A. Candy Bar Questionnaire

Name_____

Product #_____

Candy Bar

■ Please rinse your mouth before starting.

■ Evaluate the product in front of you by looking at it and tasting it.

■ Considering *ALL* characteristics (*APPEARANCE, FLAVOR,* and *TEXTURE*) indicate your overall opinion by checking one box [√].

☐ ☐ ☐ ☐ ☐ ☐ ☐ ☐ ☐ ☐ ☐
Dislike   Neither   Like
extremely   like nor   extremely
                       dislike
                       (nl/nd)

■ **Comments:** Please indicate WHAT in particular you liked or disliked about this product. (USE WORDS NOT SENTENCES.)

LIKED   DISLIKED

_____   _____
_____   _____
_____   _____
_____   _____

1. Candy Bar Liking Questions

Please retaste the product as needed and indicate how much you LIKE or DISLIKE the following. *Check* the box that represents your response[✔].

Overall appearance

☐ ☐ ☐ ☐ ☐ ☐ ☐ ☐ ☐ ☐
Dislike   nl/nd   Like
extremely   extremely

Overall flavor

☐ ☐ ☐ ☐ ☐ ☐ ☐ ☐ ☐ ☐
Dislike   nl/nd   Like
extremely   extremely

Overall texture

☐ ☐ ☐ ☐ ☐ ☐ ☐ ☐ ☐ ☐
Dislike   nl/nd   Like
extremely   extremely

2. Candy Bar Specific Evaluation

Retaste the product as needed and *check* the box for your response [√] for both questions (LIKING and INTENSITY LEVEL) for each characteristic.

	Liking	Intensity/Level
Appearance		
Color	Dislike extremely □ □ □ □ nl/nd □ □ □ Like extremely	Light □ □ □ □ □ □ □ □ Dark
Color uniformity	Dislike extremely □ □ □ □ nl/nd □ □ □ Like extremely	Non-uniform □ □ □ □ □ □ □ □ Uniform
Amount of broken blisters	Dislike extremely □ □ □ □ nl/nd □ □ □ Like extremely	None □ □ □ □ □ □ □ □ Many
Flavor		
Chocolate flavor	Dislike extremely □ □ □ □ nl/nd □ □ □ Like extremely	None □ □ □ □ □ □ □ □ High
Peanut flavor	Dislike extremely □ □ □ □ nl/nd □ □ □ Like extremely	None □ □ □ □ □ □ □ □ High
Roasted/toasted flavor	Dislike extremely □ □ □ □ nl/nd □ □ □ Like extremely	None □ □ □ □ □ □ □ □ Many
Sweetness	Dislike extremely □ □ □ □ nl/nd □ □ □ Like extremely	None □ □ □ □ □ □ □ □ High

2. Candy Bar Specific Evaluation (continued)

Texture

Firmness of whole bar
☐ ☐ ☐ ☐ ☐ ☐ ☐ ☐ ☐
Dislike extremely　　　　nl/nd　　　Like extremely

☐ ☐ ☐ ☐ ☐ ☐ ☐ ☐ ☐
Soft　　　　　　　　　　　　　　　　Firm

Crunchiness of nuts
☐ ☐ ☐ ☐ ☐ ☐ ☐ ☐ ☐
Dislike extremely　　　　nl/nd　　　Like extremely

☐ ☐ ☐ ☐ ☐ ☐ ☐ ☐ ☐
Not crunchy　　　　　　　　　　　Crunchy

Melt down
☐ ☐ ☐ ☐ ☐ ☐ ☐ ☐ ☐
Dislike extremely　　　　nl/nd　　　Like extremely

☐ ☐ ☐ ☐ ☐ ☐ ☐ ☐ ☐
Slow melt　　　　　　　　　　　Fast melt

Chalky mouth coating
☐ ☐ ☐ ☐ ☐ ☐ ☐ ☐ ☐
Dislike extremely　　　　nl/nd　　　Like extremely

☐ ☐ ☐ ☐ ☐ ☐ ☐ ☐ ☐
None　　　　　　　　　　　　　　Chalky

Raise your hand when finished. Thank you!

B. Paper Napkins Questionnaire

Name_____

Product #_____

Paper Table Napkins

■ Please be sure your hands are clean before starting.

■ Evaluate the product in front of you.

■ LOOK at this napkin, OPEN AND FEEL it, and answer the following questions.

Overall opinion

Please indicate how much you LIKED or DISLIKED this product overall (considering ALL APPEARANCE, TACTILE/FEEL CHARACTERISTICS). *Circle* one of the numbers below⊗ to express your overall opinion.

0	1	2	3	4	5	6	7	8	9	10
Dislike extremely					Neither like nor dislike (nl/nd)					Like extremely

■ **Comments:**   Please indicate WHAT in particular you liked or disliked about this product. (USE WORDS NOT SENTENCES.)

LIKED   DISLIKED

_____   _____

_____   _____

_____   _____

1. Paper Table Napkins Liking Questions

Please retest the product as needed and indicate how much you LIKE or DISLIKE the following. *Circle* the number that represents your response ⊗.

Overall appearance

0 1 2 3 4 5 6 7 8 9 10
Dislike   nl/nd   Like
extremely   extremely

Overall texture

0 1 2 3 4 5 6 7 8 9 10
Dislike   nl/nd   Like
extremely   extremely

2. Paper Table Napkins Specific Evaluation

Retest the product as needed and *circle* your response ⊗ for both questions (LIKING and INTENSITY LEVEL) for each characteristic.

Characteristic	Liking	Intensity/level
Surface gloss	Dislike extremely 0 1 2 3 4 5 6 7 8 9 10 Like extremely (nl/nd)	Dull finish 0 1 2 3 4 5 6 7 8 9 10 Glossy finish
Color/whiteness	Dislike extremely 0 1 2 3 4 5 6 7 8 9 10 Like extremely (nl/nd)	Gray color 0 1 2 3 4 5 6 7 8 9 10 Bright color
Surface embossing	Dislike extremely 0 1 2 3 4 5 6 7 8 9 10 Like extremely (nl/nd)	Not embossed 0 1 2 3 4 5 6 7 8 9 10 Very embossed
Specks in surface	Dislike extremely 0 1 2 3 4 5 6 7 8 9 10 Like extremely (nl/nd)	No specks 0 1 2 3 4 5 6 7 8 9 10 Many specks
Stiffness	Dislike extremely 0 1 2 3 4 5 6 7 8 9 10 Like extremely (nl/nd)	Not stiff 0 1 2 3 4 5 6 7 8 9 10 Very stiff
Smoothness of Surface	Dislike extremely 0 1 2 3 4 5 6 7 8 9 10 Like extremely (nl/nd)	Rough/not smooth 0 1 2 3 4 5 6 7 8 9 10 Very smooth
Body	Dislike extremely 0 1 2 3 4 5 6 7 8 9 10 Like extremely (nl/nd)	Flimsy 0 1 2 3 4 5 6 7 8 9 10 Full bodied
Soft	Dislike extremely 0 1 2 3 4 5 6 7 8 9 10 Like extremely (nl/nd)	Not soft 0 1 2 3 4 5 6 7 8 9 10 Very soft

Indicate to the test supervisor that you have completed this questionnaire. Thank you!

Appendix 12.2
Protocol Design for Consumer Studies

A. PROTOCOL DESIGN FORMAT WORKSHEETS

---

1. Product Screening

1. Test objective
   _____
   _____
   _____

2. Sample selection
   a. Variables_____
      _____
      _____

   b. Products/brands _____
      _____
      _____
      _____

3. Reasons_____
   _____
   _____

---

2. Sample Information

Sample conditions
1. Sample source _____
   _____
   _____

   Age   _____

   Place   _____

   Code   _____

   Packaging condition _____
   _____

2. Sample holding
   _____
   _____
   _____

3. Other
   _____
   _____

3. Sample Preparation

Total amount _____

Other ingredients _____

Temperature (storage or preparation) _____

Preparation/reconstitution time _____

Holding time_____

Containers _____

Other _____

_____

Special instructions _____

_____

4. Sample Presentation

Amount _____

Containers/utensils _____

Coding _____

_____

Serving size_____

Temperature _____

Presentation procedure_____

_____

Order _____

_____

_____

5. Subjects

Age range _____

Sex _____

Product usage _____

_____

Frequency of product consumption _____

Availability_____

B. PROTOCOL DESIGN EXAMPLE: CANDY BARS

---

1. Product Screening

1. Test objective
   To determine the relative acceptance and attribute diagnostics for candy bars with different chocolate to peanut ratios and with some roast differences in peanuts
2. Sample selection
   a. Variables *Amount of standard 1050 coating on bar; amount of peanuts by weight; degree of roast color in peanuts*
   b. Products/brands *Screen 18 to 22 prototypes (experimental design) and 2 competitors; have descriptive data available to identify products with little or no differences from one another; choose 12 to 15 bars to test*
3. Reasons
   14 selected samples demonstrate differences in peanut/chocolate balance and roast flavor intensity and crunchiness of nut pieces

---

2. Sample Information

Sample Conditions
1. Sample source *Trial run prototype samples (3 oz); competitors from same age carefully stored lots*
   Age *3 months old*
   Place *Lancaster production; competitors from Midwest distribution*
   Code *Ours L432-439; competition A 419Q, 7425S*
   Packaging condition *All samples overwrapped in white foil wrappers (732 equipment Lancaster)*
2. Sample Holding
   Hold all foil wrapped samples for 3 weeks prior to test in boxes of 24 overwrapped in cellophane, at 65°, in 50% RH storeroom prior to shipping to test site
3. Other
   Ship all samples by truck in styrofoam chests to Indianapolis and Syracuse for test

---

3. Sample Preparation

Total amount *250 bars of each to each test site (150 needed)*
Other ingredients *None*
Temperature (storage or preparation) *Keep at 65 to 75°F*
Preparation/reconstitution time *None*
Holding time *None*
Containers *Use plastic plates*
Other *Leave bars wrapped until just before presentation to subject; discard any broken, split, or pitted samples*
Special instructions *Do not handle bars any more than a few seconds to prevent melting and damage*

4. Sample Presentation

Amount   *Each subject to get one full bar of each product*

Containers/utensils   *Plastic plates*

Coding   *Three-digit codes; see attached sheets for codes for each subject*

Serving size   *One bar per subject*

Temperature   *65 to 75°F*

Presentation procedure   *Place sample in middle of coded 6 in. plastic plate*

Order   *See attached sheet for codes and order for each subject [Such a sheet is not included here, but should be prepared based on the experimental design used.]*

5. Subjects

Age range   *50% 12 to 25 years; 50% 25 to 55 years*

Sex   *50% male; 50% female*

Product usage   *Has eaten a chocolate coated candy bar within the last month*

Frequency of product consumption   *5 or more bars/year*

Availability   *Afternoons—3 to 5 or evenings—7 to 9*

13 Basic Statistical Methods

CONTENTS

I. INTRODUCTION

The goal of applied statistics is to draw some conclusion about a population based on the information contained in a sample from that population. The types of conclusions fall into two general categories, estimates and inferences. Further, the size and manner in which a sample is drawn from a population affects the precision and accuracy of the resulting estimates and inferences. These issues are addressed in the experimental design of a sensory study. This chapter presents the concepts and techniques of estimation, inference, and experimental design as they relate to some of the more fundamental statistical methods used in sensory evaluation. The topics are presented with a minimum of theoretical detail. Those interested in pursuing this area further are encouraged to read Gacula and Singh (1984), Smith (1988), and O'Mahony (1986) or, for more theoretically advanced presentations, Snedecor and Cochran (1980) and Cochran and Cox (1957).

Several definitions presented at this point will make the discussion that follows easier to understand. A *population* is the entire collection of elements of interest. The population of interest in sensory analysis varies from study to study. In some cases the population may be people (e.g., consumers of a particular food) while in other cases it may be products (e.g., batches of corn syrup). An *element* or *unit* from the population might be a particular consumer or a particular batch of syrup. Measurements taken on elements from a population may be discrete, that is, take on only specific values (e.g., preference for brand A), or continuous, that is, take on any value on a continuum (e.g., the intensity of sweetness). The values that the measurements take on are0 governed by a probability *distribution,* usually expressed in the form of a mathematical equation, that relates the occurrence of a specific value to the probability of that occurrence. Associated with the distribution are certain fixed quantities called *parameters.* The values of the parameters provide information about the population. For continuous distributions, for instance, the *mean* (μ) locates the center of the measurements. The *standard deviation* (σ) measures the dispersion or "spread" of the measurements about the mean. For discrete distributions, the proportion of the population that possesses a certain characteristic is of interest. For example, the *population proportion* (p) of a binomial distribution might summarize the distribution of preferences for two products.

Only in the rarest of circumstances is it possible to conduct a census of the population and directly compute the exact values of the population parameters. More typically, a subset of the elements of the population, called a *sample,* is collected, and the measurements of interest are made on each element in the sample. Mathematical functions of these measurements, called *statistics,* are used to approximate the unknown values of the population parameters. The value of a statistic is called an *estimate.*

Often a researcher is interested in determining if a population possesses a specific characteristic (e.g., more people prefer Product A than Product B). There are risks associated with drawing conclusions about the population as a whole when the only information available is that contained in a sample. Formal procedures, called tests of hypotheses, set limits on the probabilities of drawing incorrect conclusions. Then, based on the actual outcome of an experiment, the researcher's risks are constrained within these known limits. Tests of hypotheses are a type of *statistical inference* that give sensory researchers greater assurance that correct decisions will be made.

The amount of information required to draw sound statistical conclusions depends on several factors (e.g., the level of risk the researcher is willing to assume, the required precision of the information, the inherent variability of the population being studied, etc.). These issues need to be addressed and a plan of action, called the *experimental design,* developed before a study is undertaken. The experimental design, based on both technical and common sense principles, will

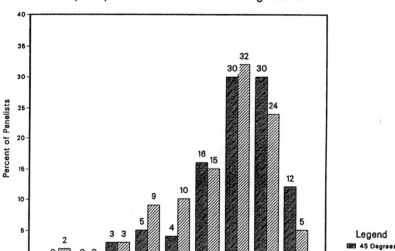

FIGURE 13.1   Histogram (with frequencies) of the overall liking scores for two samples of salad dressing.

insure that the experimental resources are focused on the critical issues in the study, that the correct information is collected, and that no excessive sampling of people or products occurs.

The remainder of the chapter is devoted to the further development of the ideas just presented. Section II presents some basic techniques for summarizing data in tabular and graphical forms. Section III combines estimation with some fundamental concepts of probability to present some methods for testing statistical hypotheses. Section IV covers the most commonly used experimental designs in sensory studies, including techniques for improving the sensitivity of panels for detecting differences among products. The basic techniques for calculating probabilities from some common distributions are presented in an appendix (see Section V).

II. SUMMARIZING SENSORY DATA

The data from sensory panel evaluations should be summarized in both graphs and tables before formal statistical analyses (i.e., tests of hypotheses, etc.) are undertaken. Examination of the graphs and tables may reveal features of the data that would be lost in the computation of test statistics and probabilities. In fact, features revealed in the tables and graphs may indicate that standard statistical analysis procedures would be inappropriate for the data at hand.

Whenever a reasonably large number of observations are available, the first step of any data analysis should be to develop the frequency distribution of responses (see Figure 13.1). Then a basic set of summary statistics should be calculated. Included in the basic set would be the arithmetic or *sample mean,* $\bar{x}$ for estimating the center (or central tendency) of the distribution of responses and the *sample standard deviation, s,* for estimating the spread (or dispersion) of the data around the mean. The sample mean is calculated as:

$$\bar{x} = \left(\sum_{i=1}^{n} x_i \right) \Big/ n$$

$$= \left(x_1 + x_2 + \ldots + x_n \right) \Big/ n$$

(13.1)

Σ in Equation 13.1 is the sum function. The subscript ($i = 1$) and superscript (n) indicate the range over which the summing is to be done. Equation 13.1 indicates that the sum is taken over all n elements in the sample. The sample standard deviation is calculated as:

$$s = \sqrt{\left[\sum_{i=1}^{n} x_i^2 - \left(\sum_{i=1}^{n} x_i\right)^2 \Big/ n\right] \Big/ (n-1)} \tag{13.2}$$

These basic statistics can sometimes be misleading. Instances where they should be used with caution include cases where the data are multimodal (i.e., several groups of data clustered at different locations on the response scale) or where there are extreme values (i.e., outliers) in the data.

Multimodal data may be indicating the presence of several subpopulations with different mean values. In such situations the sample mean of all the data may be meaningless, and, as a result, so might the sample standard deviation (since s measures the spread around the mean). Multimodal data should be examined further to determine if there is a way to break up the entire set into unimodal subgroups (e.g., by sex, age, geography, plant, batch, etc.). Separate sets of summary statistics could then be calculated within each subgroup. If it is not possible to break up the entire set, then the researcher has to determine which summary statistics are still meaningful. For instance, the *median* divides the data in half with 50% of the observations falling below the median and 50% falling above it. This may be a meaningful way to identify the center of a set of multimodal data. Similarly, the spread of the data might be measured by the difference between the *first and third quartiles* of the responses (i.e., the points that 25% and 75% of the values fall below, respectively). This difference is called the *interquartile range*.

The sample mean, $\bar{x}$, is sensitive to the presence of extreme values in the data. The median is less sensitive to extreme values, so it could again be used in place of the sample mean as the summary measure of the center of the data. Another option is a robust estimator of central tendency called the *trimmed mean*. The trimmed mean is calculated in the same way as the sample mean but after a specific proportion (e.g., 5%) of the highest and lowest data values have been eliminated. Various computerized statistical analysis packages routinely compute a variety of measures of central tendency and dispersion.

Many statistical analysis procedures assume that the data are normally distributed. If the raw data used to calculate $\bar{x}$ are normally distributed, then so is $\bar{x}$. In fact, even if the raw data are not distributed as normal random variables, $\bar{x}$ is still approximately normal providing the sample size is greater than 25 or so. The mean of the distribution of $\bar{x}$ is the same as the mean of the distribution of the raw data, i.e., μ, and if σ is the standard deviation of the raw data, then $\sigma/\sqrt{n}$ the standard deviation of $\bar{x}$. $\sigma/\sqrt{n}$ is called the *standard error of the mean*. Notice that as the sample size n increases, the standard error of the mean decreases. Therefore, as the sample size becomes larger, $\bar{x}$ is more and more likely to take on a value close to the true value of μ. The standard error (*SE*) of the mean is estimated by $SE = s/\sqrt{n}$, where s is the sample standard deviation calculated in Equation 13.2.

A. SUMMARY ANALYSIS OF DATA IN THE FORM OF RATINGS

The overall liking responses of 30 individuals in each of four cities are presented in Table 13.1. The frequency distributions of the responses are presented tabularly in Table 13.2 and graphically, using simple dot-plots, in Figure 13.2. There is no strong indication of multimodal behavior within a city. The summary statistics for these data are presented in Table 13.3. The box-and-whisker plots (see Danzart, 1986) in Figure 13.3 provide additional information about the distribution of ratings from city to city and, possibly, some minor concern about extreme observations.

TABLE 13.1
Data from a Multicity Monadic Consumer Test

Respondent	Attribute: Overall Liking[a]			
	Atlanta	Boston	Chicago	Denver
1	12.6	10.4	7.9	10.3
2	9.8	10.4	7.8	11.7
3	8.6	8.9	6.3	11.5
4	9.8	8.0	11.1	9.9
5	15.0	10.4	5.5	11.7
6	12.7	11.0	6.5	10.3
7	12.8	7.4	8.8	11.6
8	9.5	10.5	5.2	12.1
9	12.4	9.2	7.8	11.6
10	9.6	9.2	7.6	12.3
11	9.2	9.8	6.3	12.4
12	7.1	9.1	7.1	10.5
13	9.9	9.7	8.0	12.4
14	12.4	10.3	5.7	14.4
15	8.7	9.1	5.5	11.1
16	11.9	10.3	5.2	9.9
17	9.9	11.7	7.2	11.9
18	11.3	9.8	8.0	8.8
19	10.4	10.2	9.1	12.3
20	11.8	9.5	8.4	8.6
21	11.5	12.4	4.0	11.9
22	8.9	9.5	6.9	9.3
23	11.4	12.9	6.6	10.0
24	6.9	11.1	7.4	10.2
25	8.8	13.3	7.3	10.8
26	11.6	12.9	7.5	12.7
27	11.3	11.4	9.1	11.1
28	9.7	9.0	6.9	11.9
29	10.0	10.1	8.4	10.2
30	11.2	11.2	6.1	10.1

[a] Measured on a 15-cm unstructured line scale.

B. ESTIMATING THE PROPORTION OF A POPULATION THAT POSSESSES A PARTICULAR CHARACTERISTIC

The statistic used to estimate the population proportion p of a binomial distribution is $\hat{p}$ (p-hat), where

$$\hat{p} = \text{(Number of "successes")/(number of trials)} \qquad (13.3)$$

Suppose that 150 consumers participate in a preference test between two samples, A and B. Further, suppose that 86 of the participants say that they prefer Sample A. Preference for Sample A was defined as a success before the test was conducted, so from Equation 13.3, $\hat{p} = 86/150 = 0.573$. That is, 0.573 or 57.3% of consumers preferred Sample A. If a multicity test had been conducted, the estimated preferences for Sample A could be represented graphically using a bar chart such as in Figure 13.4.

TABLE 13.2
Frequency Distributions from the Multicity
Consumer Test Data in Table 13.1

Attribute: Overall Liking

Category midpoint[a]	Frequencies in			
	Atlanta	Boston	Chicago	Denver
1	0	0	0	0
2	0	0	0	0
3	0	0	0	0
4	0	0	1	0
5	0	0	2	0
6	0	0	6	0
7	2	1	8	0
8	0	1	9	0
9	5	6	3	3
10	9	12	0	8
11	4	5	1	4
12	6	2	0	13
13	3	3	0	1
14	0	0	0	1
15	1	0	0	0

[a] For example, in Atlanta nine people responded with an overall liking rating between 9.5 and 10.4.

C. Confidence Intervals on μ and p

The single valued statistics, called *point estimates,* calculated previously, provide no information as to their own precision. Confidence intervals supply this missing information. A confidence interval is a range of values within which the true value of a parameter lies with a known probability. Confidence intervals allow the researcher to determine if the point estimates are sufficiently precise to meet the needs of an investigation.

Three types of confidence intervals are presented: the one-tailed upper confidence interval, the one-tailed lower confidence interval, and the two-tailed confidence interval. The equations for calculating these intervals for both μ and p are presented in Table 13.4. In general, two-tailed confidence intervals are most useful, but if the analyst is only interested in an average value that is either "too big" or "too small," then the appropriate one-tailed confidence interval should be used.

The quantities $t_{\alpha,n-1}$ and $t_{\alpha/2,n-1}$ in Table 13.4 are *t*-statistics. The quantity α measures the level of confidence. For instance, if $\alpha = 0.05$, then the confidence interval is a $100(1 - \alpha)\% = 95\%$ confidence interval. The quantity $(n - 1)$ in Table 13.4 is a parameter associated with the *t*-distribution called degrees of freedom. The value of t depends on the value of α and the number of degrees of freedom $(n - 1)$. Critical values of t are presented in Table T3.

The quantity z in Table 13.4 is the critical value of a standard normal variable. (The standard normal distribution has mean $\mu = 0$ and standard deviation $\sigma = 1$.) Critical values of z for some commonly used levels of α are presented in the last row of Table T3 (i.e., the row corresponding to ∞ degrees of freedom).

Consider the overall liking data presented in Table 13.1. The sample mean intensity for Atlanta was $\bar{x} = 10.56$ and the sample standard deviation of the data was $s = 1.79$. To construct a lower

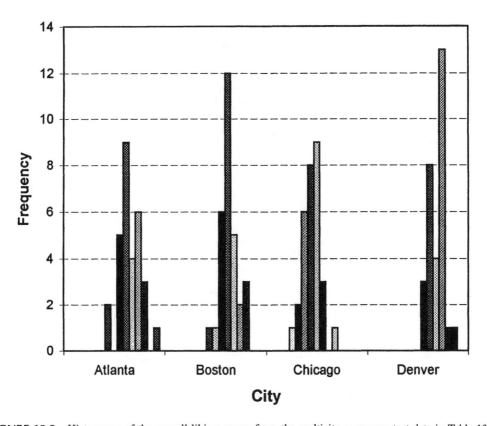

FIGURE 13.2 Histograms of the overall liking scores from the multicity consumer test data in Table 13.1.

TABLE 13.3
Summary Statistics from the Multicity Consumer Test Data in Table 13.1

			Attribute: Overall Liking				
	City	n	Mean	Median	Trimmed mean	Std. dev	Std. error
Overall liking	Atlanta	30	10.557	10.200	10.573	1.793	0.327
	Boston	30	10.290	10.250	10.273	1.401	0.256
	Chicago	30	7.173	7.250	7.146	1.448	0.264
	Denver	30	11.117	11.300	11.115	1.276	0.233
	City	Min	Max	Q1	Q3		
Overall liking	Atlanta	6.900	15.000	9.425	11.825		
	Boston	7.400	13.300	9.200	11.125		
	Chicago	4.000	11.100	6.250	8.000		
	Denver	8.600	14.400	10.175	11.950		

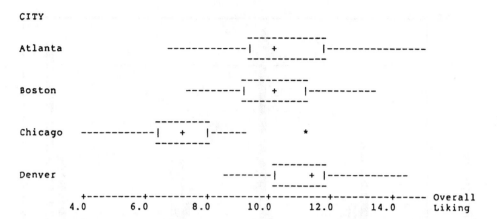

```
CITY

Atlanta              ------------|   +     |----------------

Boston                   --------|   +   |-----------

Chicago  ------------|   +   |------          *

Denver                    --------|    +  |------------
         +---------+---------+---------+---------+---------+----- Overall
        4.0       6.0       8.0      10.0      12.0      14.0   Liking
```

FIGURE 13.3 Box-and-whisker plots of the overall liking scores from the multicity consumer test data in Table 13.1.

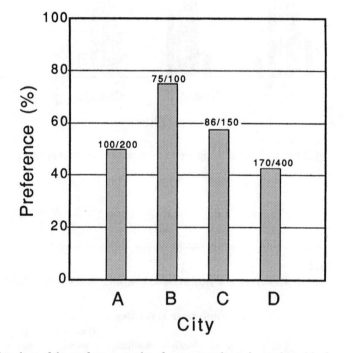

FIGURE 13.4 Bar chart of the preference results of a two-sample study conducted in four cities showing the relative difference from city to city. Actual preference results and total respondent base are included for each city.

one-tailed 95% confidence interval on the value of the population mean one uses Table 13.4 and Table T3 to obtain:

$$\bar{x} - t_{\alpha,n-1} S/\sqrt{n}$$

where $\alpha = 0.05$ and $n = 30$, so $t_{\alpha,n-1}$ is $t_{0.05,29} = 1.699$, yielding

$$10.56 - 1.699(1.79)/\sqrt{30}$$

$$= 10.56 - 0.56$$

$$= 10.0$$

TABLE 13.4
Computational Forms for Confidence Intervals

	Parameter	
Type of interval	μ	p
One-tailed upper	$\bar{x} + t_{\alpha,n-1}\,s/\sqrt{n}$	$\hat{p} + z_{\alpha}\sqrt{\hat{p}(1-\hat{p})/n}$
One-tailed lower	$\bar{x} - t_{\alpha,n-1}\,s/\sqrt{n}$	$\hat{p} - z_{\alpha}\sqrt{\hat{p}(1-\hat{p})/n}$
Two-tailed	$\bar{x} \pm t_{\alpha/2,n-1}\,s/\sqrt{n}$	$\hat{p} \pm z_{\alpha/2}\sqrt{\hat{p}(1-\hat{p})/n}$

The limit is interpreted to mean that the researcher is 95% sure that the true value of the mean overall liking rating in Atlanta is no less than 10.0.

A two-tailed 95% confidence interval on the mean is calculated as:

$$\bar{x} \pm t_{\alpha/2,n-1}\,s/\sqrt{n}$$

where $\alpha = 0.05$ and $n = 30$, so $t_{\alpha/2,n-1}$ is $t_{0.025,29} = 2.045$, yielding:

$$10.56 \pm 2.045(1.79)/\sqrt{30}$$

$$= 10.56 \pm 0.67$$

or

$$(9.89, 11.23)$$

That is, the researcher is 95% sure that the true value of the mean overall liking rating in Atlanta lies somewhere between 9.89 and 11.23. In Figure 13.5, the sample means and their associated 95% confidence intervals are presented for the overall liking data of each of the four cities presented in Table 13.1. The analyst can now begin to formulate some ideas about differences in average overall liking that may exist among the cities.

Consider the consumer preference test discussed before where 86 of the 150 ($\hat{p} = 0.573$) consumers preferred Sample A. To construct a 95% confidence interval (two-tailed) on the true value of the population proportion p one uses Table 13.4 and Table T3 to obtain:

$$\hat{p} \pm z_{\alpha/2}\sqrt{\hat{p}(1-\hat{p})/n}$$

where $n = 150$, $\alpha = 0.05$, so $z_{\alpha/2} = t_{\alpha/2,\infty} = 1.96$, yielding:

$$0.573 \pm 1.96\sqrt{(0.573)(0.427)/150}$$

or

$$(0.494, 0.652)$$

The researcher may conclude, with 95% confidence, that the true proportion of the population that prefers Sample A lies between 49.4% and 65.2%. Confidence intervals on proportions can also be

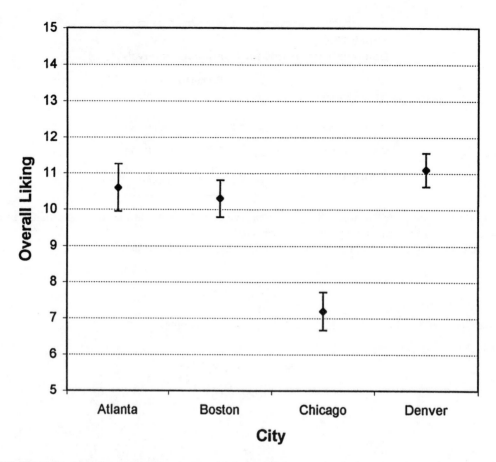

FIGURE 13.5 Average overall liking scores with 95% confidence intervals from the multicity consumer test data in Table 13.1. Note the large degree of overlap among Atlanta, Boston, and Denver, compared to the much lower average value for Chicago.

depicted graphically as in Figure 13.6, where 95% two-tailed confidence intervals have been added to the data summarized in Figure 13.4.

D. OTHER INTERVAL ESTIMATES

Confidence intervals state a range of values that have a known probability of containing the true value of a population parameter. The researcher may not always want to draw such a conclusion. There are other types of statistical interval estimates.

For instance, a *prediction interval* is a range of values that has a known probability of containing the average value of "k" future observations. The researcher may choose $k = 1$ to calculate an interval that has a known probability of containing the next observed value of some response (e.g., being 95% confident that the perceived saltiness of the next batch of potato chips will lie between 7.2 and 10.4). Two-sided prediction intervals are calculated as:

$$\bar{x} \pm t_{\alpha/2, \, n-1} s \sqrt{(1/k) + (1/n)}$$

Another statistical interval, called a *tolerance interval,* is a range of values that has a known probability of containing a specified proportion of the population. An example of a one-sided

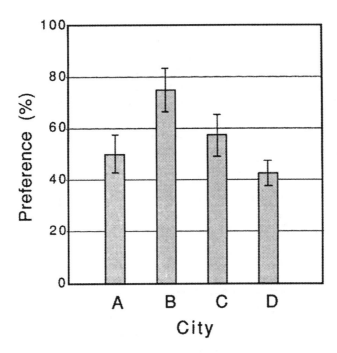

FIGURE 13.6   Bar chart of the preference results including 95% confidence intervals of a two-sample study conducted in four cities. Confidence intervals that overlap 50% indicate that no significant preference exists in that city ($\alpha = 0.05$). Confidence intervals from two cities that do not overlap indicate roughly that the two cities differ in their degree of preference for the product.

tolerance interval is that the researcher is 95% sure that 90% of all batches have firmness ratings less than 6.3. Two-sided tolerance intervals can also be computed (see Dixon and Massey, 1969).

E. DATA TRANSFORMATIONS

At times a researcher may want to transform the scale of measurement from the one used to collect the data to a more meaningful scale for presentation. This is easy to do for a kind of transformation called a linear transformation. If the original variable, x, is transformed to a new variable, y, using $y = a(x) + b$, then y is a linear transformation of x. Linear transformations are limited to multiplying the original variable by a constant, a, and/or adding a constant, b. Raising the original variable to a power, taking its logarithm, sine, inverse, etc. are all nonlinear transformations. If x has mean value μ and standard deviation σ then the mean and standard deviation of y are $a\mu + b$ and $a\sigma$, respectively. These equations for computing the mean and standard deviation of the transformed variable y apply only to linear transformations. The sample mean, $\bar{y}$ and sample standard deviation, s_y are obtained by substituting $\bar{x}$ for μ and s_x for σ.

An example of this data transformation technique occurs in tests for overall differences such as Triangle, Duo-trio, and Two-out-of-five tests where the original measurement is the proportion of correct responses, p_c. Using the Triangle test as an example, p_c can be transformed to the proportion of the population that can distinguish the samples, p_d, by using $p_d = 1.5(p_c) - 0.5$. The expression for p_d is obtained by inverting the equation for the probability of obtaining a correct answer in a Triangle test, $p_c = 1(p_d) + (\frac{1}{3})(1 - p_d)$; that is, the probability of a correct answer is the probability of selecting a distinguisher, p_d (who will always give a correct answer), plus the probability of selecting a nondistinguisher $(1 - p_d)$, and having that person guess correctly (which has a probability of $\frac{1}{3}$). Notice that when there are no perceptual differences between the samples in a Triangle test the expected proportion of correct answers is $p_c = \frac{1}{3}$ which transforms to the expected proportion of distinguisher $p_d = 0$ (i.e., everyone is guessing).

In a Triangle test involving n respondents, if x people correctly select the odd sample, then the *estimated value* of p_c is $\hat{p}_c = x/n$ and the estimated standard deviation of p_c is $s_c = \sqrt{\hat{p}_c(1-\hat{p}_c)/n}$. The estimated proportion of distinguishers is then $\hat{p}_d = 1.5(x/n) - 0.5$, with an estimated standard deviation of $s_d = 1.5s_c$. These transformations are applied in several places in Chapter 6.

These data transformations are particularly useful in the unified approach to discrimination testing discussed in Chapter 6. Confidence intervals can be constructed on the proportion of distinguishers in the population of panelists, p_d, using

$$\text{Lower Confidence Limit: } \hat{p}_d - z_\alpha s_d, \text{ and Upper Confidence Limit: } \hat{p}_d + z_\beta s_d,$$

where $\hat{p}_d$ is the estimate of the proportion of distinguishers, s_d is the sample standard deviation of the proportion of distinguishers and z_α and z_β are the α and β critical values from the standard normal distribution. The quantities $\hat{p}_d$ and s_d are obtained from $\hat{p}_c$ and s_c using the following transformations:

Method	$\hat{p}_d$	s_d
Triangle Test	$1.5\,\hat{p}_c - 0.5$	$1.5s_c$
Duo-Trio and Paired Comparison	$2\,\hat{p}_c - 1$	$2s_c$
Two-out-of-Five	$(10/9)\,\hat{p}_c - (1/9)$	$(10/9)s_c$

If the lower confidence limit is zero or less, then the null hypothesis of no perceptible difference cannot be rejected (at the $1-\alpha$ confidence level). If the lower confidence limit is greater than zero, then the samples are perceptibly different. If the upper confidence limit is less than the proportion of distinguishers that the researcher wants to be able to detect, p_{max}, then the products are sufficiently similar (at the $1-\beta$ confidence level). If the upper confidence limit is greater than p_{max}, then the samples are not sufficiently similar. (See Chapter 6 for examples using these confidence intervals.)

III. STATISTICAL HYPOTHESIS TESTING

Often the objective of an investigation is to determine if it is reasonable to assume that the unknown value of a parameter is equal to some specified value or possibly that the unknown values of two parameters are equal to each other. In the face of the incomplete and variable information contained in a sample, statistical decisions of this type are made using hypothesis testing. The process of statistical hypothesis testing is summarized by the following five steps:

1. The objective of the investigation is stated in mathematical terms, called the null hypothesis (H_0), (e.g., H_0: $\mu = 8$).
2. Based on the prior interest of the researcher, another mathematical statement, called the alternative hypothesis (H_a) is formulated (e.g., H_a: $\mu > 8$, H_a: $\mu < 8$, or H_a: $\mu \neq 8$).
3. A sample of elements from the population is collected and the measurement of interest is taken on each element of the sample.
4. The value of the statistic used to estimate the parameter of interest is calculated.
5. Based on the assumed probability distribution of the measurements and the null hypothesis assumption, H_0, the probability that the statistic takes on the value calculated in Step 4 is computed. If this probability is smaller than some predetermined value (α), the null-hypothesis is rejected in favor of the alternative hypothesis.

A. STATISTICAL HYPOTHESES

In most sensory studies, statistical hypotheses specify the value of some parameter in a probability distribution, such as the mean μ or the population proportion p. The null hypothesis is determined by the objective of the investigation and serves as the baseline condition that is assumed to exist prior to running the experiment. The value specified in the null hypothesis is used to calculate the test statistic (and resulting p-value) in the hypothesis test. The alternative hypothesis is developed based on the prior interest of the investigator. For example, if a company is replacing one of the raw ingredients in its current product with a less expensive ingredient from an alternate supplier, the sensory analyst's only interest going into the study would be to determine with a high level of confidence that the product made with the less expensive ingredient is not less preferred than the company's current product. The null hypothesis and the alternative hypothesis for this investigation are

$$H_0: p_{current} = p_{less\ expensive}$$

vs.

$$H_a: p_{current} > p_{less\ expensive}$$

where p_i is the proportion of the population that prefers Product i. Both the null and the alternative hypotheses must be specified before the test is conducted. If the alternative hypothesis is formulated after reviewing the data, the results of the statistical tests are biased in favor of rejecting the null hypothesis too often.

B. One-Sided and Two-Sided Hypotheses

There are two types of alternative hypotheses, one-sided alternatives and two-sided alternatives. Some examples of situations leading to one-sided and two-sided alternatives are:

One-sided

Confirm that a test brew is more bitter.
Confirm that a test product is preferred to the control.
In general, whenever H_a has the form:
  A is more (less) than B, where both A and B are specified.

Two-sided

Decide which test brew is more bitter.
Decide which test product is preferred.
In general, whenever H_a has the form:
  A is different from B.

Researchers often have trouble deciding whether the alternative hypothesis is one-sided or two-sided. Rules of thumb that work for one person may misguide others. There are no statistical criteria for deciding if an alternative hypothesis should be one-sided or two-sided. The form of the alternative hypothesis is determined by the prior interest of the researcher. If the researcher is only interested in determining if two samples are different, then the alternative hypothesis is two-sided. If, on the other hand, the researcher wants to test for a specific difference between two samples, that is, one sample (specified) is more preferred, or more sweet, etc. than another sample, then the alternative hypothesis is one-sided. Most alternatives are two-sided, unless the researcher states that a specific type of difference is of interest before the study is conducted.

A point of confusion may arise regarding one-sided vs. two-sided alternatives because in several common sensory testing situations one-tailed tests statistics are used to test two-sided alternatives. For example, in a Triangle test the null hypothesis is only rejected for large numbers of correct selections (i.e., a one-tailed test criterion). However, the alternative hypothesis is two-sided (i.e., H_a: The samples are perceivably different). Similar situations arise when χ^2 and F-tests are performed.

In practice, researchers should express their interests (i.e., the null and alternative hypotheses) in their own words. If the researcher's interests are clearly stated, it is easy to decide whether the alternative hypothesis is one-sided or two-sided. If not, then further probing is necessary. The

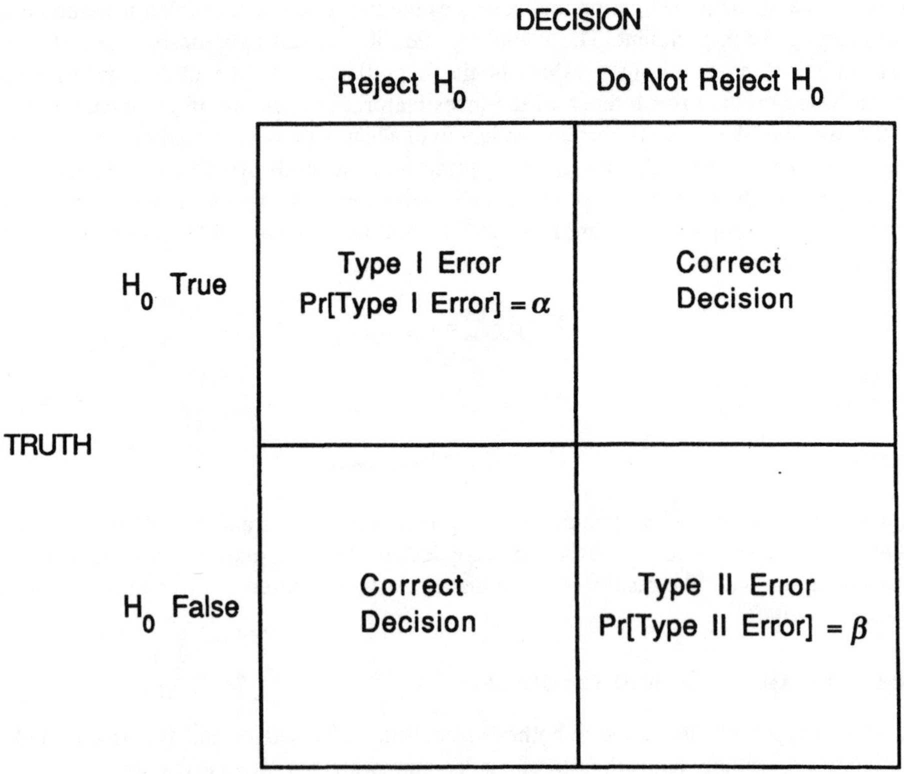

DECISION

	Reject H_0	Do Not Reject H_0
H_0 True	Type I Error Pr[Type I Error] = α	Correct Decision
H_0 False	Correct Decision	Type II Error Pr[Type II Error] = β

TRUTH

FIGURE 13.7   Type I and Type II errors of size α and β.

sensory analyst should report the results of the study in terms of the researcher's stated interests (one-sided or two-sided) regardless of whether the statistical method is one-tailed or two-tailed.

C. TYPE I AND TYPE II ERRORS

In testing statistical hypotheses some conclusion is drawn. The conclusion may be correct or incorrect. There are two ways in which an incorrect conclusion may be drawn. First, a researcher may conclude that the null hypothesis is false when, in fact, it is true (e.g., that a difference exists when it does not). Such an error is called a Type I error. Second, a researcher may conclude that the null hypothesis is true, or more correctly that the null hypothesis cannot be rejected, when, in fact, it is false (e.g., failing to detect a difference that exists). Such an error is called a Type II error (see Figure 13.7). The practical implications of Type I and Type II errors are presented in Figure 13.8.

The probabilities of making Type I and Type II errors are specified before the investigation is conducted. These probabilities are used to determine the required sample size for the study (see, for example, Snedecor and Cochran, 1957, p. 102). The probability of making a Type I error is equal to α. The probability of making a Type II error is equal to β. Although α and β are probabilities (i.e., numbers), it is currently a common practice to use Type I error and α-error (as well as Type II error and β-error) interchangeably. This somewhat casual use of terminology causes little confusion in practice.

The complementary value of Type II Error, i.e., $1 - \beta$, is called the "power" of the statistical test. Power is simply the probability that the test will detect a given sized departure from the null hypothesis (and, therefore, correctly reject the false null hypothesis). In discrimination testing, for example, the null hypothesis is H_0: $p_d = 0\%$. Departures from the null hypothesis are measured as

a) In Testing for a Difference

Truth	Reject H_0	Do Not Reject H_0
H_0 is True	**Type I Error** • Substitution takes place when it should not. • New product promotion done on same product as before. • Franchise in trouble due to loss of consumer confidence.	Correct Decision
H_0 is False	Correct Decision	**Type II Error** • Substitution does not take place when it should. • Candidate sample is missed. • Money, effort and time are lost. • We 'missed the boat."

b) In Testing for Similarity

Truth	Reject H_0	Do Not Reject H_0
H_0 is True	**Type I Error** • Substitution does not take place when it should. • Candidate sample is missed. • Money, effort and time are lost. • We 'missed the boat."	Correct Decision
H_0 is False	Correct Decision	**Type II Error** • Substitution takes place when it should not. • New product promotion done on same product as before. • Franchise in trouble due to loss of consumer confidence.

FIGURE 13.8   The practical implications of Type I and Type II errors.

values of $p_d > 0\%$. Suppose a researcher is conducting a Duo-trio test with 40 assessors and is testing at the $\alpha = 0.05$ level of significance. If the true proportion of distinguishers in the population of assessors is $p_d = 25\%$, then the power of the test is $1 - \beta = 0.44$ — that is, the test, as designed, has a 44% chance of rejecting the null hypothesis at the $\alpha = 0.05$ level when 25% of the population can distinguish the samples. The power of a statistical test is affected by the size of the departure from the null hypothesis (i.e., p_d), the size of the Type I error (α-risk) and the number of assessors, n.

D. Examples: Tests on Means, Standard Deviations, and Proportions

This section presents procedures for conducting routine tests of hypotheses on means and standard deviations of normal distributions and on the population proportion (or probability of success) from binomial distributions.

1. Example 13.1: Testing that the Mean of a Distribution is Equal to a Specified Value

Suppose in the consumer test example in Section II.A that the sensory analyst wanted to test if the average overall liking of the sample for Chicago was six or greater than six. The mathematical forms of the null hypothesis and alternative hypothesis are

$$H_0: \mu = 6$$

vs.

$$H_a: \mu > 6$$

The alternative hypothesis is one-sided.

The statistical procedure used to test this hypothesis is a *one-tailed, one-sample t-test.** The form of the test statistic is

$$t = \left(\bar{x} - \mu_{H_0}\right)/\left(s/\sqrt{n}\right) \tag{13.4}$$

The values of $\bar{x}$ and s are calculated in Table 13.3. Substituting in Equation 13.4 yields:

$$t = (7.17 - 6)/\left(1.45/\sqrt{30}\right) = 4.42 \tag{13.5}$$

This value of t is compared to the upper-α critical value of a t-distribution with $(n-1)$ degrees of freedom (denoted as $t_{\alpha,n-1}$). The value of $t_{\alpha,n-1}$ marks the point in the t-distribution (with $(n-1)$ degrees of freedom) for which the probability of observing any larger value of t is α. If the value obtained in Equation 13.5 is greater than $t_{\alpha,n-1}$, then the null hypothesis is rejected at the α-level of significance. Suppose the sensory analyst decides to control the Type I error at 5% (i.e., $\alpha = 0.05$). Then, from the row of Table T3 corresponding to 29 degrees of freedom the value of $t_{0.05,29} = 1.699$, so the sensory analyst rejects the null hypothesis assumption that $\mu = 6$ in favor of the alternative hypothesis that $\mu > 6$ at the 5% significance level.

If this alternative hypothesis had been $H_a: \mu \neq 6$ (i.e., a two-sided alternative), then the null hypothesis would be rejected for absolute values of t (in Equation 13.5) greater than $t_{\alpha/2,n-1}$, that is, reject if $|t| > t_{0.025,29} = 2.045$ (from Table T3).

---

* Tests on means can be performed using standard normal z values if the standard deviation, σ, is known. The t-tests are based on Student's t-distribution, which is similar to the normal distribution. The t-distribution is used when σ is estimated by s because the t-distribution takes into account that the estimated value of s may deviate slightly from σ.

TABLE 13.5
Data and Summary Statistics for the
Paired *t*-Test Example 13.2

Judge	Sample 1	Sample 2	Difference
1	7.3	5.7	1.6
2	8.4	5.2	3.2
3	8.7	5.9	2.8
4	7.6	5.3	2.3
5	8.0	6.1	1.9
6	7.1	4.3	2.8
7	8.0	5.7	2.3
8	7.5	3.8	3.7
9	6.9	4.5	2.4
10	7.4	5.0	2.4
			$\bar{\delta} = 2.54$
			$s_\delta = 0.61$

---

2. Example 13.2: Comparing Two Means — Paired-Sample Case

Sensory analysts often compare two samples by having a single panel evaluate both samples. When each member of the panel evaluates both samples, the paired *t*-test is the appropriate statistical method to use. In general, the null hypothesis can specify any difference of interest (i.e., H_0: $\delta = \mu_1 - \mu_2 = \delta_0$; setting $\delta_0 = 0$ is equivalent to testing H_0: $\mu_1 = \mu_2$) The alternative hypothesis can be two-sided (i.e., H_a: $\delta \neq \delta_0$) or one-sided (H_a: $\delta > \delta_0$ or H_a: $\delta < \delta_0$). In any case, the form of the paired *t*-statistic is

$$t = \frac{\bar{\delta} - \delta_0}{s_\delta / \sqrt{n}} \tag{13.6}$$

where $\bar{\delta}$ is the average of the differences between the two samples and s_δ is the sample standard deviation of the differences. Consider the data in Table 13.5 that summarizes the scores of the panel on a single attribute. The analyst wants to test whether the average rating of Sample 1 is more than two units greater than the average rating for Sample 2. The null hypothesis in this case is H_0: $\delta \leq 2$ vs. the alternative hypothesis H_a: $\delta > 2$. The test statistic is calculated as:

$$t = \frac{2.54 - 2.00}{0.61 / \sqrt{10}} = 2.79$$

($n = 10$ is used as the sample size because there are 10 judges, each contributing one difference to the data set.) The null hypothesis is rejected if this value of *t* exceeds the upper-α critical value of the *t*-distribution with $(n - 1)$ degrees of freedom (i.e., $t_{\alpha,n-1}$).

The analyst decides to set $\alpha = 0.05$ and finds in Table T3 that $t_{0.05,9} = 1.833$. The value of $t = 2.79$ is greater than 1.833, so the analyst rejects the null hypothesis and concludes at the 5% significance level that the average rating for Sample 1 is more than two units greater than the average rating for Sample 2.

TABLE 13.6
Data and Summary Statistics for the
Two-Sample *t*-Test Example 13.3

Group 1		Group 2	
Judge	Score	Judge	Score
1	6.2	1	6.7
2	7.5	2	7.6
3	5.9	3	6.3
4	6.8	4	7.2
5	6.5	5	6.7
6	6.0	6	6.5
7	7.0	7	7.0
		8	6.9
		9	6.1

$n_1 = 7$ $n_2 = 9$
$\bar{x}_1 = 6.557$ $\bar{x}_2 = 6.778$
$s_1 = 0.580$ $s_2 = 0.460$

3. Example 13.3: Comparing Two Means — Independent (or Two-Sample) Case

Suppose that a sensory analyst has trained two descriptive panels at different times and that the analyst now wants to merge the two groups. The analyst wants a high level of confidence that the two groups score samples with equivalent ratings before merging the groups and treating them as one panel.

The sensory analyst conducts several attribute panels to ensure that the two groups are similar. For each attribute considered the analyst presents samples of the same product to all panelists and records their scores and the group they belong to. The data from one of the studies is presented in Table 13.6. The null hypothesis for this test is $H_0: \mu_1 = \mu_2$ (or, equivalently, $H_0: \mu_1 - \mu_2 = 0$). The alternative hypothesis is $H_a: \mu_1 \neq \mu_2$ (i.e., a two-sided alternative).

The test statistic used to test the hypothesis is a two-sample *t*-test. The form of the test statistic is

$$t = \frac{(\bar{x}_1 - \bar{x}_2) - \delta_0}{\sqrt{\frac{(n_1-1)s_1^2 + (n_2-1)s_2^2}{n_1 + n_2 - 2}} \sqrt{\frac{1}{n_1} + \frac{1}{n_2}}} \tag{13.7}$$

where δ_0 is the difference specified in the null hypothesis ($\delta_0 = 0$ in the present example). Substituting the values from Table 13.6 into Equation 13.7 yields:

$$t = \frac{(6.557 - 6.778) - 0}{\sqrt{\frac{(7-1)(0.580)^2 + (9-1)(0.460)^2}{7+9-2}} \sqrt{\frac{1}{7} + \frac{1}{9}}}$$

$$= \frac{-0.221}{\sqrt{0.265}\sqrt{0.254}} = -0.85$$

The value of $t = -0.85$ is compared to the critical value of a t-distribution at the $\alpha/2$ significance level (because the alternative hypothesis is two-sided) with $(n_1 + n_2 - 2)$ degrees of freedom. For the present example (using $\alpha = 0.05$) $t_{0.025,14} = 2.145$ from Table T3. The null hypothesis is rejected if the absolute value (i.e., disregard the sign) of t is greater than 2.145. Since the absolute value of $t = -0.85$ (i.e., $|t| = 0.85$) is less than $t_{0.025,14} = 2.145$, the sensory analyst does not reject the null hypothesis and concludes that on the average the two groups report similar ratings for this attribute.

4. Example 13.4: Comparing Standard Deviations from Two Normal Populations

The sensory analyst in Example 13.3 should also be concerned that the variabilities of the scores of the two groups are the same. To test that the variabilities of the two groups are equal the analyst compares their standard deviations. The null hypothesis for this test is H_0: $\sigma_1 = \sigma_2$. The alternative hypothesis is H_a: $\sigma_1 \neq \sigma_2$ (i.e., a two-sided alternative). The test statistic used to test this hypothesis is

$$F = \frac{s^2_{Larger}}{s^2_{Smaller}} \tag{13.8}$$

where s^2_{Larger} is the square of the larger of the two sample standard deviations and $s^2_{Smaller}$ is the square of the smaller sample standard deviation. In Table 13.6 Group 1 has the larger sample standard deviation, so $s^2_{Larger} = s^2_1$ and $s^2_{Smaller} = s^2_2$ for this example. The value of F in Equation 13.8 is then:

$$F = (0.58)^2/(0.46)^2 = 1.59$$

The value of F is compared to the upper $\alpha/2$ critical value of an F distribution with $(n_1 - 1)$ and $(n_2 - 1)$ degrees of freedom (the numerator degrees of freedom are $(n_1 - 1)$ because $s^2_{Larger} = s^2_1$ for this example. (If s^2_{Larger} had been s^2_2, then the degrees of freedom would be $(n_2 - 1)$ and $(n_1 - 1)$.) Using a significance level of $\alpha = 0.05$ the value of $F_{0.025,6,8}$ is found in Table T6 to be 4.65. The null hypothesis is rejected if $F > F_{\alpha/2,(n_1-1),(n_2-1)}$. Since $F = 1.59 < F_{0.025,6,8} = 4.65$, the null hypothesis is not rejected at the 5% significance level. The sensory analyst concludes that there is not sufficient reason to believe the two groups differ in the variability of their scoring on this attribute.

This is another example of a two-sided alternative that is tested using a one-tailed statistical test. The criterion for two-sided alternatives is to reject the null hypothesis if the value of F in Equation 13.8 exceeds $F_{\alpha/2,df_1,df_2}$ where df_1 and df_2 are the numerator and denominator degrees of freedom, respectively. Equation 13.8 is still used for one-sided alternatives (i.e., H_a: $\sigma_1 > \sigma_2$), but the criterion becomes "reject the null hypothesis if $F > F_{\alpha,df_1,df_2}$."

5. Example 13.5: Testing that the Population Proportion is Equal to a Specified Value

Suppose that two samples (A and B) are compared in a preference test. The objective of the test is to determine if either sample is preferred by more than 50% of the population. The sensory analyst collects a random sample of $n = 200$ people, presents the two samples to each person in a balanced, random order, and asks each person which sample they prefer.* For those respondents who refuse to state a preference, the "no preference" responses are divided equally among the two samples. It is found that 125 of the people said they preferred Sample

---

* Note that this example is quite similar to the paired t-test in Example 13.2, but in this case the data are ranks (i.e., either 0 or 1).

TABLE 13.7
Results of a Two Region Preference
Test Example 13.6

| | Preference | | |
Region	Product A	Product B	Total
1	125	75	200
2	102	98	200
Total	227	173	400

A. The estimated proportion of the population that prefer Sample A is then $\hat{p}_A = 125/200 = 62.5\%$ by Equation 13.3.

The sensory analyst arbitrarily picks "preference for Sample A" as a "success" and tests the hypothesis $H_0: p_A = 50\%$ vs. the alternative $H_a: p_A \neq 50\%$. The analyst chooses to test this hypothesis at the $\alpha = 0.01$ significance level, using the appropriate z-test, of the form:

$$z = \frac{p - p_0}{\sqrt{(p_0)(1 - p_0)/n}} \quad \text{for } \hat{p} \text{ and } p_0 \text{ proportions}$$

or   (13.9)

$$z = \frac{\hat{p} - p_0}{\sqrt{(p_0)(100 - p_0)/n}} \quad \text{for } \hat{p} \text{ and } p_0 \text{ percentages}$$

where $\hat{p}$ and p_0 are the observed and hypothesized values of p, respectively. Substituting the observed and hypothesized values into Equation 13.9 yields:

$$z = (62.5 - 50.0)/\sqrt{(50)(100 - 50)/200} = 3.54$$

This value of z is compared to the critical value of a standard normal distribution. For two-sided alternatives the absolute value of z is compared to $z_{\alpha,2} = t_{\alpha/2,\infty}$ (for one-sided alternatives, the value of z is compared to $z_\alpha = t_{\alpha,\infty}$) using Table T3. The value of $z_{0.005} = t_{0.005,\infty} = 2.576$. Since $z = 3.54$ is greater than 2.576, the null hypothesis is rejected and the analyst concluded at the 1% significance level that Sample A is preferred by more than 50% of the population.

6. Example 13.6: Comparing Two Population Proportions

Let us extend Example 13.5 to take into account regional preferences. Suppose a company wishes to introduce a new product (A) into two regions and wants to know if the product is equally preferred over its prime competitor's product (B) in both regions. The sensory analyst conducts a 200-person preference test in each region and obtains the results shown in Table 13.7.

The null hypothesis in this example is $H_0: p_1 = p_2$ vs. the alternative hypothesis $H_a: p_1 \neq p_2$, where p is defined as the proportion of the population in region i that prefers Product A. This hypothesis is tested using a χ^2-test of the form:

$$\chi^2 = \sum_{i=1}^{r} \sum_{j=1}^{c} (O_{ij} - E_{ij})^2 / E_{ij}$$ (13.10)

where r and c are the numbers of rows and columns in a data table like Table 13.7. O_{ij} is the observed value in row i and column j of a data table. E_{ij} is the "expected" value for the entry in the i'th-row and j'th-column of the data table. The E_{ij} are calculated as:

$$E_{ij} = \text{(total for row i)(total for column j)/(grand total)}$$

Substituting the values from Table 13.7 into Equation 13.10 we obtain:

$$\chi^2 = \frac{(125-(200)(227)/400)^2}{(200)(227)/400} + \frac{(75-(200)(173)/400)^2}{(200)(173)/400}$$

$$+\frac{(102-(200)(227)/400)^2}{(200)(227)/400} + \frac{(98-(200)(173)/400)^2}{(200)(173)/400}$$

$$= \frac{(125-113.5)^2}{113.5} + \frac{(75-86.5)^2}{86.5} + \frac{(102-113.5)^2}{113.5} + \frac{(98-86.5)^2}{86.5}$$

$$= 5.39$$

The value of χ^2 in Equation 13.10 is compared to the upper-α critical value of a χ^2-distribution with $(r-1)(c-1)$ degrees of freedom. If the analyst chooses $\alpha = 0.10$ (i.e., 10% significance level), then the critical value $\chi^2_{0.10,1} = 2.71$ (from Table T5). Since $\chi^2 = 5.39 > \chi^2_{0.10,1} = 2.71$, the analyst concludes at the 10% significance level that Product A is not equally preferred over Product B in both regions. Regional formulations may have to be considered.

E. CALCULATING SAMPLE SIZES IN DISCRIMINATION TESTS

The sample size required for a discrimination test is a function of the test sensitivity parameters, α, β, and p_d, or in the case of directional difference tests, p_{max}. Tables T7, T9, T11, and T13 can be used to find sample sizes for commonly chosen values of the parameters. Alternatively, researchers can use a spreadsheet to perform the necessary calculations. The "Test Sensitivity Analyzer" has been developed in Microsoft Excel to allow researchers to study how various choices of α, β, and p_d (or p_{max}), affect the sample size and the number of correct responses necessary to claim that a difference exists or that the samples are similar (see Figure 13.9). The Test Sensitivity Analyzer does this indirectly by letting the researcher choose values for the same size, n, the number of correct responses, x, and the maximum allowable proportion of distinguishers, p_d. (Although p_d is not meaningful in a directional difference test, the value of p_{max} is computed based on the value entered for p_d.) The Test Sensitivity Analyzer then computes values for α and β. By adjusting the values of n, x, and p_d, the researcher can find the set of values that provides the best compromise between test sensitivity and available resources.

The binomial distribution, upon which discrimination tests are based, is a discrete probability distribution. Only integer values for the sample size, n, and the number of correct responses, x, are valid. Small changes in n and x can have large impacts on the probabilities α and β, particularly for small values of n. Generally, it is not possible to select values of n, x, and p_d (or p_{max}) that yield values for α and β that are exactly equal to their target values. Instead, the researcher must select values for n, x, and p_d (or p_{max}) that yield values for α and β that are close to their targets. As illustrated in Figure 13.9, the researcher wants to conduct a Duo-trio test for similarity with the following target sensitivity values: $\alpha = 0.25$, $\beta = 0.10$, and $p_d = 25\%$. Strictly speaking, the values of n and x should be chosen so that both α and β are no greater than their target values. However, the researcher only has access to 60 assessors. Setting $n = 60$, the researcher finds that $x = 33$ correct responses yields values for α and β that are quite close to their targets, although the value

INPUTS				OUTPUT			
Number of Respondents	Number of Correct Responses	Probability of a Correct Guess	Proportion Distinguishers	p_{max} or Probability of a Correct Response @ p_d	TYPE I Error	TYPE II Error	Power
n	x	p_o	p_d	p_{max}	α-risk	β-risk	$1-\beta$
60	33	0.50	0.25	0.625	0.2595	0.0923	0.9077

Interpretation:

33   or more correct responses is evidence of a difference at the α = 0.26 level of significance.

32   or fewer correct responses indicates that you can be 91% sure that no more than
        25% of the panelists can detect a difference -- that is, evidence of similarity relative to
                p_d= 25% at the β = 0.09 level of significance.

Instructions:
1   Make entries in Row 4 of Columns A through D ONLY!
    a.   Enter the number of respondents (n) in Cell A4.
    b.   Enter the number of correct responses (x) in Cell B4.
    c.   Enter the probability of a correct guess (p_o) in Cell C4.
          (e.g., for the Triangle test C4=1/3, for the Duo-trio test C4=1/2, etc.)
    d.   Enter the proportion of distinguishers (p_d) you want to be able to detect in
          D4.

2   Evaluate the results in Row 4, Columns F through H to decide if the test has adequate
     sensitivity.
    a.   In testing for difference, choose small values for α-risk.
          (Adjust n and x to achieve the desired sensitivity (i.e., α-risk).)
    b.   In testing for similarity, choose small values for β-risk.
          (Adjust n, x and p_d to achieve the desired sensitivity (i.e., balance between
          p_d and β-risk).)
          (Do not choose values of x that are less than what would be expected by
          chance alone (e.g., $n/3$ for a Triangle test, $n/2$ for a Duo-trio, etc.).
          Increase n when necessary.)
    c.   When testing for difference and similarity simultaneously, choose
          acceptably small values for both α-risk and β-risk. (Adjust n, x and p_d to
          achieve the desired sensitivity (i.e., balance between p_d, β-risk, and α-risk)).
    d.   When using a two-sided directional difference test, double the computed
          value of α-risk to account for the two-sided nature of the test.

FIGURE 13.9   Test Sensitivity Analyzer illustrating the values of n, x, and p_d for a Duo-trio test with target values of $\alpha = 0.25$ and $\beta = 0.10$. Note that the α-risk is slightly greater than the target value specified.

for $\alpha = 0.26$ is slightly larger than desired. By adjusting n and x, the researcher finds that $n = 67$ assessors with $x = 37$ correct responses would be needed to yield values for α and β that are both at or below their targets. The researcher decides that the 60-assessor test is adequate, given that is the maximum number of assessors available and that the α-risk is only 1% greater than the target value.

The Test Sensitivity Analyzer is a useful tool for planning discrimination tests. Researchers quickly can run a variety of scenarios with different values of n, x, and p_d (or p_{max}) to observe the resulting impacts on α-risk and β-risk, selecting the values that offer the best compromise solution. The Test Sensitivity Analyzer can be programmed in Excel by making the entries in the cells indicated in Table 13.8. The explanatory text is entered in the appropriate cells, using fonts and sizes necessary to achieve the desired visual effect.

In testing for similarity or in the unified approach, the number of correct responses, x, should not be chosen to be less than the number that would be expected by chance alone (e.g., $n/3$ in a Triangle test, $n/2$ in a Duo-trio test, etc.). Such values correspond to negative values for the proportion of distinguishers ($p_d < 0$). This is a logical impossibility which, therefore, should not be used as the decision criterion in a test.

TABLE 13.8
EXCEL Programming Information for Test Sensitivity Analyzer

Cell	Entry
E4	=D4+C4*(1-D4) *(Can be Hidden if Desired)*
F4	=1-BINOMDIST(B4-1,A4,C4,TRUE)
G4	=BINOMDIST(B4-1,A4,E4,TRUE)
H4	=1-G4
A7	=B4
F7	=F4
A9	=B4-1
F9	=H4 *(Using % Format)*
B10	=D4 *(Using % Format)*
D11	=B10 *(Using % Format)*
F11	=G4

In using the Test Sensitivity Analyzer for two-sided directional difference tests, researchers must remember to double the computed value for α-risk to account for the two-sided nature of the test.

IV. THE STATISTICAL DESIGN OF SENSORY PANEL STUDIES

In this section experimental designs that are commonly used in sensory evaluation are presented. The discussion is structured to avoid much of the confusion that often surrounds this topic. In Section IV. A, independent replications of an experiment are distinguished from multiple observations of a single sample. It is shown that confusing replications with multiple observations, which result directly from failing to recognize the population of interest, can lead to the incorrect use of measurement error in place of experimental error in the statistical analysis of sensory data. When this happens samples are often declared to be significantly different when they are not. In Section IV. B, the most commonly used designs for sensory panel studies are presented. These include randomized (complete) block designs, balanced incomplete block designs, Latin square designs, and split-plot designs.

A. SAMPLING: REPLICATION VS. MULTIPLE OBSERVATIONS

The fundamental intent of the statistical analysis of a designed experiment is to generate an accurate and precise estimate of the experimental error. All tests of hypotheses and confidence statements are based on this. Experimental error is the unexplainable, natural variability of the population being studied. Experimental error is expressed quantitatively as the variance or as the standard deviation of the population. One measurement taken on one unit from a population provides no means for estimating experimental error. In fact, multiple observations of the same unit provide no means to estimate experimental error either. The differences among the multiple observations taken on a single unit result from measurement error. Several units from the same population need to be sampled in order to develop a valid estimate of experimental error. The measurements taken on different units are called replications. It is the unit-to-unit (or "rep-to-rep") differences that contain the information about the variability of the population (i.e., experimental error).

A common objective of sensory studies is to differentiate products based on differences in the perceived intensities of some attributes. If only a single sample (batch, jar, preparation, etc.) of each product is evaluated, there is no way to estimate the experimental error of the population of products. Often measurement error, that is, judge-to-judge variability, is substituted for experimental error in the statistical analysis of sensory panel data. This is a very dangerous mistake because ignoring experimental error and replacing it with measurement error can lead to falsely concluding that significant differences exist among the products when, in fact, no such differences exist. Evaluating a single batch of product ignores the batch-to-batch differences that may contribute substantially to product variability. Just as repeated measurements of one individual's height tell us nothing about person-to-person differences, repeated evaluations of a single sample (regardless of the size of the panel) tell us nothing about product batch-to-batch variability.

Measurement error is real (as sensory professionals are well aware). However, measurement error cannot be casually substituted for experimental error without incurring the large risk of obtaining misleading results from statistical analyses. If in a taste test the contents of one jar of mayonnaise are divided into 20 servings and presented to panelists, or a single preparation of a sweetener solution is poured into 20 cups and served, then the results of the test are equivalent to the repeated measurements of an individual's height. The variability estimate obtained from the study estimates measurement error. It is not a measure of the product variability (the valid experimental error) because the independent replicates (e.g., different batches) of the product were not presented. The only legitimate conclusion that could be drawn from such a study is whether the panelists were able to detect differences among the particular *samples* they evaluated. This is not the same as concluding that the *products* are different because there is no way to assess how constant any of the observed differences would be in future evaluations of different batches of the same products.

To avoid confusing independent replications of a treatment with multiple observations the sensory analyst must have a clear understanding of what population is being studied. If the objective of a study is to compare several brands of a product, then several units from each brand need to be evaluated. If an ingredient is known to be extremely uniform, then at the very least, separate preparations of samples with that ingredient should be served to each judge. (For extremely uniform products the major source of variability may well be the preparation-to-preparation differences.)

Suggesting that only one sample be taken from each jar of product or that each serving be prepared separately is undeniably more inconvenient than taking multiple observations on a single jar. However, the sensory analyst must compare this inconvenience to the price paid when, for instance, a new product fails in the market because a prototype formulation was falsely declared to be significantly superior to a current formulation based on the evaluation of a single batch of each product.

B. BLOCKING AN EXPERIMENTAL DESIGN

The blocking structure of an experimental design is a description of how the treatments are applied to the experimental material. To understand blocking structure two concepts must be understood: the "block" and the "experimental unit."* A block is simply a group of homogeneous experimental

---

* Many terms used in the area of experimental design originated from agricultural experiments. "Block" originally meant a block of land in a field that had the same fertility (or that received the same level of irrigation) throughout, while other blocks of land in the field might have had different fertility (or irrigation) levels. Individual plots of land were the experimental units.

TABLE 13.9
ANOVA Table for a Completely
Randomized Design for the Multicity
Consumer Test Data in Table 13.1

Source of variability	df	Sum of squares	Mean square	F
Total	119	541.56		
City	3	283.34	94.45	42.43
Error	116	258.22	2.23	

material. Theoretically any unit within a block will yield the same response to the application of a given treatment. The level of the response may vary from block to block, but the difference between any two treatments applied within a block is constant for all blocks. The experimental material within a block is divided into small groups called "experimental units." An experimental unit is that portion of the total experimental material to which a treatment is independently applied.

The sensitivity of a study is increased by taking into account the block-to-block variability that is known to exist prior to running the experiment. If the treatments are applied appropriately, the block effects can be separated from the treatment effects and from the experimental error, thus providing "clean" reads of the treatment effects while simultaneously reducing the unexplained variability in the study.

In more familiar terms, in sensory tests the experimental material is the large group of evaluations performed by the judges. The evaluations are typically arranged into blocks according to judge, in recognition of the fact that, due to differing thresholds for instance, judges may use different parts of the rating scale to express their perceptions. It is assumed that the size of the perceived difference between any two samples is the same from judge to judge. Within each judge (i.e., block) a single evaluation is the experimental unit. The treatments, which can be thought of as products at this point, must be independently applied at each evaluation. This is accomplished through such techniques as randomized orders of presentation, sequential monadic presentations, and wash-out periods of sufficient duration to allow the respondent to return to some baseline level of perception (constant for all evaluations).

1. Completely Randomized Designs

The simplest blocking structure is the completely randomized design (CRD). In a CRD all of the experimental material is homogeneous; that is, a CRD consists of one large block of experimental units. CRDs are used, for example, when a single product is being evaluated at several locations by distinct groups of respondents (e.g., a monadic, multicity consumer test). In such cases the significance of the differences due to location is determined in light of the variability that occurs within each location.

The overall liking data in Table 13.1 conform to a CRD. The box-and-whisker plots presented in Figure 13.3 and the confidence intervals in Figure 13.5 suggest that some city-to-city differences may exist. The average liking response can be used to summarize city-to-city differences. Analysis of variance (ANOVA)* is used to determine if the observed differences in average liking are statistically significant. The ANOVA table for these data is presented in Table 13.9.

---

* ANOVA is the statistical method used to compare more than two means in a single study. The computational forms for obtaining the test statistics in ANOVA are sufficiently complicated to make manual calculations impractical (see Poste et al., 1991). It is recommended that computerized statistical data analysis packages be used instead.

TABLE 13.10
Average Overall Liking Scores from the Multicity
Monadic Consumer Test Data in Table 13.1

City	Atlanta	Boston	Chicago	Denver
Mean Rating	10.56BC	10.29B	7.17A	11.12C

Note: Means not followed by the same letter are significantly
different at the 5% level.

The *F*-ratio for cities in Table 13.9 is highly significant ($F_{0.01,3,116}$ = 3.95), indicating that at least some of the observed differences among cities are real. In order to determine which of the averages are significantly different another statistical method, called a multiple comparisons procedure, must be applied. For the present example the multiple comparison technique called Fisher's least significant difference (LSD) is used. In general, the LSD value used to compare two averages $\bar{x}_i$ and $\bar{x}_j$ is calculated as:

$$\text{LSD}_\alpha = t_{\alpha/2,df_E}\sqrt{MS_E}\ \sqrt{(1/n_i)+(1/n_j)} \tag{13.11}$$

where $t_{\alpha/2,df_E}$ is the upper-$\alpha/2$ critical value of a *t*-distribution with df_E degrees of freedom (i.e., the degrees of freedom for error from the ANOVA), MS_E is the mean square for error from the ANOVA, and n_i and n_j are the number of observations that went into the calculation of $\bar{x}_i$ and $\bar{x}_j$, respectively. If the sample sizes are the same for all $\bar{x}$'s, then Equation 13.11 reduces to:

$$\text{LSD}_\alpha = t_{\alpha/2,df_E}\sqrt{2MS_E/n} \tag{13.12}$$

where *n* is the common sample size. In the example, *n* = 30 so the $\text{LSD}_{0.05}$ = $1.96\sqrt{2(2.23)/30}$ = 0.76. Any two samples whose means differ by more than 0.76 are significantly different at the 5% level. As shown in Table 13.10, Chicago has a significantly lower average value than the other three cities, and Boston has a significantly lower average value than Denver.

Completely randomized designs are seldom used in multisample studies involving sensory panels because it is inefficient to have each panelist evaluate only a single sample, and yet it is recognized that different panelists might use different parts of the rating scales to express their perceptions. More elaborate panel designs are needed for such studies. Four of the most commonly used designs for sensory panels are discussed in the remainder of this section.

C. RANDOMIZED (COMPLETE) BLOCK DESIGNS

If the number of samples is sufficiently small so that sensory fatigue is not a concern, then a randomized (complete) block design is appropriate. Panelists are the "blocks"; samples are the "treatments." Each panelist evaluates (either by rating or ranking) all of the samples (hence the term "complete block").

A randomized block design is effective when the sensory analyst is confident that the panelists are consistent in rating the samples but recognizes that panelists might use different parts of the scale to express their perceptions. The analysis applied to data from a randomized block design takes into account this type of judge-to-judge difference, yielding a more accurate estimate of experimental error and thus more sensitive tests of hypotheses than would otherwise be available.

TABLE 13.11
Data Table for a Randomized (Complete) Block Design

Blocks (judges)	Samples 1	2	•	•	•	t	Row total
1	x_{11}	x_{12}	•	•	•	x_{1t}	$x_{1.} = \sum_{j=1}^{t} x_{1j}$
2	x_{21}	x_{22}	•	•	•	x_{2t}	$x_{2.} = \sum_{j=1}^{t} x_{2j}$
•	•	•				•	•
•	•	•				•	•
•	•	•				•	•
b	X_{b1}	X_{b2}	•	•	•	X_{bt}	$x_{b.} = \sum_{j=1}^{t} x_{bj}$
Column total	$x_{.1} = \sum_{i=1}^{b} x_{i1}$	$x_{.2} = \sum_{i=1}^{b} x_{i2}$	•	•	•	$x_{.t} = \sum_{i=1}^{b} x_{it}$	

TABLE 13.12
ANOVA Table for Randomized Block Designs Using Ratings

Source of variability	Degrees of freedom	Sum of squares	Mean square	F
Total	$bt - 1$	SS_T		
Blocks (judges)	$b - 1$	SS_J		
Samples	$df_s = t - 1$	SS_s	$MS_s = SS_s/df_s$	MS_s/MS_E
Error	$df_E = (b - 1)(t - 1)$	SS_E	$MS_E = SS_E/df_E$	

Independently replicated samples of the test products are presented to the panelists in a randomized order (using a separate randomization for each panelist). The data obtained from the panelists' evaluations can be arranged in a two-way table as in Table 13.11.

1. Randomized Block Analysis of Ratings

Data in the form of ratings from a randomized block design are analyzed by ANOVA. The form of the ANOVA table appropriate for a randomized block design is presented in Table 13.12. The null hypothesis is that the mean ratings for all of the samples are equal (H_0: $\mu_i = \mu_j$ for all samples i and j) vs. the alternative hypothesis that the mean ratings of at least two of the samples are different (H_a: $\mu_i \neq \mu_j$ for some pair of distinct samples i and j). If the value of the F-statistic calculated in Table 13.12 exceeds the critical value of an F with $(t - 1)$ and $(b - 1)(t - 1)$ degrees of freedom (see Table T6), then the null hypothesis is rejected in favor of the alternative hypothesis.

If the F-statistic in Table 13.12 is significant, then multiple comparison procedures are applied to determine which samples have significantly different average ratings. Fisher's LSD for randomized (complete) block designs is

$$\text{LSD} = t_{\alpha/2, df_E} \sqrt{2\, MS_E/b} \tag{13.13}$$

where b is the number of blocks (typically judges) in the study and $t_{\alpha/2,df_E}$ and MS_E are as defined previously.

2. Randomized Block Analysis of Rank Data

If the data from a randomized block design are in the form of ranks, then a nonparametric analysis is performed using a Friedman-type statistic. The data are arranged as in Table 13.11, but instead of ratings, each row of the table contains the ranks assigned to the samples by each judge. The column totals at the bottom of Table 13.11 are the rank sums of the samples.

The Friedman-type statistic for rank data, which takes the place of the F-statistic in the analysis of ratings, is

$$T = \left([12/bt(t+1)] \sum_{j=1}^{t} x_{.j}^2 \right) - 3b(t+1) \tag{13.14}$$

where b = the number of panelists, t = the number of samples, and $x_{.j}$ = the rank sum of sample j (i.e., the column total for sample j in Table 13.11). The "dot" in $x_{.j}$ indicates that summing has been done over the index replaced by the dot, that is, $x_{.j} = \sum_{i=1}^{b} x_{ij}$.

The test procedure is to reject the null hypothesis of no sample differences at the α-level of significance if the value of T in Equation 13.14 exceeds $\chi^2_{\alpha,t-1}$ and to accept H_0: otherwise, where $\chi^2_{\alpha,t-1}$ is the upper-α percentile of the χ^2 distribution with $t-1$ degrees of freedom (see Table T5). The procedure assumes that a relatively large number of panelists participate in the study. It is reasonably accurate for studies involving 12 or more panelists.

If the χ^2-statistic is significant, then a multiple comparison procedure is performed to determine which of the samples differ significantly. The nonparametric analog to Fisher's LSD for rank sums from a randomized (complete) block design is

$$\begin{aligned} \text{LSD}_{\text{rank}} &= z_{\alpha/2}\sqrt{bt(t+1)/6} \\ &= t_{\alpha/2,\infty}\sqrt{bt(t+1)/6} \end{aligned} \tag{13.15}$$

Two samples are declared to be significantly different at the α-level if their rank sums differ by more than the value of LSD_{rank} in Equation 13.15.

If the panelists are permitted to assign equal ranks or ties to the samples, then a slightly more complicated form of the test statistic T' must be used (see Hollander and Wolfe, 1973). Assign the average of the tied ranks to each of the samples that could not be differentiated. For instance, in a four sample test if the middle two samples (normally of ranks 2 and 3) could not be differentiated, then assign both the samples the average rank of 2.5. Replace T in Equation 13.14 with:

$$T' = \frac{12 \sum_{j=1}^{t} \left(x_{.j} - G/t \right)^2}{bt(t+1) - [1/(t-1)] \sum_{i=1}^{b} \left[\left(\sum_{j=1}^{g_i} t_{i,j}^3 \right) - t \right]} \tag{13.16}$$

TABLE 13.13
Data Table for a Balanced Incomplete Block Design

$$(t = 7, k = 3, b = 7, r = 3, \lambda = 1, p = 1)$$

Sample Block	1	2	3	4	5	6	7	Block total
1	X	X		X				B_1
2		X	X		X			B_2
3			X	X		X		B_3
4				X	X		X	B_4
5	X				X	X		B_5
6		X				X	X	B_6
7	X		X				X	B_7
Treatment total	R_1	R_2	R_3	R_4	R_5	R_6	R_7	G

Note: X = an individual observation, B_i = the sum of the observations in row i, R_j = the sum of the observations in column j, G = the sum of all of the observations.

---

where $G = bt(t + 1)/2$, g_i is the number of tied groups in block i and $t_{i,j}$ is the number of samples in the jth tied group in block i. (Nontied samples are each counted as a separate group of size $t_{i,j} = 1$.)

D. BALANCED INCOMPLETE BLOCK DESIGNS

Balanced incomplete block (BIB) designs allow sensory analysts to obtain consistent, reliable data from their panelists even when the total number of samples in the study is greater than the number that can be evaluated before sensory fatigue sets in. In BIB designs the panelists evaluate only a portion of the total number of samples (notationally, each panelist evaluates k of the total of t samples, $k < t$). The specific set of k samples that a panelist evaluates is selected so that in a single repetition of a BIB design every sample is evaluated an equal number of times (denoted by r), and all pairs of samples are evaluated together an equal number of times (denoted by λ). The fact that r and λ are constant for all the samples in a BIB design ensures that each sample mean is estimated with equal precision and that all pair-wise comparisons between two sample means are equally sensitive. The number of blocks required to complete a single repetition of a BIB design is denoted by b. Table 13.13 illustrates a typical BIB layout. A list of BIB designs, such as the one presented by Cochran and Cox (1957), is very helpful in selecting a specific design for a study.

In order to obtain a sufficiently large number of total replications, the entire BIB design (b blocks) may have to be repeated several times. The number of repeats or repetitions of the fundamental design is denoted by p. The total number of blocks is then pb, yielding a total of pr replications for every sample, and a total of $p\lambda$ for the number of times every pair of samples occurs in the total BIB design.

Experience with 9-point category scales and unstructured line scales has shown that the total number of replications (pr) should be at least 18 in order to yield sufficiently precise estimates of the sample means. This is a rule of thumb, suggested only to provide a starting point for determining how many panelists are required for a study. The total number of replications needed to ensure that meaningfully large differences among the samples are declared statistically significant is influenced by many factors: the products, panelist acuity, level of training, etc. Only experience and trial and error can answer the question of how many replications are needed for any given study.

There are two general approaches for administering a BIB design in a sensory study. First, if the number of blocks is relatively small (four or five, for example), it may be possible to have a

TABLE 13.14
ANOVA Tables for Balanced Incomplete Block Designs

a) Each of p Panelists Evaluates All b Blocks

Source of variability	Degrees of freedom	Sum of square	Mean square	F
Total	$tpr - 1$	SS_T		
Panelists	$p - 1$	SS_P		
Blocks (within panelists)	$p(b - 1)$	$SS_{B(P)}$		
Samples (adj. for blocks)	$df_S = t - 1$	SS_S	$MS_S = SS_S/df_S$	MS_S/MS_E
Error	$df_E = tpr - t - pb + 1$	SS_E	$MS_E = SS_E/df_E$	

b) Each of pb Panelists Evaluates One Block

Source of variability	Degrees of freedom	Sum of square	Mean square	F
Total	$tpr - 1$	SS_T		
Blocks	$pb - 1$	SS_B		
Samples (adj. for blocks)	$df_S = t - 1$	SSs	$MS_S = SS_S/df_S$	MS_S/MS_E
Error	$df_E = tpr - t - pb + 1$	SS_E	$MS_E = SS_E/df_E$	

small number of panelists (p in all) return several times until each panelist has completed an entire repetition of the design. (The order of presentation of the blocks should be randomized separately for each panelist, as should be the order of presentation of the samples within each block.) Second, for large values of b, the normal practice is to call upon a large number of panelists (pb in all) and to have each evaluate the samples in a single block. The block of samples that a particular panelist receives should be assigned at random. The order of presentation of the samples within each block should again be randomized in all cases.

1. BIB Analysis of Ratings

ANOVA is used to analyze BIB data in the form of ratings (see Table 13.14). As in the case of a randomized (complete) block design, the total variability is partitioned into the separate effects of blocks, samples, and errors. However, the formulas used to calculate the sum of squares in a BIB analysis are more complicated than for a randomized (complete) block analysis. The sensory analyst should ensure that the statistical package used to perform the analysis is capable of handling a BIB design. Otherwise a program specifically developed to perform the BIB analysis is required.

The form of the ANOVA used to analyze BIB data depends on how the design is administered. If each panelist evaluates every block in the fundamental design, then the "panelist effect" can be partitioned out of the total variability (see Table 13.14a). If each panelist evaluates only one block of samples, then the panelist effect is confounded (or mixed-up) with the block effect (see Table 13.14b). The panelist effect is accounted for in both cases, thus providing an uninflated estimate of experimental error regardless of which approach is used.

If the F-statistic in Table 13.14 exceeds the critical value of an F with the corresponding degrees of freedom, then the null hypothesis assumption of equivalent mean ratings among the samples is rejected. Fisher's LSD for BIB designs has the form:

$$\text{LSD} = t_{\alpha/2, df_E} \sqrt{2MS_E/pr} \sqrt{[k(t-1)]/[(k-1)t]} \qquad (13.17)$$

where t is the total number of samples, k is the number of samples evaluated by each panelist during a single session, r is the number of times each sample is evaluated in the fundamental design (i.e., in one repetition of b blocks), and p is the number of times the fundamental design is repeated. MS_E and $t_{\alpha,2,df_E}$ are as defined before.

2. BIB Analysis of Rank Data

A Friedman-type statistic is applied to rank data arising from a BIB design. The form of the test statistic is

$$T = \left[12/p\lambda t(k+1)\right]\sum_{j=1}^{t} R_j^2 - 3(k+1)pr^2/\lambda \tag{13.18}$$

where t, k, r, λ, and p were defined previously and R_j is the rank sum of the j'th sample (i.e., the value for sample j in the last row of Table 13.13) (see Durbin, 1951). Tables of critical values of T in Equation 13.18 are available for selected combinations of $t = 3$ to 6, $k = 2$ to 5, and $p = 1$ to 7 (see Skillings and Mack, 1981). However, in most sensory studies the total number of blocks exceeds the values in the tables. For these situations, the test procedure is to reject the assumption of equivalency among the samples if T in Equation 13.18 exceeds the upper-α critical value of a χ^2-statistic with $(t - 1)$ degrees of freedom (see Table T5).

If the χ^2-statistic is significant, then a multiple comparison procedure is performed to determine which of the samples differ significantly. The nonparametric analog to Fisher's LSD for rank sums from a BIB design is

$$\begin{aligned} \mathrm{LSD}_{\mathrm{rank}} &= z_{\alpha/2}\sqrt{p(k+1)(rk-r+\lambda)/6} \\ &= t_{\alpha/2,\infty}\sqrt{p(k+1)(rk-r+\lambda)/6} \end{aligned} \tag{13.19}$$

E. LATIN SQUARE DESIGNS

In randomized block and balanced incomplete block designs a single source of variability (i.e., judges) is recognized and compensated for before the sensory panel study is conducted. When two sources of variability are known to exist before the panel is run, then a Latin square design should be used. For example, it is commonly recognized that panelists can vary in how they perceive attributes from session to session (i.e., a session effect) and according to the order in which they evaluate the samples (i.e., a context effect). A Latin square design can be used to compensate for these two sources of variability and, as a result, yield more sensitive comparisons of the differences among the samples.

The number of samples must be small enough so that all of them can be evaluated in each session ($t \leq 5$, typically). Further, each panelist must be able to return repeatedly for a number of sessions equal to the number of samples in the study (i.e., t represents the number of samples and the number of sessions in a Latin square design). For each panelist, each sample is presented once in each session. Across the t-sessions each sample is presented once in each serving position. As can be seen in Figure 13.10, the equal allocation of the samples to each serving order/session combination can be displayed in a square array, thus giving rise to the term "Latin square."

A separate randomization of serving orders is used for each judge. This can be done, for example, by first randomly assigning the codes S_1, S_2, S_3, S_4, and S_5 in Figure 13.10 to the samples for each judge separately. Then, again for each judge, a particular order (i.e., row of Figure 13.10) is randomly selected for each session.

SERVING ORDER

SESSION	1	2	3	4	5
1	S_1	S_2	S_3	S_4	S_5
2	S_5	S_1	S_2	S_3	S_4
3	S_4	S_5	S_1	S_2	S_3
4	S_3	S_4	S_5	S_1	S_2
5	S_2	S_3	S_4	S_5	S_1

FIGURE 13.10 The Latin square arrangement of serving orders by sessions for one judge and a five-sample study. The sample codes S1, S2, . . . , S5 are assigned randomly to the samples in the study. The serving orders and session orders are randomly permuted for each judge individually.

ANOVA is used to analyze data from a Latin square design (see Table 13.15). The total variability is partitioned into the separate effects of judges, panel sessions, order of evaluations, samples, and error. If the F statistic in Table 13.15 exceeds the critical value of an F with the corresponding degrees of freedom then the null hypothesis assumption of equivalent mean ratings among the samples is rejected. Fisher's LSD for Latin square designs has the form:

$$\text{LSD} = t_{\alpha/2, df_E} \sqrt{2MS_E / pt} \tag{13.20}$$

where p is the number of panelists.

TABLE 13.15
ANOVA Table for Latin Square Designs Using Ratings

Source of variability	Degrees of freedom	Sum of squares	Mean square	F
Total	$pt^2 - 1$	SS_T		
Judges	$p - 1$	SS_J		
Sessions (within judge)	$p(t - 1)$	SS_P		
Order (within judge)	$p(t - 1)$	SS_O		
Samples	$df_S = t - 1$	SS_S	$MS_S = SS_S/df_S$	MS_S/MS_E
Error	$df_E = (pt - p - 1)(t - 1)$	SS_E	$MS_E = SS_E/df_E$	

F. SPLIT-PLOT DESIGNS

In randomized block and balanced incomplete block designs panelists are treated as a blocking factor; that is, it is assumed that the panelists are an identifiable source of variability that is known to exist before the study is run and, therefore, should be compensated for in the design of the panel. In ANOVA the effects of environmental factors (e.g., judges) and treatment factors (e.g., products) are assumed to be additive. In practice this assumption implies that while panelists may use different parts of the sensory rating scales to express their perceptions, the size and direction of the differences among the samples are perceived and reported in the same way by all of the panelists. Of course, the data actually collected in a study diverge slightly from the assumed pattern due to experimental error. Another way of stating this assumption is that there is no "interaction" between blocks and treatments (e.g., judges and samples) in a randomized block or BIB design. For a group of highly trained, motivated, and "calibrated" panelists the assumption of no interaction between judges and samples is reasonable. However, during training, for instance, the sensory analyst may doubt the validity of this assumption. Split-plot designs are used to determine if a judge-by-sample interaction is present.

In split-plot designs judges are treated as a second experimental treatment along with the samples. A group of b panelists are presented with t samples (in a separately randomized order for each panelist) in each of at least two panels ($p \geq 2$). The p panels are the blocks or "replicates" of the experimental design. Randomly selected batches or independent preparations of the samples are used for each panel. This is the first layer of randomization in a split-plot study. Then the panelists receive their specific sets of samples (arranged in a randomized order based on their arrival times at each panel). Due to the sequential nature of the randomization scheme, where first one treatment factor (samples) is randomized within replicates and then a second treatment factor (judges) is randomized within the first treatment factor (i.e., samples), a split-plot design is appropriate.

1. Split-Plot Analysis of Ratings

A special form of ANOVA is used to analyze data from a split-plot design (see Table 13.16). The sample effect is called the "whole-plot effect." Judges and the judge-by-sample interaction are called the "subplot effects." Separate error terms are used to test for the significance of whole-plot and subplot effects (because of the sequential nature of the randomization scheme described previously).

The whole-plot error term (Error(A) in Table 13.16) is calculated in the same way as a panel-by-sample interaction term would be, if one existed. The F_1-statistic in Table 13.16 is used to test for a significant sample effect. If the value of F_1 is larger than the upper-α critical value of the F distribution with $(t-1)$ and $(p-1)(t-1)$ degrees of freedom, then it is concluded that there are significant differences among the average values of the samples.

The F_2 and F_3 statistics in Table 13.16 are used to test for the significance of the subplot effects, judges and the judge-by-sample interaction, respectively. The denominator of both F_2 and F_3 is the subplot error term $MS_{E(B)}$. If F_3 exceeds the upper-α critical value of the F-distribution with $(b-1)(t-1)$ and $t(p-1)(b-1)$ degrees of freedom, then a significant judge-by-sample interaction exists. The significance of the interaction indicates that the judges are expressing their perceptions of the differences among the samples in different ways. Judge-by-sample interactions result from insufficient training, confusion over the definition of an attribute, or lack of familiarity with the rating technique. When a significant judge-by-sample interaction exists, it is meaningless to examine the overall sample effect (tested by F_1 in Table 13.16) because the presence of an interaction indicates that the pattern of differences among the samples depends on which judge or

TABLE 13.16
ANOVA Table for Split-Plot Designs Using Ratings

Source of variability	Degrees of freedom	Sum of squares	Mean square	F
Total	$pbt - 1$	SS_T		
Panel	$p - 1$	SS_P		
Samples	$t - 1$	SS_S	$MS_S = SS_S/(t - 1)$	$F_1 = MS_S/MS_{E(A)}$
Error(A)	$df_{E(A)} = (p - 1)(t - 1)$	$SS_{E(A)}$	$MS_{E(A)} = SS_{E(A)}/df_{E(A)}$	
Judges	$b - 1$	SS_J	$MS_J = SS_J/(b - 1)$	$F_2 = MS_S/MS_{E(B)}$
Judge-by-sample	$df_{js} = (b - 1)(t - 1)$	SS_{JS}	$MS_{JS} = SS_{JS}/df_{JS}$	$F_3 = MS_{JS}/MS_{E(B)}$
Error(B)	$df_{E(B)} = t(p - 1)(b - 1)$	$SS_{E(B)}$	$MS_{E(B)} = SS_{E(B)}/df_{E(B)}$	

judges are being considered. Tables of individual judges' mean ratings and plots of the judge-by-sample means should be examined to determine which judges are causing the interaction to be significant (see Chapter 9, p. 146).

If F_3 is not significant but F_2 is, then an overall judge effect is present. A significant judge effect confirms that the judges are using different parts of the rating scale to express their perceptions. This is not of as great a concern as a significant judge-by-sample interaction. However, depending on the magnitude of the differences among the judges, it may indicate that the panel needs to be re-calibrated through the use of references.

If F_3 is not significant but F_1 is, then an overall sample effect is present. To determine which of the samples differ significantly use Fisher's LSD for split-plot designs:

$$LSD = t_{\alpha/2,df_{E(A)}} \sqrt{2MS_{E(A)}/p} \tag{13.21}$$

where p is the number of independently replicated panels and $df_{E(A)}$ and $MS_{E(A)}$ are the degrees of freedom and mean square for Error(A), respectively.

G. A SIMULTANEOUS MULTIPLE COMPARISON PROCEDURE

Thus far we have used only Fisher's LSD multiple comparison procedure to determine which samples differ significantly in a designed sensory panel study. There are, in fact, two classes of multiple comparison procedures. The first class, including Fisher's LSD, controls the comparison-wise error rate; that is, the Type I error (of size α) applies each time a comparison of means or rank sums is made. Procedures that control the comparison-wise error rate are called one-at-a-time multiple comparison procedures. The second class controls the experiment-wise error rate; that is, the Type I error applies to all of the comparisons among means or rank sums simultaneously. Procedures that control the experiment-wise error rate are called simultaneous multiple comparison procedures.

Tukey's honestly significant difference (HSD) is a simultaneous multiple comparison procedure. Tukey's HSD can be applied regardless of the outcome of the overall test for differences among the samples. The general form of Tukey's HSD for the equal sample-size case for ratings data is

$$HSD = q_{\alpha,t,df_E} \sqrt{MS_E/n} \tag{13.22}$$

where q_{α,t,df_E} is the upper-α critical value of the studentized range distribution with df_E degrees of freedom (see Table T4) for comparing t sample means. As with the LSD, df_E and MS_E are the degrees of freedom and the mean square for error from the ANOVA, respectively (Error(A) in the split-plot ANOVA); n is the sample size common to all the means being compared. For randomized (complete) block designs $n = b$; for split-plot designs $n = p$. Tukey's HSD for BIB designs has the form:

$$\text{HSD} = q_{\alpha,t,df_E} \sqrt{MS_E/pr} \sqrt{k(t-1)/(k-1)t} \qquad (13.23)$$

The nonparametric analog to Tukey's HSD for rank sums is

$$\text{HSD}_{\text{rank}} = q_{\alpha,t,\infty} \sqrt{bt(t+1)/12} \qquad (13.24)$$

for randomized (complete) block designs, and:

$$\text{HSD}_{\text{rank}} = q_{\alpha,t,\infty} \sqrt{p(k+1)(rk-r+\lambda)/12} \qquad (13.25)$$

for BIB designs.

V. APPENDIX ON PROBABILITY

The purpose of this section is to present the techniques for calculating probabilities based on some commonly used probability distributions. The techniques are the foundation for statistical estimation and inference that were discussed in Sections II and III, as well as for the more advanced topics discussed in Chapter 14.

A. THE NORMAL DISTRIBUTION

The *normal distribution* is among the most commonly used distributions in probability and statistics. The form of the normal distribution function is

$$f(x) = \left[1/\sqrt{2\pi}\sigma\right] \exp\left(-[x-\mu]^2/2\sigma^2\right)$$

where exp is the exponential function with base e. The parameters of the normal distribution are the mean μ $(-\infty < \mu < \infty)$ and the standard deviation σ $(\sigma > 0)$. The normal distribution is symmetric about μ, that is, $f(x - \mu) = f(\mu - x)$. The mean μ measures the central location of the distribution. The standard deviation σ measures the dispersion or "spread" of the normal distribution about the mean. For small values of σ the graph of the distribution is narrow and peaked; for large values of σ the graph is wide and flat (see Figure 13.11). As with all continuous probability distributions, the total area under the curve is equal to one, regardless of the values of the parameters.

Let x be a random variable having a normal distribution with mean μ and standard deviation σ (often abbreviated as $x \sim \text{n}(\mu, \sigma)$). Define the variable z as:

$$z = (x - \mu)/\sigma \qquad (13.26)$$

The random variable z also has a normal distribution. The mean of z is 0 and its standard deviation is 1 (i.e., $z \sim \text{n}(0, 1)$). z is said to have a *standard normal distribution,* or oftentimes z is called a *standard normal deviate.* Given the values of μ and σ for a normal random variable x and a table of standard normal probabilities (see Table T2), it is possible to calculate various probabilities of interest.

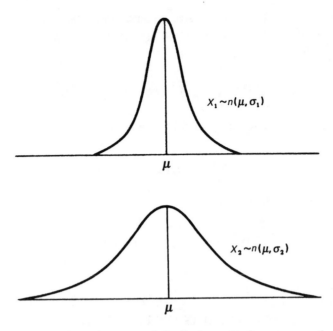

FIGURE 13.11   A comparison of two normal distributions with the same mean but with $\sigma_1 < \sigma_2$.

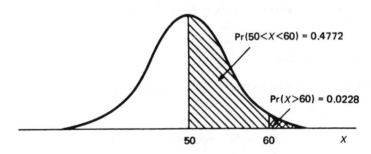

FIGURE 13.12   A graphical depiction of calculating normal probabilities on an interval and in the tail of the distribution.

1. Example 13.7: Calculating Normal Probabilities on an Interval

Consider the problem of calculating the probability that a normal random variable x with mean $\mu = 50$ and standard deviation $\sigma = 5$ takes on a value between 50 and 60 (notationally, $\Pr[50 < x < 60]$). The first step in solving the problem is to "standardize" x using (Equation 13.26):

$$\Pr[50 < x < 60] = \Pr\left[(50-50)/5 < (x-50)/5 < (60-50)/5\right]$$

$$= \Pr[0 < z < 2]$$

Table T2 gives the probabilities of a standard normal deviate taking on a value from zero (i.e., its mean) to some specified number. Therefore, entering Table T2 in the row corresponding to 2.0 and the column corresponding to 0.00 we find that the probability sought is equal to 0.4772 (see Figure 13.12).

Next consider the problem of finding $\Pr[45 < x < 50]$, where, as before $x \sim n(50, 5)$. Standardizing, we find:

$$\Pr[45 < x < 50] = \Pr\big[(45-50)/5 < (x-50)/5 < (50-50)/5\big]$$

$$= \Pr[-1 < z < 0]$$

Because the standard normal distribution is symmetric about its mean, 0, it follows that $\Pr[-c < z < 0] = \Pr[0 < z < c]$ for any constant c. Therefore, by Table T2:

$$\Pr[-1 < z < 0] = \Pr[0 < z < 1]$$

$$= 0.3413$$

(13.27)

Therefore, we find that $\Pr[45 < x < 50] = 0.3413$.

Finally, consider $\Pr[45 < x < 60]$ for the same random variable $x \sim n(50, 5)$. This problem is solved as follows:

$$\Pr[45 < x < 60] = \Pr[-1 < z < 2] \quad \big(\text{Standardizing by } [13.26]\big)$$

$$= \Pr[-1 < z < 0] + \Pr[0 < z < 2]$$

$$= \Pr[0 < z < 1] + \Pr[0 < z < 2] \quad \big(\text{by } [13.27]\big)$$

$$= 0.3413 + 0.4772 \quad \quad (\text{from Table T2})$$

$$= 0.8185$$

2. Example 13.8: Calculating Normal Tail Probabilities

Tail probabilities are associated with the areas under the probability curve at the extremes of the distribution (see Figure 13.13). Notationally, tail probabilities are stated as $\Pr[x > c]$ or $\Pr[x < c]$ for some constant c. Tail probabilities are widely used in testing statistical hypotheses.

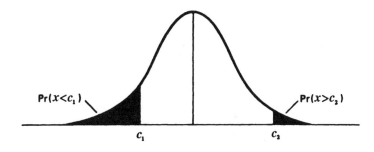

FIGURE 13.13   Tail probabilities of a normal distribution.

Consider the problem of finding $\Pr[x > 60]$, where $x \sim n(50, 5)$. Noting that the total area (i.e., probability) under any probability curve is one, it follows from the symmetry of the normal distribution that $\Pr[x < \mu] = \Pr[x > \mu] = 0.50$. Therefore:

$$\Pr[x > 60] = \Pr[(x-50)/5 > (60-50)/5]$$

$$= \Pr[z > 2] \qquad (\text{by } [13.26])$$

$$= 0.50 - \Pr[0 < z < 2]$$

$$= 0.50 - 0.4772 \qquad (\text{from Example } 13.7)$$

$$= 0.0228$$

(See the crosshatched area in Figure 13.12 for an understanding of the third step.)

B. THE BINOMIAL DISTRIBUTION

The binomial distribution function is

$$\Pr[x = k] = b[k] = \binom{n}{k} p^k (1-p)^{n-k} \tag{13.28}$$

for $k = 0, 1, 2,..., n$; $n > 0$ and an integer; and $0 \le p \le 1$.

The parameters of the binomial distribution are n = the number of trials and p = the probability of "success" on any trial. The choice of what constitutes a success on each trial is arbitrary. For instance, in a two-sample preference test (A vs. B) preference for A could constitute a success or preference for B could constitute a success. Regardless, $k_i = 1$ for $i = 1, 2, ..., n$ if the result of the i'th trial is a success; $k_i = 0$, otherwise, in Equation 13.28 $k = \sum_{i=1}^{n} k_i$ is the total number of successes in n trials. Exact binomial probabilities can be calculated using spreadsheet functions such as Excel's BINOMDIST. Approximate binomial probabilities can be calculated using the normal approximation to the binomial.

1. Example 13.9: Calculating Exact Binomial Probabilities

Suppose $n = 16$ assessors participate in a Two-out-of-five difference test. The probability of correctly selecting the two odd samples from among the five follows a binomial distribution with probability of success, $p = 0.10$ (when there is no perceptible difference among the samples). To find the probability that exactly two ($k = 2$) of the assessors make the correct selections (i.e., exactly 2 successes in 16 trials), enter the following in a cell in an Excel spreadsheet,* = BINOMDIST(2, 16, 0.10, FALSE). The response displayed will be 0.2745, which is the desired probability.

To find the probability that between two and six assessors (inclusive) make the correct selections, one notes that

$$\Pr[2 \le x \le 6] = \Pr[x \le 6] - \Pr[x \le 1], \tag{13.29}$$

---

* Other spreadsheets have similar probability functions. Consult the user manuals of your spreadsheet for the appropriate syntax.

which is computed in Excel by entering the following in the spreadsheet:

1. In cell A1 enter: = BINOMDIST(6, 16, 0.10, TRUE)

2. In cell A2 enter: = BINOMDIST(1, 16, 0.10, TRUE)

3. In cell A3 enter: = A1-A2

The desired probability, 0.4848, is displayed in cell A3.

There are two approaches for calculating tail probabilities using spreadsheet functions, depending on whether you want to compute the probability in the lower or upper tail of the distribution. To compute probabilities in the lower tail of the distribution, for example that less than three assessors make the correct selections, use the following technique:

$$\Pr[x < 3] = \Pr[x \le 2] \tag{13.30}$$

so enter the following in a cell in an Excel spreadsheet: = BINOMDIST(2, 16, 0.10, TRUE), and the resulting probability, 0.7892, will be displayed. On the other hand, consider the probability that at least three (i.e., three or more) assessors make the correct selections — that is,

$$\Pr[x \ge 3] = 1 - \Pr[x < 3] = 1 - \Pr[x \le 2] \tag{13.31}$$

so enter the following in a cell in an Excel spreadsheet: = 1-BINOMDIST(2, 16, 0.10, TRUE), and the resulting probability, 0.2108, will be displayed.

2. Example 13.10: The Normal Approximation to the Binomial

When a computerized spreadsheet with a binomial probability function is not available, approximate binomial probabilities can be calculated using the normal distribution. To use the methods of Section V.A (immediately preceding), one needs to know the values of μ and σ. For the number of successes, these are

$$\mu = np$$
$$\sigma = \sqrt{np(1-p)} \tag{13.32}$$

Let $n = 36$ and $p = \frac{1}{3}$ and consider the problem of calculating the probability of at least 16 successes. For Equation 13.32 one computes:

$$\mu = (36)(1/3) = 12$$
$$\sigma = \sqrt{36(1/3)(1-1/3)}$$
$$= \sqrt{8} = 2.828$$

Therefore, using the methods of Example 13.8:

$$\Pr[x \geq 16] = \Pr[(x-12)/2.828 \geq (16-12)/2.828]$$

$$= \Pr[z \geq 1.41]$$

$$= 0.5 - \Pr[0 < z < 1.41]$$

$$= 0.5 - 0.4207$$

$$= 0.0793$$

One can also use the normal approximation to the binomial to calculate probabilities associated with the proportion of successes. For this case:

$$\mu = p$$

$$\sigma = \sqrt{p(1-p)/n} \tag{13.33}$$

In most sensory evaluation tests the number of trials is large enough so that the normal approximation gives adequately accurate results. A rule of thumb often used is that the normal approximation to the binomial is sufficiently accurate if both $np > 5$ and $n(1-p) > 5$; that is, for the normal approximation to be reasonably accurate, the sample size n should be sufficiently large so that one would expect to see at least five successes and at least five failures in the sample results.

REFERENCES

Cochran, W.G. and Cox, G.M., 1957, *Experimental Designs,* 2nd ed. John Wiley & Sons, New York.

Danzart, M., 1986. Univariate Procedures In: *Statistical Procedures in Food Research*, Piggott, J.R., Ed. Elsevier Applied Science Publishers Ltd., Essex, England.

Dixon, W.J. and Massey, F.J., 1969. *Introduction to Statistical Analysis.* McGraw-Hill, New York.

Durbin, J., 1951. Incomplete Blocks in Ranking Experiments. *Br. J. of Stat. Psychol.* 4, 85.

Gagula, M.C. and Singh, J., 1984. *Statistical Methods in Food and Consumer Research.* Academic Press, Orlando, FL.

Hollander, M. and Wolfe, D.A., 1973. *Nonparametric Statistical Methods.* John Wiley & Sons, New York.

O'Mahony, M., 1986. *Sensory Evaluation of Food: Statistical Methods and Procedures.* Marcel Dekker, New York.

Poste, L.M., Mackie, D.A., Butler, G., and Larmond, E., 1991. Laboratory Methods for Sensory Evaluation of Food. Publication 1864, Agriculture Canada, Ottawa, 91 pp.

Skillings, H.H. and Mack, G.A., 1981. On the Use of a Friedman Type Statistic in Balanced and Unbalanced Block Designs. *Technometrics* 23, 171.

Smith, G.L., 1988. Statistical Analysis of Sensory Data, In: *Sensory Analysis of Foods*, Piggott, J.R., Ed. Elsevier Applied Science Publishers Ltd., Essex, England.

Snedecor, G.W. and Cochran, W.G., 1980. *Statistical Methods.* Iowa State University Press, Ames, IA.

14 Advanced Statistical Methods

CONTENTS

I. INTRODUCTION

The basic statistical techniques presented in Chapter 13 are all that would be required to analyze the results of most sensory tests. However, when the objectives of the study go beyond simple estimation or discrimination, then more sophisticated statistical methods may need to be applied. This chapter presents some of the more common of these advanced techniques. The computational complexity of the methods makes hand calculation impractical. It is assumed that the reader has access to computer resources capable of performing the necessary calculations.

Sensory studies seldom include only a single response variable. More often, many variables are measured on each sample and often one of the goals of the study is to determine how the different "multivariate" measurements relate to each other. Approaches for studying multivariate data relationships are presented in Section II. First, correlation analysis, principal components analysis (PCA), and cluster analysis are discussed. These techniques are used to study sets of multivariate data in which all of the variables are of equal status. Second, regression analysis, principal component regression, partial least squares, and discriminant analysis are presented. These methods apply when the variables in the data set can be classified as being either independent or dependent, with the goal of the analysis being to predict the value of the dependent variables using the independent variables. In Section III, experimental plans for systematically studying the combined effects of more than one experimental variable are presented. The discussion includes factorial experiments, fractional factorials (or "screening studies") and response surface methodology (or "product optimization studies").

II. DATA RELATIONSHIPS

The need to determine if relationships exist among different variables often arises in sensory evaluation. The manner and degree that different descriptive attributes increase or decrease together, the similarity among consumers' liking patterns for various products, and the ability to predict the value of a perceived attribute based on the age of a product are three examples of this type of problem.

The various statistical methods that exist for drawing relationships among variables can be divided into two groups. The first group of methods handles data sets in which all of the variables are independent, in the sense that they are all equally important with no one or few variables being viewed as being driven by the others (e.g., a set of descriptive flavor attributes). The second group of methods is applied to data sets that contain both dependent and independent variables. These are data sets in which one or more of the variables are of special or greater interest relative to some others (e.g., overall liking vs. descriptive attribute ratings). Because the methods in both of these groups deal with more than one variable at a time, they are members of the class of "multivariate" statistical methods.

A. ALL INDEPENDENT VARIABLES

When all of the variables are viewed as being equally important, the goal of the statistical analyses is to determine the nature and degree of relationships among the variables, to determine if groups of related variables exist, or to determine if distinct groups of observations exist.

1. Correlation Analysis

The simplest of multivariate techniques, correlation analysis, is used for measuring the strength of the linear relationship between two variables. The strength of the relationship between attributes X and Y, for instance, is summarized in the correlation coefficient r, where:

$$r = \frac{\Sigma(y_i - \bar{y})(x_i - \bar{x})}{\sqrt{\Sigma(y_i - \bar{y})^2 \Sigma(x_i - \bar{x})^2}}$$

The value of r lies between -1 and $+1$. A value of -1 indicates a perfect inverse linear relationship (i.e., one variable decreases as the other increases) while a value of $+1$ indicates a perfect direct linear relationship (i.e., both variables either increase or decrease together). A value near zero implies that little linear relationship exists between the two variables. A strong correlation does not imply causality, that is, neither variable can automatically be assumed to be "driving" the other, but rather that the two co-vary to some degree.

Correlation coefficients are summary measures. An analyst should always examine the scatterplots of the paired variables before deciding if the value of r is an adequate summary of the relationship. A strong linear trend among relatively evenly spaced observations, as in Figure 14.1(A) is safely summarized by the correlation coefficient, as is an unpatterned spread of observations spanning the ranges of both variables, as in Figure 14.1(B). Some relationships may have high values of r but are clearly better summarized in nonlinear terms, as in

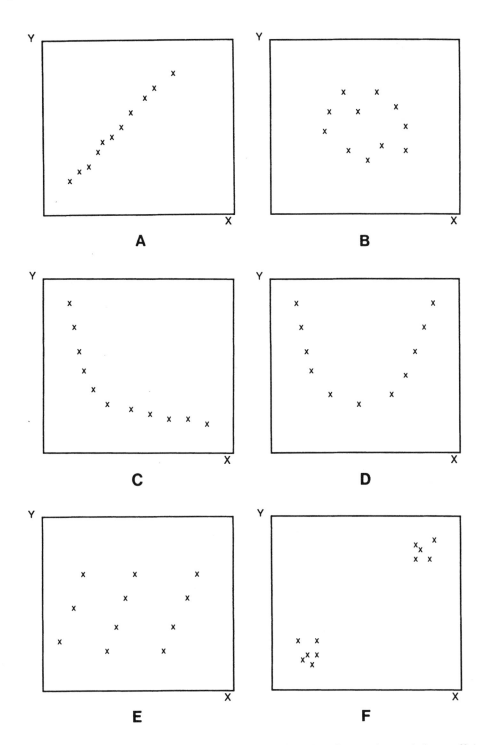

FIGURE 14.1 Scatterplots of two variables, X and Y, showing when the sample correlation coefficient, r, is a good summary measure (i.e., plots [A] and [B]) and when it is not (i.e., plots [C] through [F]).

Figure 14.1(C). In other cases patterns that are clearly apparent visually may have very low values of r, as in Figure 14.1(D) and Figure 14.1(E). Conversely, the correlation coefficient may be misleadingly large when distinct groups of observations (with no internal correlation) are present, as in Figure 14.1(F). A scatterplot matrix (see Figure 14.2) allows all of the pairwise plots of the data to be displayed in a compact format.

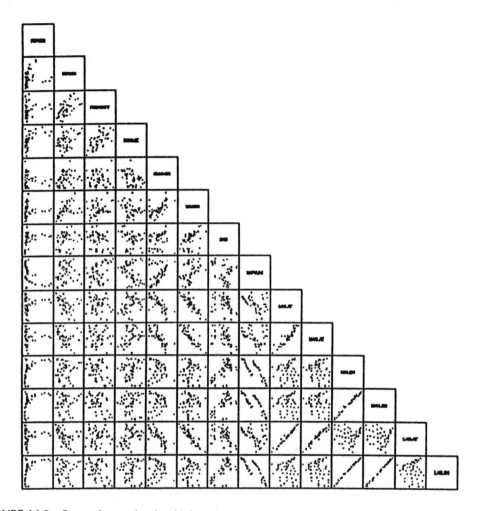

FIGURE 14.2   Scatterplot matrix of multiple responses useful for identifying correlated or otherwise related pairs of variables.

Correlation analysis can be used to identify groups of responses that vary in similar ways, possibly distinct from other such groups. Also, correlation analysis can be used to determine the strength of the relationship between data arising from different sources (e.g., consumer ratings and descriptive data from a trained panel, descriptive attribute ratings, and instrumental measurements, etc.). Chapter 12, p. 249, contains examples of these types of relationships.

Patterns of correlation may vary across product categories, regionally, or from one market segment to another. Care should be taken to ensure that the data to which correlation analysis is applied arise from a single population and not a blend of heterogeneous ones.

2. PRINCIPAL COMPONENTS ANALYSIS

An initial correlation analysis might identify one or more groups of variables that are highly correlated with each other (and not highly correlated with variables from other groups). This suggests that variables in each group contain related information and that possibly a smaller number of unobserved (or "latent") variables would provide an adequate summary of the total variability. Principal components analysis (PCA) is the statistical technique used to identify the smallest number of latent variables, called "principal components," that explain the greatest amount of observed variability. It is often possible to explain as much as 75% to 90% of the total variability in a data set consisting of 25 to 30 variables with as few as 2 to 3 principal components.

Computer programs that extract the principal components from a set of multivariate data are widely available so theoretical and computational details are not included in the following discussion. Those interested in a more analytical discussion of PCA are referred to Piggott and Sharman (1986).

PCA analyzes the correlation structure of a group of multivariate observations and identifies the axis along which the maximum variability in the data occurs. This axis is called the first principal component. The second principal component is the axis along which the greatest amount of remaining variability lays subject to the constraint that the axes must be perpendicular (at right angles) to each other (i.e., orthogonal or uncorrelated). Each additional principal component is selected to be orthogonal to all others and such that each successive principal component explains as much of the remaining unexplained variability as possible. The number of principal components can never be larger than the number of observed variables and, in practice, is often much less. The process of extracting principal components ends either when a prespecified amount of the total variability has been explained (this quantity is always included in the computer output of a PCA) or when extraction of another principal component would add only trivially to the explained variability. This situation is depicted graphically by the flattening out of the Scree plot (1966) in Figure 14.3.

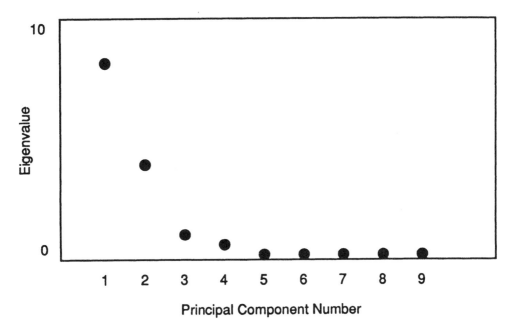

FIGURE 14.3   A Scree plot of the variability explained by each principal component used to determine how many principal components should be retained in a study. Only those principal components that are extracted before the plot flattens out are retained.

The direction of the axis defined by each principal component, y_i, is expressed as a linear combination of the observed variables, x_j, as in:

$$y_i = a_{i1}x_1 + a_{i2}x_2 + \ldots + a_{ip}x_p \tag{14.1}$$

The coefficients, a_{ij}, are called weights or loading factors. They measure the importance of the original variables on each principal component. Like correlation coefficients, a_{ij} takes on values between −1 and +1. A value close to −1 or +1 indicates that the corresponding variable has a large influence on the value of the principal component; values close to 0 indicate that the corresponding variable has little influence on the principal component. Typically, groups of highly correlated observed variables segregate themselves into nonoverlapping groups predominantly associated with a specific principal component.

Examination of the loading factors reveals how the observed variables group together and may lead to a meaningful interpretation of the type of variability being summarized by each principal component. To further aid in interpretation, the principal component axes are sometimes rotated to increase their alignment with the axes of the original variables. After rotation the first principal component will no longer lie in the direction of maximum variability, followed by the second, etc., but the advantage gained by having a small number of interpretable latent variables offsets this effect.

In addition to depicting the associations among the original variables, PCA can be used to display the relative "locations" of the samples. A plot of the principal component scores for a set of products can reveal groupings and polarizations of the samples that would not be as readily apparent in an examination of the larger number of original variables. Cooper et al. (1989) used PCA to depict in two dimensions the relationship of 16 orange juice products originally evaluated on ten attributes (see Figure 14.4). In their analysis the first two principal components explained 79% of the original variability. Note that no attempt is made to interpret the principal components themselves. Rather, it is recommended that the relationship among the original attributes be dis-

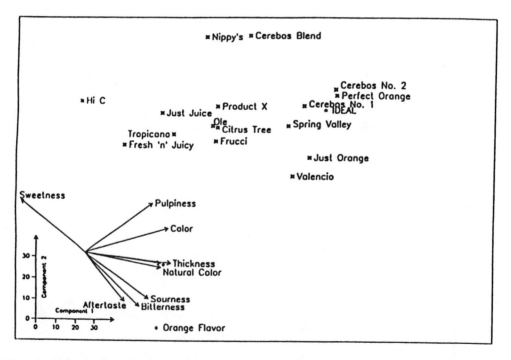

FIGURE 14.4 Results of a PCA on orange drinks (Cooper et al., 1989) showing both the relationships of the products to each other and the associations among the original descriptive attributes. The plot of the products is offset from the plot of the attributes to aid the visual presentation while maintaining the relative directions and magnitudes. For example, Hi C is high in sweet and low in sour and bitter, while Cerebos No. 2 is less sweet and relatively high in pulp, color. (Copyright ASTM. Reprinted with permission.)

played graphically as in Figure 14.4. Piggott and Sharman (1986) present additional examples. Powers (1988) presents numerous references of the application of PCA in descriptive analysis.

PCA provides a way to summarize data collected on a large number of variables in fewer dimensions. It is tempting to ask if it is necessary to continue to evaluate all of the original variables as opposed to only a few "representative" ones. The number of original variables studied should not be reduced based on PCA results. As seen in Equation 14.1, each of the original variables is included in the computation of each principal component. Retaining only a small group of representative variables on a sensory ballot ignores the multivariate nature of the effects of the original variables, would not allow for future verification of the stability of the principal components, and could lead to misleading results in future evaluations.

3. Cluster Analysis

In the same spirit that PCA identifies groups of attributes based on their degree of correlated behavior, the multivariate statistical method cluster analysis identifies groups of observations based on the degree of similarity among their ratings. The ratings may be different attributes collected on a single sample or a single attribute collected on a variety of samples. There are a large number of cluster analysis algorithms in common use at present so no fair treatment of the computational details of cluster analysis could be presented in a general discussion such as this. Interested readers are referred to Jacobsen and Gunderson (1986) for their discussion of applied cluster analysis that includes a step-by-step example, a list of food science applications, and a list of texts and computer programs on the topic. Godwin et al. (1978) present an interesting application of cluster analysis in which sensory attributes and instrumental measurements are grouped based on their relation to concomitantly collected hedonic responses. Although not entirely proper statistically (attributes are not randomly sampled observations from some extant population), their approach is an interesting numerical technique for studying data relationships and should not be overlooked.

There are two classes of cluster analysis algorithms: the hierarchical and the nonhierarchical methods. The practical distinction between them is that once an observation is assigned to a cluster by a hierarchical method it can never be moved to another cluster, while moving an observation from one cluster to another is possible in nonhierarchical methods.

Hierarchical methods proceed in one of two directions. In the more common approach each observation is initially considered to be a cluster of size one and the analysis successively merges the observations (or intermediate clusters of observations) until only one cluster exists. Alternatively, the analysis may begin by treating all the observations as belonging to one cluster and then proceed to break groups of observations apart until only single observations remain. The successive mergers or divisions are depicted graphically in a dendrogram (or tree-diagram) (see Figure 14.5). The dendrogram charts the hierarchical structure of the observations, measures the degree of change in the clustering criterion, and is used to decide how many clusters truly exist.

The general difference between hierarchical cluster analysis algorithms is the way in which the distance (or linkage) between two clusters is measured. Commonly used algorithms include average linkage, centroid linkage, median linkage, furthest neighbor (or complete) linkage, nearest neighbor (or single) linkage, and Ward's minimum variance linkage (see SAS, 1989). Ward's method uses ANOVA-type sum of squares as a "distance" measure. Each approach has its advantages and disadvantages. None has emerged as a clear favorite for general use.

Nonhierarchical methods include the k-means method (MacQueen, 1967) and the fuzzy objective function (or FCV) method (Bezdek, 1981). Iterative mathematical techniques are used in both. For both, the user must indicate the number of clusters that are believed to exist. In the k-means method each observation is assigned to a cluster based on its (Euclidean) distance from the center of the cluster. As more observations are added to a cluster the center moves; thus the assignments of the observations must be repeated until no further changes occur. FCV replaces

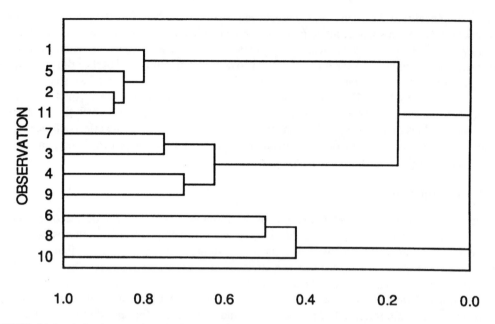

FIGURE 14.5 A dendrogram from a cluster analysis showing which observations are grouped together and the degree of separation among the clusters.

the concept of "cluster membership" with "degree of membership." The method assigns a membership weight, between 0 and 1, to each observation for every one of the prespecified number of clusters. Instead of reassigning observations to different clusters, adjustment of the membership weights continues until the convergence criteria are met (e.g., minimal shift in the locations of the centers of the clusters).

FCV offers the advantage of distinguishing observations that are strongly linked to a particular cluster (i.e., with membership weights close to +1.0) from those observations that have some association with more than one cluster (i.e., with membership weights nearly equal for two or more clusters). In addition, Jacobsen and Gunderson (1986) present a discussion of some approaches for determining the discriminatory importance of the original variables using an FCV clustering example of Norwegian beers based on gas chromatographic data.

An application of cluster analysis particularly important in sensory acceptance testing is that of identifying groups of respondents that have different patterns of liking across products. While some respondents may favor an increasing intensity of some flavor note, others may find it objectionable. Merging such distinct groups may lead to a misunderstanding of the acceptability of a product because, in statistical terms, failing to recognize the clusters leads to computing the mean of a multimodal set of data. The center of such a set of observations may not represent any real group of respondents and, thus, is an inappropriate summary measure of overall liking (see Figure 14.6). Performing cluster analyses to discriminate patterns of liking may uncover groups of respondents that cross over demographic boundaries known to exist prior to running the study. As such, cluster analysis has an advantage over this classical approach to "segmentation."

Once clusters of respondents are identified, correlation analysis, PCA, and/or regression analyses (to be discussed in the next section) can be performed to determine the similarity and differences among the clusters in how the perceived attributes relate to liking. Multiple products/varieties, "niche" marketing, or line extensions may be indicated. Lastly, demographic summaries of each cluster could be performed to determine if the members form a targetable population for marketing purposes.

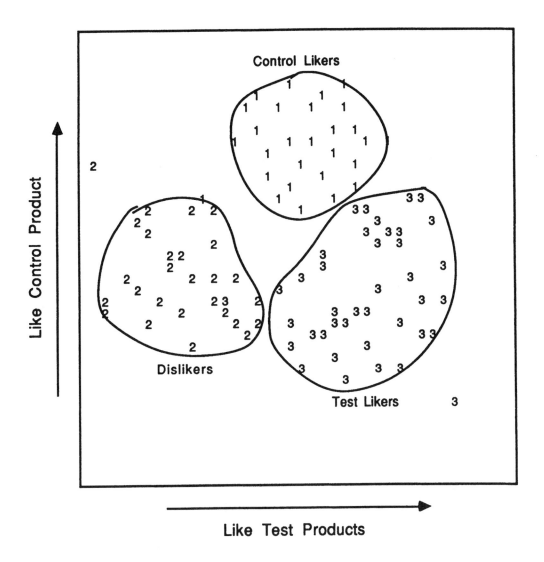

FIGURE 14.6   A plot of three clusters of respondents grouped by their patterns of overall liking for a group of yogurt products. The plot shows how the overall mean of the three groups would be a poor summary of the "average" liking for a product.

B. DEPENDENT AND INDEPENDENT VARIABLES

For this set of methods the values of some variable(s) are viewed as being dependent on the values of the other ("independent") variables in the set. The statistical methods for such data either use the independent variables to predict the value of a continuous dependent variable or they use the independent variables to group observations into particular categories of a discrete dependent variable. Even when both dependent and independent variables are present, a researcher should first apply the methods described in the previous section to uncover fundamental relationships among all the variables (using both correlation and PCA) and to determine if all the observations can be analyzed as a single group or if clusters exist that display distinctly different patterns of relationships (via cluster analysis). These preliminary analyses help to ensure that a meaningful and complete summary of the information contained in the data is obtained.

1. Regression Analysis

Predicting the value of one variable based on the values of one or more other variables has become commonplace. Consumer acceptability has been predicted by descriptive data or by formula and process values. Descriptive data values have been predicted by instrumental results. The perceived intensity of various responses has been predicted based on the intensity (or concentration) of a stimulus using either psychophysical models (e.g., Stevens' law) or by kinetic models (e.g., the Michaelis-Menten/Beidler equation). All of these examples use regression analysis to relate the value of a continuous dependent variable to the values of one or more independent variables.

Regression can be used simply to predict the value of a response or, not so simply, to determine what and how changes in one variable cause changes in another. By itself, regression analysis does not yield causal relationships. If a researcher comes prepared with hypotheses about the dynamics of a system, then regression analysis can be used to test the validity of the hypothesized relationships. In general, however, a highly accurate predictive model obtained by regression analysis is only just that. A highly accurate model does not imply that the independent variables drive the dependent variable. The researcher must provide the meaning behind data. It cannot be obtained from the numerical analysis procedures used to analyze the data.

Plotting data is essential to a successful regression analysis. For the same reasons noted in the discussion of correlation analysis, researchers could easily be misled by blindly applying computer programs to perform regression computations without first examining plots of the dependent variable(s) vs. the independent variable(s) (see Figure 14.1). Other plots that are useful in determining the quality of the regression model are presented in the following.

a. Simple Linear Regression

In simple linear regression the value of a single dependent variable, y, is predicted using the value of a single independent variable, x, using a linear model of the form:

$$y = \beta_0 + \beta_1 x + \varepsilon \tag{14.2}$$

where β_0 and β_1 are parameters of the regression equation that will be estimated in the analysis and ε is the unexplained deviation between the observed value of y and its predicted value, called a "residual."

The original units of measure do not have to be retained in simple linear regression. If examination of a plot of y vs. x reveals a nonlinear relationship, it is often possible to transform either x or y, or both, to obtain a straight-line relationship. These transformed values of y and x can then be substituted into Equation 14.2 to obtain estimates of β_0 and β_1 (on the transformed scales). For example, the data in Figure 14.1(C) might be linearized by taking the logarithm of y.

The coefficients β_1 and β_0 are estimated by:

$$b_1 = \frac{\Sigma(x_i - \bar{x})(y_i - \bar{y})}{\Sigma(x_i - \bar{x})^2} \tag{14.3}$$

and

$$b_0 = \bar{y} - b_1\bar{x} \tag{14.4}$$

Based on the estimated regression coefficients, the predicted (or expected) value of the dependent variable, y, is

$$\hat{y}_i = b_0 + b_1 x_i \qquad (14.5)$$

The estimates in Equations 14.3 and 14.4 are "best" in the sense that they minimize the sum of the squared differences between the observed and predicted values of y, that is, they minimize the sum of the squared residuals:

$$SS_{Res} = \Sigma\left(y_i - \hat{y}_i\right)^2 = \Sigma\hat{\varepsilon}_i^2 \qquad (14.6)$$

This is what is meant when it is said that the regression equation was fit to the data using the "method of least squares."

A fundamental criterion used to assess the quality of the regression equation is to determine if the fitted line results in a substantial reduction in the variability of the dependent variable. The variability of y around the line (i.e., vs. $\hat{y}$) is compared with the variability of y around its sample mean $\bar{y}$ (which is the original "expected" value of y) (see Figure 14.7). This notion is formalized statistically by adding the assumption that the residuals of the regression analysis are normally distributed, independent of each other, all with the mean value of 0 and the same variance, σ^2, i.e., $\varepsilon \sim n(0,\ \sigma^2)$. ANOVA can then be used to determine if a significant reduction in unexplained variability is obtained by using least-squares estimates to predict y based on x. The F-ratio in the ANOVA table for simple linear regression, such as in Table 14.1, actually tests H_0: $\beta_1 = 0$ vs. H_A:

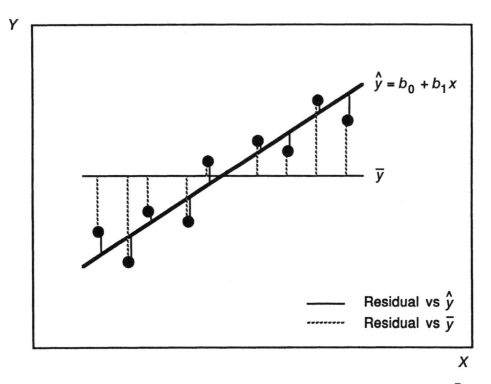

FIGURE 14.7   A comparison of the residuals from a fitted regression line with the residuals from $\bar{y}$ used to determine if the fitted line significantly reduces the amount of unexplained variability in the response. The reduction in the size of the residuals (i.e., distance from the "expected" value) between $\hat{y}$ vs. $\bar{y}$ shows that the regression line is a better summary of the data.

$\beta_1 \neq 0$, which is equivalent to the reduction in variability argument stated previously (if $\beta_1 = 0$ then the line is horizontal and $\hat{y} = \overline{y}$, so no reduction in variability could occur).

TABLE 14.1
ANOVA Table for Simple Linear Regression

Source of variation	Degrees of freedom	Sum of squares	Mean square	F
Total	$n - 1$	SS_T		
Regression	1	SS_{Reg}	$MS_{Reg} = SS_{Reg}$	MS_{Reg}/MS_E
Error	$df_E = n - 2$	SS_{Res}	$MS_{Res} = SS_E/df_E$	

Other criteria are used to assess the quality of the regression equation. The coefficient of determination:

$$R^2 = 1 - \frac{SS_{Res}}{SS_{Tot}} \tag{14.7}$$

summarizes the proportion of the total variability that is explained by using x to predict y. SS_{Res} and SS_{Tot} in Equation 14.7 are the residual and total ANOVA sums of squares from Table 14.1, respectively. In sensory evaluation, values of $R^2 > 0.75$ are generally considered to be acceptable. However, whether this is true depends on the intended use of the regression equation. Other criteria may be more informative. A confidence interval on β_1 can be constructed using:

$$b_1 \pm t_{\alpha/2, n-2}\sqrt{MS_E/SS_x} \tag{14.8}$$

where $SS_x = \sum(x_1 - \overline{x})^2$ and $t_{\alpha/2, n-2}$ is the upper-$\alpha/2$ critical value of Student's t-distribution with $n - 2$ degrees of freedom. The F-ratio from ANOVA only tells you if β_1 is different from zero. The confidence interval in Equation 14.8 tells you if β_1 is estimated with sufficient precision to be useful in applications. The idea of a confidence interval on β_1 can be extended to confidence bands on the predicted value of y using:

$$\hat{y}_0 \pm t_{\alpha/2, n-2}\sqrt{MS_E\left[(1/n) + (x_0 - \overline{x})^2/SS_x\right]} \tag{14.9}$$

The confidence bands can be plotted along with the predicted values to provide a visual assessment of the quality of the fit (see Figure 14.8). If the confidence bands are too wide, then regardless of the F-ratio test or the value of R^2 the fitted simple linear regression equation is not good enough.

Several possibilities exist to explain a poor fitting regression equation. Most of these possibilities can be studied with plots of the residuals from the regression. The residuals, $\hat{\varepsilon}$, should be plotted vs. the predicted values, $\hat{y}$, and vs. the independent variable, x. The residuals should be randomly dispersed across the range of both the predicted values and the independent variable (see Figure 14.9a). Any apparent trends indicate that a simple linear regression is not sufficient and that a more complex relationship exists between x and y. Higher order terms (e.g., x^2) may be needed (see Figure 14.9b) or data transformations may need to be performed (see Figure 14.9c). An individual point falling far from the rest in either the vertical or horizontal direction may be an

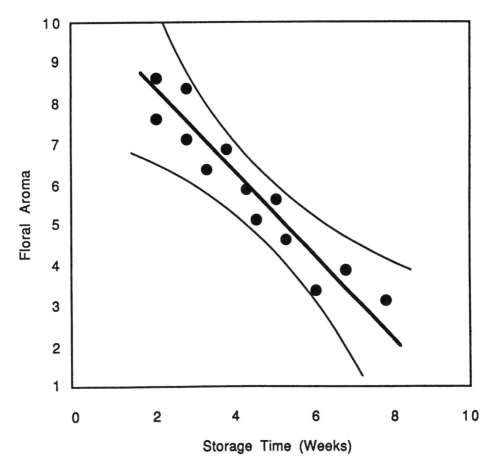

FIGURE 14.8   A fitted regression line with 95% confidence bands. The width of the bands provides a visual assessment of the quality of the fitted line. Narrow bands such as these indicate that the data are well fitted by the line, while wide bands indicate that a large amount of unexplained variability remains.

outlier that is having an unreasonably large influence on the fit of the model (see Figure 14.9d). Such observations should be examined and, when appropriate, eliminated from the data.

b. Multiple Linear Regression

Sometimes more than one independent variable is needed to obtain an acceptable prediction of a response, y. It may be that a polynomial in a single variable, x, is needed because the relationship between x and y is not a straight line, such as in:

$$y = \beta_0 + \beta_1 x + \beta_2 x^2 + \beta_3 x^3$$

In other cases, the response may be influenced by more than one independent variable such as in:

$$y = \beta_0 + \beta_1 x_1 + \beta_2 x_2 + \beta_3 x_3$$

or a combination of both cases may exist. Regardless, multiple regression analysis is a straight-forward extension of simple linear regression that allows multiple independent variables to be

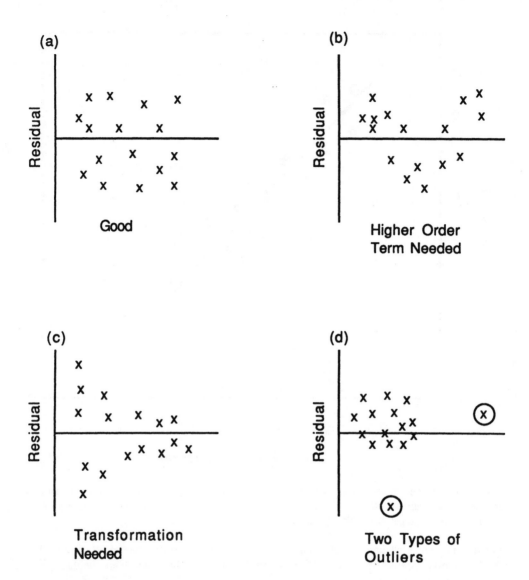

FIGURE 14.9 Plots of the residuals from a simple linear regression showing the desired, random arrangement in (a) and several undesirable patterns, i.e., plots (b), (c), and (d), along with their interpretation.

included in the regression equation for y. An integral part of the analysis involves the assessment of the value of each term considered for inclusion in the model.

A pitfall associated with multiple regression is that now any relationships that exist among the independent variables will influence the resulting regression equation. Multicolinearity is the term used to describe situations in which two or more independent variables are highly correlated with each other. This mutual correlation will influence the values of the estimated coefficients, b_i's, and could lead to incorrect conclusions about the importance of each term in the model. It is important that the correlation structure of the independent variables be studied before undertaking a multiple linear regression analysis. The correlation analysis will be more meaningful if it is accompanied by a scatterplot matrix of the independent (and dependent) variables, like the one in Figure 14.2. A designed approach to multiple regression, called response surface methodology (RSM), avoids the problems of multicolinearity. RSM will be discussed in Section III.

Not all of the independent variables that could be included in a multiple linear regression model may be needed. Some of the independent variables may be poor predictors of the response, or due to multicolinearity, two or more independent variables may explain the same variability in the dependent variable. Several approaches for selecting variables to include in the model are available (see Draper and Smith, 1981). Most computer packages include more than one.

One "brute force" approach is the "All Possible Regressions" method. As the name implies, all possible subsets of the independent variables are considered, starting with all of the one-variable models, then all of the possible two-variable models, etc. Computer output typically only presents a small number of the best models from each size group. Several criteria are used to determine which models are best in all possible regressions. The multiple R^2 (from Equation 14.7) is a common measure, with larger values being preferred. Another criterion is the size of the residual mean square, MS_{Res}, from the ANOVA. MS_{Res} is the estimated residual variance, so smaller values are desirable.

Associated with the residual mean square criteria is the adjusted R^2, where:

$$R_{adj}^2 = 1 - \frac{(n-1)MS_{Res}}{SS_{Tot}} \qquad (14.10)$$

R_{adj}^2 is interpreted in the same way as R^2. In multiple regression, the residual mean square will initially decrease with the addition of new terms into the model. R^2 will increase. Some studies reach a point where the further additions of new terms will result in an increase in the residual mean square (a bad sign), but R^2 will continue to increase (a good sign). When the residual mean square begins to increase, the adjusted R^2 will begin to decrease so that the two statistics agree qualitatively.

A final criterion commonly used in all possible regressions is Mallow's C_p statistic (1973).

$$C_p = \frac{SS_{Res(p)}}{MS_{Full}} + 2p - n \qquad (14.11)$$

where $SS_{Res(p)}$ is the residual sum of squares from a model containing p terms and MS_{Full} is the residual mean square from the model containing all of the independent variables. Unlike R^2 and MS_{Res}, C_p considers how good the full model is and uses this as a base of comparison to gauge the quality of a model containing only a subset of the independent variables. One drawback to C_p is that it can only be calculated when the number of observations in the data set is greater than the number of independent variables. For instance, if the number of descriptive attributes is greater than the number of samples, then Mallow's C_p criterion could not be used.

Another group of variable selection procedures used in multiple regression is the forward inclusion, backward elimination, and stepwise selection procedures. These use similar criteria for deciding if an independent variable should be included in the model or not. Forward inclusion starts by adding to the model the independent variable that maximizes the reduction in the unexplained variability, measured by the residual sum of squares (SS_{Res}). Additional terms are added based on the additional reduction in SS_{Res} that occurs as a result of their inclusion. Computer packages use an "F-to-enter" statistic to determine if adding a particular term will result in a statistically significant reduction in the unexplained variability. The independent variable with the largest F-to-enter value is the next term added to the model. The F-to-enter value can be set to correspond to the analyst's desired significance level. For example, a value of 4.0 corresponds roughly to $\alpha = 0.05$ for data sets containing 30 to 50 observations. The forward inclusion procedure continues to add terms until none of the F-to-enter values are large enough (compared to the value set in the program).

Backward elimination starts with all of the independent variables in the model and proceeds to eliminate terms based on how little of the variability in the dependent variable they explain. Computer packages use an "F-to-remove" statistic to measure how unimportant a particular term is. The term with the smallest F-to-remove is the next one to be excluded from the model. Once again, the analyst can select the value for F-to-remove, and the procedure will continue to remove terms until none of the F-to-removes are too small (4.0 is a good initial value for F-to-remove, also).

Stepwise selection is the recommended variable selection procedure for building multiple linear regression models. Stepwise selection combines forward inclusion with backward elimination by allowing for either the addition or removal of a term at each step of the procedure (starting after two terms have been added). Terms are initially added to the model using the F-to-enter criteria from forward selection. Because of multicolinearity, the importance of each term in the model changes depending on which other terms are also in the model. A term in the model may become redundant as others are added. Stepwise selection allows for such a term to be removed, using the F-to-remove criteria from backward elimination. The values of F-to-enter and F-to-remove are recomputed for each independent variable at each step of the procedure. The analysis ends when none of the statistics satisfy the values set by the analyst.

All of the diagnostics used to assess the quality of the model in simple linear regression should be used to determine how good the multiple linear regression model is. These include R^2 (now also including R^2_{adj}), MS_{Res}, confidence intervals on the individual coefficients, and most importantly, *plots of the residuals.* The potential for missing nonlinear relationships or outliers is higher in multiple linear regression because of the difficulty of visualizing in more than three dimensions. Plots of the residuals vs. the independent variables is the easiest way to explain problems and/or to determine if further improvements in the model are possible.

2. Principal Component Regression

A weakness of multiple linear regression is the manner in which it deals with correlated predictor variables (i.e., the x variables), a problem called multicolinearity. As noted in the previous section, if two highly correlated predictor variables are included in the regression model, the size and even the sign of the slope coefficients can be misleading. The problem is overcome to some degree by using one of the variable selection procedures, such as stepwise regression. However, the regression model that results from a stepwise procedure does not tell the whole story when it comes to identifying all of the predictor variables, x's, that are related to the predicted variable, y. For example, when using attribute intensity data from a descriptive panel to predict consumer acceptance, a particular attribute, such as sweet taste, may be highly related to acceptance, but it may not appear in the regression model because another attribute which is highly correlated with sweet taste, such as sweet aftertaste, may already be in the model. The stepwise procedure will not include both sweet taste and sweet aftertaste in the regression model precisely because they are highly correlated with each other (and, thus, are explaining the same variability in acceptance). This gives the researcher the incorrect impression that only sweet aftertaste, and not sweet taste, is important to acceptance. A more correct interpretation is that the term in the regression model, e.g., sweet aftertaste, is, in effect, representing all of the descriptive attributes with which it is highly correlated. However, the regression model does not reveal which attributes are correlated. Thus, a stepwise regression procedure applied to correlated predictor variables does not, by itself, uncover all of the attributes that are "driving" acceptance.

Principal Components Regression (PCR) is a method that overcomes this weakness. PCR is a straightforward combination of principal components analysis (PCA) and regression. Continuing the example of using descriptive data to predict acceptance, a PCA is performed on the average attribute profiles of the samples in the study. A set of factor scores is obtained for each sample. The factor scores are used as the predictor variables, i.e., the x's, in a regression analysis to predict consumer acceptance, y. The factors obtained from the PCA represent the underlying dimensions of sensory variability in the samples and are typically easy to interpret based on the factor loadings. Attributes with large positive or negative loadings (i.e., close to +1 or –1) on a single factor are the attributes that define the factor, so if the factor is found to be a significant driver of acceptance, the researcher knows that all of the attributes associated with that factor are, as a group, influencing acceptance. In addition, the factors obtained from PCA are not correlated with each other. Thus, the problem of multicolinearity is avoided. Popper, Heymann, and Rossi (1997) present an excellent example of PCR applied to the prediction of consumer acceptance of twelve honey mustard salad dressings based on their descriptive profiles.

The factors identified with PCA are ordered according to how much of the variability in the original data each explains. The first factor accounts for the greatest amount of variability, the second factor accounts for the next greatest amount, etc. It may not be the case that the factor that explains the most variability in the original x variables is the factor that is most highly related to y. In fact, some factors may not be related to y at all and, therefore, do not need to be included in the PCR model. Stepwise regression can be used to generate a PCR model that includes only those factors which are statistically significant predictors of y. In the descriptive analysis example, the stepwise approach to PCR allows the researcher to identify which under-lying dimensions of sensory variability (and all of their associated attributes) "drive" acceptance and which do not.

The PCR approach can be further extended by recognizing that the factor scores may be related to the response variable, y, in nonlinear and interactive ways. Nonlinear and interactive relationships can be accounted for in PCR by including the squares and the cross-products of the factor scores, respectively, as variables in the regression, as is done in response surface methodology (see Section III.C, pp. 330–334). This allows the researcher to identify the levels of the factor scores that are associated with, for example, the best liked product — that is, an optimum or "ideal" point.

3. Partial Least Squares Regression

As noted above, PCA generates factors which may not be related to the independent variable of the regression, y. Partial Least Squares (PLS) regression (see Martens and Martens, 1986) is a multivariate technique related to PCR that overcomes this weakness. Where PCA concentrates only on explaining the variability exhibited by the correlated predictor variables (x-variables), PLS derives factors (i.e., linear combinations of the x-variables) that 1) explain large portions of the variability in the x-variables and simultaneously 2) correlate to as great a degree as possible with the dependent predictor variable y. While the PLS factors may not explain as much of the variability in the x's as would the PCA factors, PLS ensures that each factor identified has maximal predictive power on y.

PLS has two other important advantages over PCR. First, it generates graphical output which clearly illustrates the relationships both among the predictor variables, x's, and between the predictor variables and y. Second, PLS readily extends to predicting more than one dependent variable simultaneously. When a single dependent variable is predicted the analysis is called PLS1. When several dependent variables are predicted the analysis is called PLS2. The graphical relationships in PLS2 can be used to illustrate, for example, how consumer vocabulary relates to descriptive attribute ratings (see Figure 14.10; Muñoz and Chambers, 1993; and Popper, Heymann, and Rossi, 1997).

The computations performed in a PLS regression are beyond the scope of this discussion. Several computer programs that perform PLS analyses are available (Pirouette, 1991; SAS, 1994; and The Unscrambler, 1993). The programs and the growing number of publications with examples of PLS applied to sensory data make this technique increasingly accessible to interested researchers.

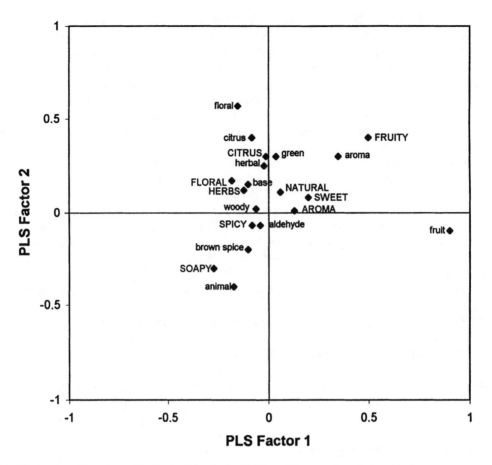

FIGURE 14.10 Use of graphical relationships in PLS2 to illustrate the relationship between consumer vocabulary and descriptive attribute ratings. Terms in CAPITALS are consumer responses. Terms in lower case are descriptive attributes. Note that consumer and descriptive vocabularies do not agree in all cases.

4. Discriminant Analysis

Discriminant analysis is a multivariate technique that is used to classify items into preexisting categories (defined by a discrete dependent variable). A mathematical function is developed using the set of continuous independent variables that best discriminate among the categories from which the items arise. For instance, descriptive attribute data might be used to classify a finished product as being "acceptable" or "unacceptable" from a quality control perspective or descriptive and/or instrumental measures might be used to determine the source (e.g., country or manufacturer) of a raw ingredient.

Discriminant analysis is similar to several of the multivariate techniques that have already been discussed. In one sense, discriminant analysis is similar to regression in that a group of continuous independent variables is used to predict the "value" of a dependent variable. In regression analysis the value is the magnitude of a continuous dependent variable, predicted using a "regression equation." In discriminant analysis the value is the category of a discrete dependent variable, predicted using a "discriminant function." In another sense, discriminant analysis is similar to PCA in that the correlated nature of the independent variables is considered in developing new axes (i.e., weighted linear combinations of the original variables). In PCA the axes are chosen to successively explain the maximum amount of variability. In discriminant analysis the axes are chosen to maximize the differences between the centers of the discrete categories of the dependent variable. A simple graphical depiction of discriminant analysis is presented in Figure 14.11 in which accept-

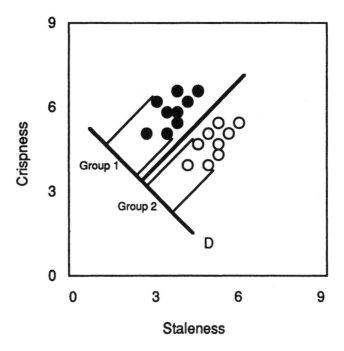

FIGURE 14.11   A graphical depiction of discriminant analysis with the samples plotted in two of the original attributes (staleness and crispness) and with the "axis of maximum discrimination," D.

able and unacceptable samples of product are displayed in a plot of two descriptive attributes, staleness and crispness. The discriminant function defines the new axis, D, on which the difference between the means of the two groups is maximized.

If the dependent variable contains only two categories, then only a single discriminant function is needed. If the dependent variable contains more than two categories, then more than one discriminant function may be needed to accurately classify the observations. The number of possible discriminant functions is one less than the number of categories. Regardless, the linear combination(s) of the original variables that best separate the categories is the one that maximizes the ratio:

$$\frac{\text{Variance between category means}}{\text{Variance within categories}}$$

In addition to this ratio, the quality of the discriminant function is measured by the proportion of the items that it correctly classifies. This evaluation can be done by using the same observations that were used to build the discriminant function. However, when sufficient data are available, it is preferable to withhold some of the observations from the model building analysis and use them only after the fact to verify that the discriminant function performs the classification task satisfactorily. As such, the verification process is more objective.

Not all of the original independent variables may be needed to accurately classify each item. As in regression analysis, there are four commonly used variable selection criteria: forward inclusion, backward elimination, stepwise, and all-possible-functions. In each, the criterion for determining the value of a variable is the degree to which it contributes to the discrimination among the categories. Powers and Ware (1986) present a summary of a stepwise discriminant analysis in which six blue cheese products were best categorized by using only 14 of the original 28 profile attributes.

The Powers and Ware reference just cited is a comprehensive discussion of applied discriminant analysis in sensory evaluation. Two alternatives to linear discriminant analysis, canonical discriminant analysis and nearest-neighbor discriminant analysis, are discussed. A variety of industrial applications, relations to other multivariate techniques, and relevant computer software concerns are also presented. Analysts interested in performing discriminant analysis are encouraged to familiarize themselves with the material presented there.

III. THE TREATMENT STRUCTURE OF AN EXPERIMENTAL DESIGN

In the experimental designs discussed in Chapter 13 the treatments (or products) were viewed as a set of qualitatively distinct objects, having no particular association among themselves. Such designs are said to have a one-way treatment structure. One-way experiments commonly occur toward the end of a research program when the objective is to decide which product should be selected for further development.

In many experimental situations, however, the focus of the research is not on the specific samples but rather on the effects of some factor or factors that have been applied to the samples. For instance, a researcher may be interested in the effects that different flour and sugar have on the flavor and texture of a specific cake recipe, or he may be interested in the effects that cooking time and temperature have on the flavor and appearance of a prepared meat. In situations such as these there are specific plans available that provide highly precise and comprehensive comparisons of the effects of the factors, while at the same time minimize the total amount of experimental material required to perform the study.

Two "multiway" treatment structures are discussed in this section. They are the factorial treatment structure (often called factorial experiments) and the response surface treatment structure (often called response surface methodology or RSM).

A. FACTORIAL TREATMENT STRUCTURES

Researchers are often interested in studying the effects that two or more factors have on a set of responses. Factorial treatment structures are the most efficient way to perform such studies. In a factorial experiment, specific levels for each of several factors are defined. A single replication of a factorial experiment consists of all possible combinations of the levels of the factors. For example, a brewer may be interested in comparing the effects of two kettle boiling times on the hop aroma of his beer. Further, if the brewer is currently using two varieties of hops, he may not be sure if the two varieties respond similarly to changes in kettle boiling time. Combining the two levels of the first factor, kettle boiling time, with the two levels of the second factor, variety of hops, yields four distinct treatment combinations that form a single replication of a factorial experiment (see Table 14.2). The experimental variables in a factorial experiment may be quantitative (e.g., boiling time) or qualitative (e.g., variety of hops). Any combination of quantitative and qualitative factors may be run in the same factorial experiment.

TABLE 14.2
Factorial Treatment Structure for Two
Factors Each Having Two Levels

		Hop variety	
		A	B
Kettle boiling	Low (1)	$T_{1A} = 6$	$T_{1B} = 13$
time	High (2)	$T_{2A} = 12$	$T_{2B} = 7$

An "effect" of a factor is the change (or difference) in the response that results from a change in the level of the factor. The effects of individual factors are called "main effects." For example, if the entries in Table 14.2 represent the average hop aroma rating of the four beer samples, the main effect due to boiling time is

$$\frac{\left(T_{1A} - T_{2A}\right) + \left(T_{1B} - T_{2B}\right)}{2} \tag{14.12}$$

Similarly, the main effect due to variety of hops is

$$\frac{\left(T_{1A} - T_{1B}\right) + \left(T_{2A} - T_{2B}\right)}{2} \tag{14.13}$$

In some studies the effect of one factor depends on the level of a second factor. When this occurs there is said to be an interaction between the two factors. Suppose for the beer brewed with hop variety A that the hop aroma rating increased when the kettle boiling time was increased, but that hop aroma decreased for the same change in boiling time when the beer was brewed with hop variety B (see Table 14.2). There is an interaction between kettle boiling time and variety of hops because the effect of boiling time depends on which variety of hops is being used.

Graphs can be used to illustrate interactions. Figure 14.12a illustrates the interaction between boiling time and variety. The points on the graph are the average hop aroma ratings of the four experimental conditions presented in Table 14.2. The interaction between the two factors is indicated by the lack of parallelism between the two lines. If there were no interaction between the two factors, the lines would be nearly parallel (deviating only due to experimental error) as in Figure 14.12b. Researchers must be very cautious in interpreting main effects in the presence of interactions. Consider the data in Table 14.2 that illustrate the "boiling time by hop variety" interaction. Applying Equation 14.12 yields an estimated main effect due to boiling time of [(6 – 12) + (13 – 7)/2 = 0, which says that boiling time has no effect. However, Figure 14.12a clearly shows that for each variety of hops there is a substantial effect due to boiling time. Because the separate variety effects are opposite, they cancel each other in calculating the main effect due to boiling time. In the presence of an interaction, the effect of one factor can only be meaningfully studied by holding the level of the second factor fixed.

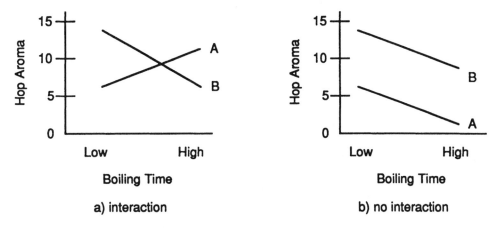

FIGURE 14.12 Plots of the mean hop aroma response illustrating (a) interaction and (b) no interaction between the factors in the study.

Researchers sometimes use an alternative to factorial treatment structures, called one-at-a-time treatment structures, in the false belief that they are economizing the study. Suppose in the beer brewing example that the brewer had only prepared three samples: the low boiling time/variety A point T_{1A}, the low boiling time/variety B point T_{1B}, and the high boiling time variety B point T_{2B}. (The high boiling time/variety A treatment combination T_{2A} is omitted.) Since only three samples are prepared, it would appear that the one-at-a-time approach is more economical than the full factorial approach. This is not true, however, if one considers the precision of the estimates of the main effects. Only one difference due to boiling time is available to estimate the main effect of boiling time in the one-at-a-time study (i.e., $T_{1B} - T_{2B}$). The same is true for the variety effect (i.e., $T_{1A} - T_{2B}$). Equations 14.12 and 14.13 show that for the factorial treatment structure two differences are available for estimating each effect. The entire one-at-a-time experiment would have to be replicated twice, yielding six experimental points, to obtain estimates of the main effects that are as precise as those obtained from the four points in the factorial experiment.

Another advantage that factorial treatment structures have over one-at-a-time experiments is the ability to detect interactions. If the high temperature/variety A observation $T_{2A} = 12$ were omitted from the data in Table 14.2 (as in the one-at-a-time study) one would observe that beer brewed at the high boiling time has less hop aroma than beer brewed at the low boiling time and that beer brewed with hop variety A has less hop aroma than beer brewed with hop variety B. The most obvious conclusion would be that beer brewed at the high boiling time using hop variety A would have the least hop aroma of all. The complete data in Table 14.2 and the plot of the interaction in Figure 14.12a show this would be an incorrect conclusion.

The recommended procedure for applying factorial treatment structures in sensory evaluation is as follows. Prepare at least two independent replications of the full factorial experiment. Submit the resulting samples for panel evaluation using the appropriate blocking structure as described in Chapter 13, p. 288. Take the mean responses from the analysis of the panel data and use them as raw data in an ANOVA. The output of the ANOVA includes tests for main effects and interactions among the experimental factors (see Table 14.3). This procedure avoids confusing the measurement error, obtained from the analysis of the panel data, with the true experimental error, which can only be obtained from the differences among the independently replicated treatment combinations.

B. FRACTIONAL FACTORIALS AND SCREENING STUDIES

Early in a research program many variables are proposed as possibly having meaningful effects on the important responses. In order to execute an efficient research plan experimenters need an approach that will allow them to screen out the influential variables from those that have little or

---

TABLE 14.3
ANOVA Table for a Factorial Experiment

Source of variation	Degrees of freedom	Sum of squares	Mean square	F
Total	$rab - 1$	SS_T		
A	$a - 1$	SS_A	$MS_A = SS_A/(a - 1)$	$F_A = MS_A/MS_E$
B	$b - 1$	SSB	$MS_B = SS_B/(b - 1)$	$F_B = MS_B/MS_E$
AB	$df_{AB} = (a - 1)(b - 1)$	SS_{AB}	$MS_{AB} = SS_{AB}/df_{AB}$	$F_{AB} = MS_{AB}/MS_E$
Error	$df_E = ab(r - 1)$	SS_E	$MS_E = SS_E/df_E$	

Note: Factor A has "a" levels, Factor B has "b" levels, and the entire experiment is replicated "r" times. The samples are prepared according to a completely randomized blocking structure.

no impact on the responses. This determination must be done with a minimum amount of work so that sufficient resources exist at the end of the program to do the necessary fine-tuning and "finishing" work on the final prototype. There are a class of experimental plans called "fractional factorials" that allow researchers to screen for the effects of many variables simultaneously with a minimum number of experimental samples.

As the number of experimental variables grows in a factorial experiment, each main effect is estimated by an increasing number of "hidden replications." For example, as noted in the previous section, in a 2×2 (or 2^2) factorial each main effect is estimated by two differences (i.e., two hidden replications). In a 2^6 factorial (i.e., six factors, each with two levels) the number of hidden replications for estimating each main effect has grown to 32. This may be excessive. A single replication of a 2^6 factorial consists of 64 experimental samples. If interest is primarily focused on identifying individual experimental variables with significant main effects, then the number of hidden replications could safely be reduced to 16 or even 8 without sacrificing too much sensitivity. The number of experimental samples would be concurrently reduced to 32 or 16, thus yielding a manageable experiment. Figure 14.13 shows that the number of samples in a 2^3 factorial can be cut in half, from eight to four, while still providing two differences for estimating each main effect.

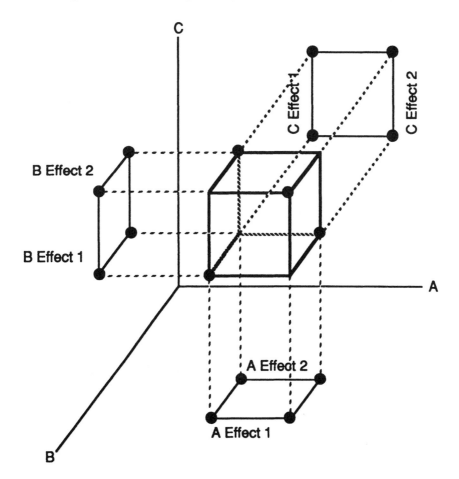

FIGURE 14.13 A graphical display of a ½-replicate fractional factorial of a 2^3 experiment showing by projection that two differences remain for estimating each main effect even though the total experiment has been reduced from eight to four samples.

1. Constructing Fractional Factorials

Most screening studies are performed by selecting two levels, low and high, for each experimental variable. The various treatment combinations of low and high levels make up the experimental design. That is, the treatment combinations define the levels of the experimental variables that should be used to produce each of the experimental samples. A convenient notation has been developed to identify the levels of the factors in each treatment combination. The high level of a variable A is denoted by the lower case a, the high level of B by b, etc. The low level of a variable is denoted by the absence of the lower case letter. For example, in a 2^3 factorial the treatment combination high A, high B, high C would be denoted as abc; the treatment combination low A, high B, high C would be denoted as bc; and the treatment combination low A, low B, high C would be denoted by c. The symbol used to represent the combination of all factors at their low levels is (1).

The eight treatment combinations that make up a single replication of a 2^3 factorial experiment are presented in the first column of Table 14.4. The remaining columns contain the signs of the coefficients that would be used to estimate each of the factorial effects. (The coefficients are either -1 or $+1$ so only the sign is needed.) The treatment combinations are grouped by the sign of the coefficient for estimating the three-way interaction ABC. The two groups formed in this way are each ½-replications of a 2^3 factorial. Either of the two groups of four treatment combinations could be selected for use in a screening study. Cochran and Cox (1957) present plans for fractional factorial experiments for both 2^n and 3^n experiments where the number of factors, n, is as large as eight.

By choosing the treatment combinations that have the same sign for the coefficients of the three-way interaction, we have given up any ability to estimate the magnitude of this effect. ABC is called the "defining contrast" because it is the criterion that was used to split the factorial into two ½-replications.

Notice in Table 14.4 that within each group there are two $+$'s and two $-$'s for estimating each effect. These are the hidden replications that remain even when only half of the full factorial is run. Suppose that the first group of four treatment combinations was selected to be run (i.e., the ABC + group). Then the main effect of variable A would (apart from a divisor of 2) be estimated by:

$$A = abc + a - b - c$$

TABLE 14.4
Factorial Effects in a 2^3 Factorial Experiment Arranged as Two ½-Replicate Fractional Factorial Experiments (ABC + and ABC –)

Treatment combination	Factorial effects						
	A	B	C	AB	AC	BC	ABC
a	+	–	–	–	–	+	+
b	–	+	–	–	+	–	+
c	–	–	+	+	–	–	+
abc	+	+	+	+	+	+	+
(1)	–	–	–	+	+	+	–
ab	+	+	–	+	–	–	–
ac	+	–	+	–	+	–	–
bc	–	+	+	–	–	+	–

However, the estimate of the two-way interaction BC is also:

$$BC = abc + a - b - c$$

The main effect of A is said to be "aliased" with the two-way interaction BC, notationally denoted as A = BC. Similarly, B = AC and C = AB. In practical terms, if two factorial effects are aliased, then it is impossible to separate their individual impacts on the responses of interest. The apparent effect of A may be really due to A or due to BC or possibly even due to a combination of the two.

The aliasing of main effects with interactions is the price paid for fractionalizing a factorial experiment. Typically, it is reasonable to assume that the magnitudes of the main effects are larger than the magnitudes of the interactions and, in such cases, fractional factorials can be used safely to screen for important experimental variables. If, however, large interactive effects are present, then a researcher may be misled into concluding that a variable has an important influence on the response when, in fact, it does not. This caution is not intended to frighten researchers away from using fractional factorials, but rather only to make them aware of the issue because it may serve to explain otherwise incongruous results that arise as a research program progresses.

2. Plackett-Burman Experiments

Fractional factorials are not the only plans that can be used to screen for influential variables. Plackett-Burman experiments (1946) are even more economical in the number of samples they require. The number of samples in a Plackett-Burman experiment is always a multiple of four. The number of experimental variables that can be screened with a Plackett-Burman experiment is at most one less than the number of samples (i.e., 4, 5, 6, or 7 variables can be screened with 8 samples; 8, 9, 10, or 11 variables can be screened with 12 samples; etc.) Box and Draper (1987) present the construction of Plackett-Burman experiments covering the range from 4 to 27 experimental factors (i.e., for studies involving 8 to 28 experimental samples).

3. Analysis of Screening Studies

Both fractional factorials and Plackett-Burman experiments can be analyzed by ANOVA. However, because of the small number of samples involved, it is sometimes impossible to compute F-ratios to test for the significance of the effects. This happens because in some screening experiments there are no degrees of freedom available for estimating experimental error. Regardless, even when the ANOVA computations can be performed the tests are not very sensitive, so that the possibility of missing a real effect (i.e., a Type II error) is relatively high.

A graphical technique for analyzing screening experiments allows the researcher more input into the decisions on which variables are affecting the response. The technique is motivated by the logic that if none of the variables have an impact on the response, then the values of their estimated effects are actually just random observations from a distribution (assumed to be normal) with a mean of zero. If these estimated effects are plotted vs. their corresponding normal random deviates, they should form a straight line (in the absence of any real effects). If, however, some of the variables affect the response, then the estimated effects are more than random observations. Real effects will fall off the line in the plot, either high and to the right (for positive effects) or low and to the left (for negative effects). The researcher can examine the "normal probability plot" of the estimated effects, such as presented in Figure 14.14, to decide which variables actually affect the response.

Constructing a normal probability plot is a four-step process:

1. Estimate the effects of the experimental variables using ANOVA.
2. Rank the estimated effects in increasing order from $i = 1$ to n.

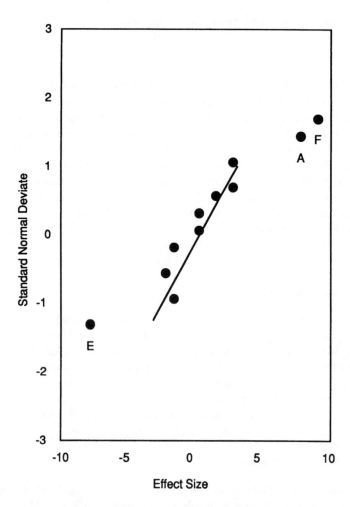

FIGURE 14.14 A normal probability plot showing the "nonsignificant" factorial effects falling on the line and the "significant" effects falling high and to the right, and low and to the left.

3. Pair the ordered estimates with the new variable $z = \Phi^{-1}[p]$, where $p = (3i - 1)/(3n + 1)$ and Φ^{-1} is the inverse of the standard normal distribution function. Many statistical computer packages contain a function for computing the value of z from p (sometimes called "PROBIT").

4. Plot z vs. the estimated effects using a standard plotting routine, fit (by eyeball) a straight line to the data, and look for points that fall high and to the right or low and to the left. These identify the "significant" variables.

C. Response Surface Methodology

The treatment structure known as response surface methodology (RSM) is essentially a designed regression analysis (see Montgomery, 1976 and Giovanni, 1983). Unlike factorial treatment structures, where the objective is to determine if (and how) the factors influence the response, the objective of an RSM experiment is to predict the value of a response variable (called the dependent variable) based on the controlled values of the experimental factors (called independent variables). All of the factors in an RSM experiment must be quantitative.

RSM treatment structures provide an economical way to predict the value of one or more responses over a range of values of the independent variables. A set of samples (i.e., experimental points) is prepared under the conditions specified by the selected RSM treatment structure. The samples are analyzed by a sensory panel, and the resulting average responses are submitted to a stepwise regression analysis. The analysis yields a predictive equation that relates the value of the response(s) to the values of the independent variables. The predictive equation can be depicted graphically in a response surface plot as shown in Figure 14.15. Alternatively, the predicted relationship can be displayed in a "contour plot" as in Figure 14.16. Contour plots are easy to interpret. They allow the researcher to determine the predicted value of the response at any point inside the experimental region without requiring that a sample be prepared at that point.

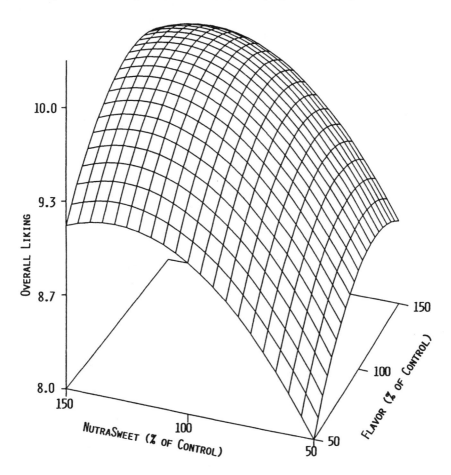

FIGURE 14.15   A response surface plot showing the predicted relationship between overall liking and the levels of NutraSweet and flavor in the product.

Several classes of treatment structures can be used as RSM experiments. The most widely used class, discussed here, is very similar to a factorial experiment. One part of the plan consists of all possible combinations of the low and high levels of independent variables. [In a two-factor RSM experiment this portion consists of the four points: (low, low), (low, high), (high, low), and (high, high).] This factorial portion of the RSM experiment is augmented by a center point [i.e., the point where all of the factors take on their average values, (low + high)/2]. Typically, the center point is replicated several times (not less than three) to provide an independent estimate of experimental error (see Figure 14.17). The regular practice in an RSM experiment is to assign the low levels of all the factors the coded value of −1; the high levels are all assigned the coded value of +1; and the center point is assigned the coded value of 0.

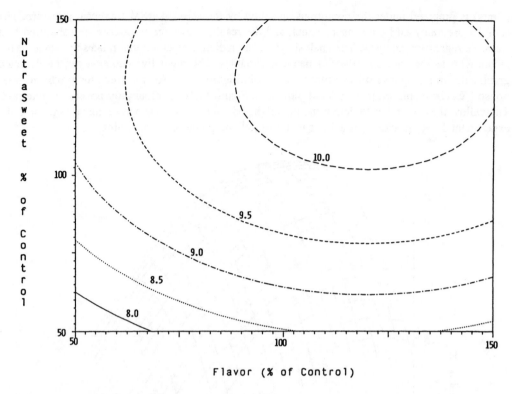

FIGURE 14.16 A contour plot of the predicted relationship between overall liking and the levels of NutraSweet and flavor in the product. Contour plots provide a quantitative assessment of the sensitivity of the product to changes in the levels of the ingredients.

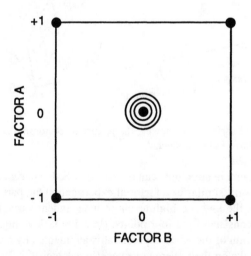

FIGURE 14.17 A two-factor, first-order RSM experiment. The figure illustrates the arrangement of the factorial and center points in an RSM experiment with two independent variables that permit estimation of a first-order regression model in Equation 14.14.

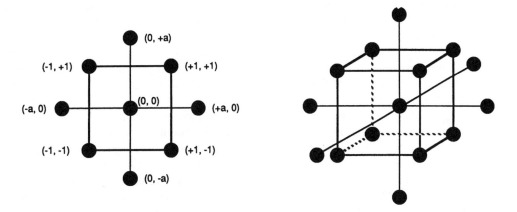

FIGURE 14.18 Central composite RSM experiments. The figures illustrate the arrangement of the factorial, axial, and center points in an RSM experiment with two and three independent variables that permit estimation of a second-order regression model as in Equation 14.15.

The treatment structure of an RSM experiment depicted in Figure 14.17 is called a first-order RSM experiment. The full regression equation that can be fit by the treatment structure has the form:

$$y = \beta_0 + \beta_1 x_1 + \beta_2 x_2 + \ldots + \beta_k x_k \tag{14.14}$$

where β_1 is the coefficient of the regression equation to be estimated and x_1, the coded level of the k-factors in the experiment. First-order RSM experiments are used to identify general trends and to determine if the correct ranges have been selected for the independent variables. The first-order models are used early in a research program to identify the direction in which to shift the levels of the independent variables to affect a desirable change in the dependent variable (e.g, increase desirable response or decrease undesirable response).

First-order models may not be able to adequately predict the response if there is a complex relationship between the dependent variable and the independent variables. A second-order RSM treatment structure is required for these situations. The full regression model that can be fit to a second-order RSM treatment structure has the form:

$$y = \beta_0 + \beta_1 x_1 + \beta_2 x_2 + \ldots + \beta_k x_k$$
$$+ \beta_{11} x_1^2 + \beta_{22} x_2^2 + \ldots + \beta_{kk} x_k^2 \tag{14.15}$$
$$+ \beta_{12} x_1 x_2 + \beta_{13} x_1 x_3 + \ldots + \beta_{k-1,k} x_{k-1} x_k$$

The addition of the squared and cross-product terms in the model allows the predicted response surface to "bend" and "flex," resulting in an improved prediction of complex relationships.

A popular class of second-order RSM experiments is the central-composite, rotatable treatment structures. Central-composite experiments are developed by adding a set of axial or "star" points to a first-order RSM treatment structure (see Figure 14.18). There are $2k$ axial points in a k-factor RSM experiment. Using the normal -1, 0, $+1$ coding for the factor levels, the axial points are $(\pm\alpha, 0,\ldots,0)$, $(0, \pm\alpha, 0,\ldots, 0),\ldots,(0, 0,\ldots, 0 \pm\alpha)$, where α is the distance from the axial point to the center of the experimental region (i.e., the center point). The value of α is $(F)^{1/4}$, where F is the number of noncenter factorial points in the first-order experiment. For example, in a two-factor experiment $F = 4$ and $\alpha = (4)^{1/4} = 1.414$.

Second-order RSM models have several advantages over first-order models. As mentioned before, the second-order models are better able to fit complex relationships between the dependent variable and the independent variables. In addition, second-order models can be used to locate the predicted maximum or minimum value of a response in terms of the levels of the independent variables.

The recommended procedure for performing an RSM experiment is as follows (also see Carr, 1989). First, the experimental samples should be prepared according to the RSM plan. Second, perform a regular BIB analysis of the samples from the RSM treatment structure, ignoring the association among the samples. (The BIB blocking structure is suggested because there are normally too many samples in an RSM experiment to evaluate at one sitting. If, however, it is possible to evaluate all of the samples together, then a randomized [complete] block design can be used.) The only output of interest from the BIB analysis is the set of adjusted sample means. The significance (or lack of significance) of the overall test statistic is of no interest. Next, submit the sample means to a stepwise regression analysis in order to develop the predictive equation that relates the value of the response to the levels of the experimental factors. The predictive equation is then used to generate a contour plot that provides a graphical depiction of the effects of the factors on the response. If there is only one response, the region where the response takes on acceptable values (or attains a minimum or maximum value) is apparent in the contour plot. When several responses are being considered, the individual contour plots can be overlayed. Hopefully, a region where all of the responses take on acceptable values can be identified as in Figure 14.19.

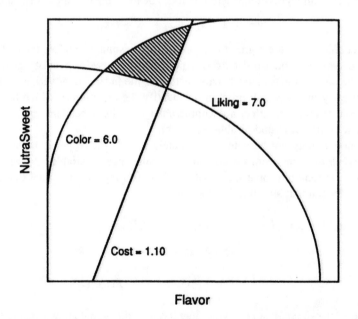

FIGURE 14.19 Overlaid contour plots of the critical limits for several response variables showing the region of formula levels predicted to satisfy all of the constraints simultaneously.

REFERENCES

Bezdek, J., 1981. *Pattern Recognition with Fuzzy Objective Function Algorithms.* Plenum Press, New York.
Box, G.E.P. and Draper, N.R., 1987. *Empirical Model Building and Response Surfaces.* John Wiley & Sons, New York.
Carr, B.T., 1989. An Integrated System for Consumer-Guided Product Optimization. In: *Product Testing with Consumers for Research Guidance*, Wu, L.S., Ed. ASTM, STP 1035, Philadelphia, PA.
Cattell, R.B., 1966. The Scree Test for the Number of Factors. *Multivar. Behav. Res.* 1, 245.

Cochran, W.G. and Cox, G.M., 1957. *Experimental Designs,* 2nd ed. John Wiley & Sons, New York.

Cooper, H.R., Earle, M.D., and Triggs, C.M., 1989. Ratio of Ideals — A New Twist on an Old Idea. In: *Product Testing with Consumers for Research Guidance*, Wu, L.S., Ed., ASTM, STP 1035, Philadelphia, PA.

Draper, N. and Smith, H., 1981. *Applied Regression Analysis,* 2nd ed. John Wiley & Sons, New York.

Giovanni, M., 1983. Response Surface Methodology and Product Optimization, *Food Technol.* 37(11), 41.

Godwin, D.R., Bargmann, R.E., and Powers, J.J., 1978. Use of Cluster Analysis to Evaluate Sensory-Objective Relations of Processed Green Beans. *J. of Food Sci.* 43, 1229.

Jacobsen, T. and Gunderson, R.W., 1986. Applied Cluster Analysis. In: *Statistical Procedures in Food Research*, Piggott, J.R., Ed. Elsevier Science Publ., Essex.

MacQueen, J.B., 1967. Some Methods for Classification and Analysis of Multivariate Observations. *Proc. 5th Berkeley Symp. Math. Stat. Prob.* 1, 281.

Mallows, C.L., 1973. Some Comments on C_p. *Technometrics* 15, 661.

Martens, M. and Martens, H., 1986. Partial Least Squares Regression. In: *Statistical Procedures in Food Research*, Piggott, J.R., Ed. Elsevier Science Publ., Essex.

Montgomery, D.C., 1976. *Design and Analysis of Experiments.* John Wiley & Sons, New York.

Muñoz, A.M. and Chambers, E., IV, 1993. Relating Sensory Measurements to Consumer Acceptance of Meat Products. *Food Technol.* 47(11), 128.

Piggott, J.R. and Sharman, K., 1986. Methods to Aid Interpretation of Multivariate Data. In: *Statistical Procedures in Food Research*, Piggott, J.R., Ed. Elsevier Science Publ., Essex.

Pirouette, 1991. Infometrix, Inc., Seattle, WA.

Plackett, R.L. and Burman, J.P., 1946. The Design of Optimum Multifactor Experiments. *Biometika* 33, 305.

Popper, R., Heymann, H. and Rossi, F., 1997. Three Multivariate Approaches to Relating Consumer to Descriptive Data. In: *Relating Consumer, Descriptive and Laboratory Data to Better Understand Consumer Responses*, Muñoz, A.M., Ed. ASTM, Manual 30, Philadelphia, PA.

Powers, J.J., 1988. Descriptive Methods of Analysis. In: *Sensory Analysis of Foods,* 2nd ed., Piggott, J.R., Ed. Elsevier Science Publ., Essex.

Powers, J.J. and Ware, G.O., 1986. Discriminant Analysis. In: *Statistical Procedures in Food Research*, Piggott, J.R., Ed. Elsevier Science Publ., Essex.

SAS, 1989. *SAS/STAT User's Guide Version 6,* 4th ed., Vol. 1. The SAS Institute, Cary, NC.

SAS, 1994. The SAS Institute, Cary, NC.

The Unscrambler, 1993. CAMO, Trondheim, Norway.

15 Guidelines for Choice of Technique

CONTENTS

I. INTRODUCTION

The five tables that follow are meant as memory joggers. They are not a substitute for study of the individual methods described in this book, but once the methods have become familiar, preferably via practical hands-on testing of most of them, the tables can be used to check whether there might be a better way to attack a given problem. Most of us tend to give preference to a few trusted favorite tests, and perhaps we bend the test objective a bit to allow their use — a dangerous habit.

To avoid this practice or find a way out of it, the authors suggest the following practical steps.

A. DEFINE THE PROJECT OBJECTIVE

Read the text in Chapter 1, then refer to Table 15.1 to classify the type of project. Review the 13 entries. Write down the project objective, then look up the test to which the table refers.

B. DEFINE THE TEST OBJECTIVE

Four tables are available for this purpose:

- Table 15.2: Difference tests — Does a sensory difference exist between samples?
- Table 15.3: Attribute difference tests — How does attribute X differ between samples?
- Table 15.4: Affective tests — Which sample is preferred? How acceptable is sample X?
- Table 15.5: Descriptive tests — Rate each of the attributes listed in the scoresheet.

Write down the test objective and list the tests required. Then meet with the project leader and others involved in the project and discuss and refine the design of the tests.

C. Reissue Project Objective and Test Objectives — Revise Test Design

In sensory testing, a given problem frequently requires appreciable thought before the appropriate practical tests can be selected (IFT, 1981). This is because the initial conception of the problem may require clarification. It is not unusual for problem and test objectives to be defined and redefined several times before an acceptable design emerges. Sensory tests are expensive, and they often give results which cannot be interpreted. If this happens, the design may be at fault. Pilot tests are often useful as a means of refining a design. It would, for example, be meaningless to carry out a consumer preference test with hundreds of participants without first having shown that a perceptible difference exists; the latter can be established with 10 or 20 tasters, using a difference test. In another example, islands of opposing preference may exist, invalidating a normal preference test; here, the solution may be a pilot study in which various types of customers receive single-sample acceptability tests.

TABLE 15.1
Types of Problems Encountered in Sensory Analysis

Type of problem	Tests applicable
1. New product development — The product development team needs information on the sensory characteristics and also on consumer acceptability of experimental products as compared with existing products in the market	All tests in this book
2. Product matching — Here the accent is on proving that no difference exists between an existing and a developmental product	Difference tests in similarity mode, Chapter 6
3. Product improvement — Step 1: define exactly what sensory characteristics need improvement; Step 2: determine that the experimental product is indeed different; Step 3: confirm that the experimental product is liked better than the control	All difference tests, Table 15.2; then Affective tests, Table 15.4; see *Note*
4. Process change — Step 1: confirm that no difference exists; Step 2: if a difference does exist, determine how consumers view the difference	Difference tests in similarity mode, Chapter 6; Affective tests, Table 15.4; see *Note*
5. Cost reduction and/or selection of new source of supply — Step 1: confirm that no difference exists; Step 2: if a difference does exist determine how consumers view the difference	Difference tests in similarity mode, Chapter 6; Affective tests, Table 15.4; see *Note*
6. Quality control — Products sampled during production, distribution, and marketing are tested to ensure that they are as good as the standard: Descriptive tests (well-trained panel) can monitor many attributes simultaneously	Difference tests, Table 15.2; Descriptive tests, Table 15.5
7. Storage stability — Testing of current and experimental products after standard aging tests; Step 1: ascertain when difference becomes noticeable; Step 2: descriptive tests (well-trained panel) can monitor many attributes simultaneously; Step 3: Affective tests can determine the relative acceptance of stored products	Difference tests, Table 15.2; Descriptive tests. Table 15.5; Affective tests, Table 15.4
8. Product grading or rating — Used where methods of grading exist which have been accepted by agreement between producer and user, often with government supervision	Grading, Chapter 5
9. Consumer acceptance and/or opinions — After laboratory screening, it may be desirable to submit product to a central-location or home-placement test to determine consumer reaction; Acceptance tests will indicate whether the current product can be marketed, or improvement is needed	Affective tests, Table 15.4; Chapter 12
10. Consumer preference — Full-scale consumer preference tests are the last step before test marketing; employee preference studies cannot replace consumer tests, but can reduce their number and cost whenever the desirability of key attributes of the product is known from previous consumer tests	Affective tests, Table 15.4; Chapter 12
11. Panelist selection and training — An essential activity for any panel; may consist of: (1) interview; (2) sensitivity tests; (3) difference tests; and (4) descriptive tests	Chapter 9
12. Correlation of sensory with chemical and physical tests — Correlation studies are needed: (1) to lessen the load of samples on the panel by replacing a part of the tests with laboratory analyses; (2) to develop background knowledge of the chemical and physical causes of each sensory attribute	Descriptive tests, Table 15.5; Attribute difference tests, Table 15.3
13. Threshold of added substances — Required: (1) in trouble shooting to confirm suspected source(s) of off-flavor(s); (2) to develop background knowledge of the chemical cause(s) of sensory attributes and consumer preferences	Chapter 8

Note: In 3, 4, and 5, if new product is different, Descriptive tests (Table 15.5) may be useful in order to characterize the difference. If the difference is found to be in a single attribute, attribute difference tests (Table 15.3) are the tools to use in further work.

TABLE 15.2
Area of Application of Difference Tests: Does a Sensory Difference Exist Between Samples?

The tests in this table are suitable for applications such as:

(1)   To determine whether product differences result from a change in ingredients, processing, packaging or storage
(2)   To determine whether an overall difference exists, where no specific attribute(s) can be identified as having been affected
(3)   To determine whether two samples are sufficiently similar to be used interchangeably
(4)   To select and train panelists and to monitor their ability to discriminate between test samples

Test	Areas of application
1. Triangle test	Two samples not visibly different; one of the most-used difference tests; statistically efficient, but somewhat affected by sensory fatigue and memory effects; generally 20–40 subjects, can be used with as few as 5–8 subjects; brief training required
2. Duo-trio tests	Two samples not visibly different; test has low statistical efficiency, but is less affected by fatigue than the Triangle test: useful where product well known to subjects can be employed as the reference; generally 30 or more subjects, can be used with as few as 12–15; brief training required
3. Two-out-of-five test	Two samples without obvious visible differences; statistically highly efficient, but strongly affected by sensory fatigue, hence use limited to visual, auditory, and tactile applications; generally 8–12 subjects, can be used with as few as 5; brief training required
4. Same/Different test	Two samples not visibly different; test has low statistical efficiency, but is suitable for samples of strong or lingering flavor, samples which need to be applied to the skin in half-face tests, and samples which are very complex stimuli and therefore confusing to the subjects; generally 30 or more subjects, can be used with as few as 12–15; brief training required
5. "A" – "Not A" test	As for No. 4, but used where one of the samples has importance as a standard or reference product, is familiar to the subjects, or essential to the project as the current sample against which all other samples are measured
6. Difference-from-control test	Two samples which may show slight visual differences such as are caused by the normal heterogeneity of meats, vegetables, salads, and baked goods; test is used where the size of the difference affects a decision about the test objective, e.g., in quality control and storage studies; generally 30–50 presentations of the sample pair; moderate amount of training required
7. Sequential tests	Used with any of the above tests 1 to 3, to determine with a minimum of testing, at a predetermined significance level, whether the two samples are perceptibly (1) identical or (2) different
8. Similarity mode	Used with tests 1 to 3 or 7, when the test objective is to prove that no perceptible difference exists between two products; used in situations such as: (1) the substitution of a new ingredient for an old one that has become too expensive or unavailable or (2) a change in processing brought about by replacement of an old or inefficient piece of equipment

TABLE 15.3
Area of Application of Attribute Difference Tests:
How Does Attribute X Differ Between Samples?

The tests in this table are used to determine whether or not, or the degree to which, two or more samples differ with respect to one defined attribute. This may be a single attribute such as sweetness, or a combination of several related attributes, such as freshness, or an overall evaluation, such as preference. With the exception of preference, panelists must be carefully trained to recognize the selected attribute, and the results are valid only to the extent that panelists understand and obey such instructions. A lack of difference in the selected attribute does not imply that no overall difference exists. Samples need not be visibly identical, as only the selected attribute is evaluated.

Test	Areas of application
1. Paired Comparison test (2-AFC test)	One of the most-used attribute difference tests; used to show which of two samples has more of the attribute under test ("Directional Difference test") or which of two samples is preferred ("Paired Preference test"); test exists in one- or two-sided applications; generally 30 or more subjects, can be used with as few as 15
2. Pairwise Ranking test	Used to rank three to six samples according to intensity of one attribute; paired ranking is simple to perform and the statistical analysis is uncomplicated, but results are not as actionable as those obtained with rating; generally 20 or more subjects, can be used with as few as 10
3. Simple Ranking test	Used to rank three to six, certainly no more than eight, samples according to one attribute; ranking is simple to perform, but results are not as actionable as those obtained by rating; two samples of small or large difference in the attribute will show the same difference in rank (i.e., one rank unit); ranking is useful to presort or screen samples for more detailed tests; generally 16 or more subjects, can be used with as few as 8
4. Rating of Several Samples	Used to rate three to six, certainly no more than eight, samples on a numerical intensity scale according to one attribute; it is a requirement that all samples be compared in one large set; generally 16 or more subjects, can be used with as few as 8; may be used to compare descriptive analyses of several samples, but note (Chapter 7, p. 111) that there will be some carryover (halo effect) between the attributes
5. Balanced Incomplete Block test	As No. 4, but used when there are too many samples (e.g., 7 to 15) to be presented together in one sitting
6. Rating of Several Samples, Balanced Incomplete Block	As No. 5, but used when there are too many samples (e.g., 7 to 15) to be presented together in one sitting

TABLE 15.4
Area of Application of Affective Tests Used in Consumer Tests and Employee Acceptance Tests

Affective tests can be divided into Preference tests in which the task is to arrange the products tested in order of preference, Acceptance tests in which the task is to rate the product or products on a scale of acceptability, and "Attribute diagnostics" in which the task is to rank or rate the principal attributes which determine a product's preference or acceptance. With regard to the statistical analysis, preference and acceptance tests can be seen as a special case of Attribute difference tests (Table 15.3) in which the attribute of interest is either preference or degree of acceptance. In theory, all tests listed in Table 15.3 can be used as Preference tests and/or as Acceptance tests. In practice, subjects in Affective tests are less experienced, and complex designs such as balanced incomplete blocks are not usable. The tests in this table are equally suitable for presentation in laboratory tests, employee acceptance tests, central location consumer tests, or home use consumer tests unless otherwise indicated.

Test	Question typically asked	Areas of application
Preference tests		
1. Paired Preference	Which sample do you prefer? Which sample do you like better?	Comparison of two products
2. Rank Preference	Rank samples according to your preference with 1 = best, 2 = next best, etc.	Comparison of three to six products
3. Multiple Paired Preference	As No. 1	Comparison of three to six products
4. Multiple Paired Preference, Selected Pairs	As No. 1	Comparison of five to eight products
Acceptance tests		
5. Simple Acceptance test	Is the sample acceptable/not acceptable?	First screening in employee acceptance test
6. Hedonic Rating	Chapter 12, Figures 12.2 and 12.3	One or more products to study how acceptance is distributed in the population represented by the subjects
Attribute diagnostics		
7. Attribute-by-Preference test	Which sample did you prefer for fragrance?	Comparison of two to six products to determine which attributes "drive" preference
8. Hedonic Rating of Individual Attributes	Rate the following attributes on the hedonic scale provided	Study of one or more products to determine which attributes, and at what level, "drive" preference
9. Intensity Rating of Individual Attributes	Rate the following attributes on the intensity scale provided, comparing with your ideal rating	Study of one or more products, in cases where groups of subjects differ in their preference

TABLE 15.5
Area of Application of Descriptive Tests

Descriptive tests are very diverse, often designed or modified for each individual application, and therefore difficult to classify in a table such as this. A classification by inventor is perhaps the most helpful.

Test	Areas of application
1. Flavor Profile (Arthur D. Little)	In situations where many and varied samples must be judged by a few highly trained tasters
2. Texture Profile (General Foods)	In situations where many and varied samples must be judged for texture by a few highly trained tasters
3. QDA® method (Tragon Corp.)	In situations such as quality assurance in a large company, where large numbers of the same kind of products must be judged day in and day out by a well-trained panel; in product development in situations where reproducibility over time and place is not required
4. Time-intensity descriptive analysis	Useful for samples in which the perceived intensity of flavor varies over time after the product is taken into the mouth, e.g., bitterness of beer, sweetness of artificial sweeteners
5. Free-Choice Profiling	In consumer testing, when it is desirable not to teach the subjects a common scale
6. Spectrum® method	A custom-design system suitable for most applications, including those under tests 1, 2, and 3; suitable where reproducibility over time and place is needed
7. Modified, short-version Spectrum® descriptive analysis	To monitor a few critical attributes of a product through shelf-life studies; to examine possible manufacturing defects and product complaints; for routine Quality Assurance

REFERENCE

IFT, 1981. Guidelines for the preparation and review of papers reporting sensory evaluation data. Sensory Evaluation Division, Institute of Food Technologist. *Food Technol.* 35(11), 50.

16 Guidelines for Reporting Results

CONTENTS

I. INTRODUCTION

For the user of sensory results, the most important consideration is how much confidence he or she can place in them. Two main factors determine this (Larmond, 1981):

1. Reliability: Would similar results be obtained if the test were repeated with the same panelists? With different panelists?
2. Validity: How valid are the conclusions? Did the test measure what it was intended to measure?

Because of the many opportunities for variability and bias resulting from the use of human subjects, reports of sensory tests must contain more detail than reports of physical or chemical measurements. It can be difficult to decide how much information to include; the recommendations below are mainly those of Prell (1976) and the Sensory Evaluation Division of the Institute of Food Technologists (1981). Application of the suggested guidelines is illustrated in the example at the end of this chapter.

II. SUMMARY

What did the test teach us? It is an important courtesy to the user not to oblige him or her to hunt through pages of text in order to discover the essence of the results. The conclusion is obvious to the sensory analyst and he or she should state it briefly and concisely in the opening summary. The summary should not exceed 110 words (Prell, loc. cit.) and should answer the four whats:

- What was the objective?
- What was done?
- What were the results?
- What can be concluded?

III. OBJECTIVE

As reiterated many times in this book, a clearly written formulation of the project objective and the test objective is fundamental to the success of any sensory experiment. The report (if directed to the project leader) should state and explain the test objective; if the report covers a complete project, it should state and explain the project objective as well as the objective of each test that formed part of the project.

In some cases, e.g., if the report is for publication, the explanation should take the form of an introduction which includes a review, with references, of pertinent previous work. This should be followed by a brief definition of the problem. It is of great importance to state the approach which was taken to solve the problem; Chapter 15, Table 15.1, p. 339, which follows the IFT (loc. cit.), should assist in this regard. If the study is based on a hypothesis, this hypothesis should be made evident to the reader in the introduction. Subsequent sections of the report should provide the test of the hypothesis.

IV. EXPERIMENTAL

The experimental section should provide sufficient detail to allow the work to be repeated. Accepted methods should be cited by adequate references. It is sometimes overlooked that subheadings in the experimental section help the reader find specific information. The section should describe the important steps in collecting the sensory data and will usually include the following:

Experimental design — Assuming that the objective was clearly stated previously, the text should now explain the "layout" of the experiment in terms of the objective. If there are major and minor objectives, the report should show how this is reflected in the design. If an advanced design is used (randomized complete block, balanced incomplete block, Latin square, etc.), it can be described by reference to the appropriate section of Cochran and Cox (1957). Next, state the measurements made (e.g., sensory, physical, chemical), sample variables and level of the variables (where appropriate), number of replications, and limitations of the design (e.g., lots available for sampling, nature and number of samples evaluated in a test session). Describe the efforts made to reduce the experimental error.

Sensory methods — When describing the methods employed, use the terminology in this book (see Chapter 15, Tables 15.2 to 15.5), which is the same as that of the International Standards Organization (1983) and the IFT (loc. cit.).

The panel — The number of panelists for each experimental condition should be stated as it influences the statistical significance of the results obtained. If too few panelists are used, large differences are required for statistical significance, whereas if too many are used (e.g., 1000 for a Triangle test), statistical significance may result when the actual difference is too small to have practical meaning. Changes in the panel during the course of the experiments should be avoided, but if they do occur, they must be fully described. The extent of previous training and the methods used to prepare the panelists for the current tests, including full description of any reference standards used, are important information needed to judge the validity of the results. The composition of the panel (age, sex, etc.) should be described if any affective tests were part of the experiment.

Conditions of the test — The physical conditions of the test area as well as the way samples are prepared and presented are important variables which influence both reliability and validity of the results. The report should contain the following information:

1. Test area — The location of the test area (booth, store, home, bus) should be stated, and any distractions present (odors, noise, heat, cold, lighting) should be described together with efforts made to minimize their influence.
2. Sample preparation — The equipment and methods of sample preparation should be described (time, temperature, any carrier used). Identify and describe raw materials and formulations if applicable.
3. Sample presentation — The description should enable the reader to judge the degree of bias likely to be contained in the results and may include any of the following capable of influencing them:

- Whether panelists work individually or as a group
- Lighting used if different from normal
- Sample quantity, container, utensils, temperature
- Order of presentation (randomized, balanced)
- Coding of sample containers, e.g., three-digit random numbers
- Any special instructions such as mouth rinsing, information about the identity of samples or variable under test; time intervals between samples; samples being swallowed or expectorated
- Any other variable which could influence the results, e.g., time of day, high or low humidity, age of samples, etc.

Statistical techniques — The manner in which the data reported were derived from actual test responses should be defined, e.g., conversion of scores to ranks. The type of statistical analysis used and the degree to which underlying assumptions (e.g., normality) are met should be discussed, as should the null hypothesis and alternate hypothesis, if not trivial.

V. RESULTS AND DISCUSSION

Results should be presented concisely in the form of tables and figures, and enough data should be given to justify conclusions. However, the same information should not be presented in both forms. Tabular data generally are more concise except for trends and interactions which may be easier to see from figures.

The results section should summarize the relevant collected data and the statistical analyses. All results should be shown, including those that run counter to the hypotheses. Reports of tests of significance (F, χ^2, t, r, etc.) should list the probability level, the degrees of freedom if applicable, the obtained value of the test statistic, and the direction of the effect.

In the discussion section, the theoretical and practical significance of the results should be pointed out and related to previous knowledge. The discussion should begin by briefly stating whether the results support or fail to support any original hypothesis. The interpretation of data should be logically organized and should follow the design of the experiment. The results should be interpreted, compared, and contrasted (with limitations indicated), and the report should end with clear-cut conclusions.

See Table 16.1 and Chapter 11, which illustrate the development of terminology and scales for a descriptive study.

TABLE 16.1

Example of Report: Hop Character in Five Beers

Summary

What was the objective?
What was done?

What were the results?
What can be concluded?

In order to choose among five lots of hops on the basis of the amount of hop character they are likely to provide, pilot brews were made with hop samples 1,2,3,4, and 5 costing $1.00, $1.20, $1.40, $1.60 and $1.80/lb; 20 trained members of the brewery panel judged each beer three times on a scale from 0 to 9. Sample 4 received a rating of 3.9, significantly higher than Samples 2 and 5, at 3.0 and 2.9. Samples 1 and 3 were significantly lowest at 2.1 and 1.4. It can be concluded that hop samples 4 and 2 deliver more hop character per dollar than the remainder.

Objectives

Project objective, test objectives, agreed before the experiment

The brewery obtained representative lot samples from several suppliers. The project objective was to choose among the lots based on their ability to provide hop character. The test objectives were: (1) to compare the five beers for degree of hop character on a meaningful scale and (2) to obtain a measure of the reliability of the results.

Experimental

Design which accomplishes objectives 1 and 2

Design — The five samples were test brewed to produce a standard bitterness level of 14 BU. The test beers were evaluated by 20 selected members of the brewery panel; the test set was tasted three times on separate days.

Describe sensory tests used

Sensory evaluation — The tasters evaluated the amount of hop character on a scale of 0 to 9; reference standards were available as follows; synthetic hop character at 1.0 mg/L = 3.0 scale units, and at 3 mg/L = 6.0 scale units.

Describe panel: number, training, etc.

The Panel — 20 panel members were selected on the basis of past performance evaluating hop character; all 20 panelists tested all three sets.

Describe conditions of test:
 Screening of samples
 Information to panel
 Panel area
 Sample presentation

Sample preparation and presentation — The test beers were stored at 12°C and evaluated 7–10 days after bottling. Samples were screened by two experienced tasters who found them representative of the type of beer with no differences in color, foam, or flavor other than hop character. Panel members were informed that samples were test brews with different hops, but the identity of individual samples was not disclosed. Members worked individually in booths and no discussion took place after the sessions. Sample portions of 70 mL were served at 12°C in clear 8-oz glasses. The five samples were presented simultaneously in balanced, random order. Samples were swallowed.

Statistical techniques

Statistical evaluation — Results were evaluated by split-plot analysis of variance.

Results and Discussion

Present results concisely

The average results for the five beers are shown in Table I and the corresponding statistical analysis in Table II. Sample 4 received a significantly higher rating for hop character (3.9) than the remaining samples.

TABLE 16.1 (continued)
Example of Report: Hop Character in Five Beers

Give enough data to justify conclusions

TABLE I
Average Hop Character Ratings for the Five Beer Samples

Sample	4	2	5	1	3
Mean	3.9[a]	3.0	2.9	2.1	1.4
Hops used, lb/bbl	0.36	0.38	0.34	0.32	0.35

[a] Samples not connected by a common underscore are significantly different at the 5% significance level.

Give probability levels, degrees of freedom, obtained value of test statistic

TABLE II
Split-Plot ANOVA of the Results

Source of variation	Degrees of freedom	Sum of squares	Mean squares	F
Total	299	975.64		
Replications	2	8.89		
Samples	4	221.52	55.38	41.88[a]
Error(A)	8	10.58	1.32	
Subjects	19	412.30	21.70	17.79[a]
Sample × subject	76	89.81	1.18	0.97
Error(B)	190	232.53	1.22	

Note: Error(A) is calculated as would be the Rep × Sample interaction. Error(B) is calculated by subtraction.

[a] Significant at the 1% level.

Interpret the data, following the design of the experiment

Samples 2 and 5 with nearly identical ratings of 3.0 and 2.9 had significantly less hop character than Sample 4, but significantly more than Samples 1 and 3. The statistical evaluation shows no significance for the subject-by-sample interaction ($F = 0.97$), so it may be assumed that the panelists were consistent in their ratings; the significance of the subject effect ($F = 17.79$) suggests that the panelists used different parts of the scale to express their perceptions; this is not uncommon; further, when there is no interaction, the subject-to-subject differences are of secondary interest. The primary concern, the difference among samples, was evaluated using an HSD multiple comparison procedure; $HSD_{5\%} = 0.7$ which results in the differences shown by underscoring in Table I. Variations in the amounts of hops used to obtain the BU level of 14 were small compared with the variations in perceived hop character intensity.

Conclusions

End with clear-cut conclusions

Of the five samples tested, Sample 4 ($1.60/lb) produced a significantly higher level of hop character. Sample 2 ($1.20) merits consideration for less expensive beers.

Note: This report covers the test described in Example 7.6, Chapter 7, p. 113.

REFERENCES

Cochran, W.G. and Cox, G.M., 1957. *Experimental Designs.* John Wiley & Sons, New York, p. 469.

IFT, 1981. Guidelines for the preparation and review of papers reporting sensory evaluation data. Sensory Evaluation Division, Institute of Food Technologists. *Food Technol.* 35(11), 50.

ISO, 1985. *International Standard ISO 6658:1985. Sensory Analysis — Methodology — General Guidance.* International Organization for Standardization. Available from American National Standards Institute, 11 West 42nd St., New York, NY 10036, or from ISO, 1 rue Varembé, CH 1211 Génève 20, Switzerland.

Larmond, E., 1981. Better reports of sensory evaluation. *Tech. Q. Master Brew. Assoc. Am.* 18, 7.

Prell, P.A., 1976. Preparation of reports and manuscripts which include sensory evaluation data. *Food Technol.* 30(11), 40.

Statistical Tables

CONTENTS

TABLE T1
Random Orders of the Digits 1 to 9: Arranged in Groups of Three Columns

Instructions

(1) To generate a sequence of three-digit random numbers, enter the table at any location, e.g., closing the eyes and pointing. Without inspecting the numbers, decide whether to move up or down the column entered. Record as many numbers as needed. Discard any numbers that are unsuitable (out of range, came up before, etc.). The sequence of numbers obtained in this manner is in random order.

(2) To generate a sequence of two-digit random numbers, proceed as in (1), but first decide, e.g., by coin toss, whether to use the first two or last two digits of each number taken from the table. Treat each three-digit number in the same manner, i.e., discard the same digit from each. If a two-digit number comes up more than once, retain only the first.

(3) Random number tables are impractical for problems such as: "place the numbers from 15 to 50 in random order." Instead, write each number on a card and draw the cards blindly from a bag or use a computerized random number generator such as PROC PLAN from SAS.®

```
862 245 458 396 522 498 298 665 635 665 113 917 365 332 896 314 688 468 663 712 585 351 847
223 398 183 765 138 369 163 743 593 252 581 355 542 691 537 222 746 636 478 368 949 797 295
756 954 266 174 496 133 759 488 854 187 228 824 881 549 759 169 122 919 946 293 874 289 452
544 537 522 459 984 585 946 127 711 549 445 793 734 855 121 885 595 152 237 574 611 145 784
681 829 614 547 869 742 822 554 448 813 976 688 959 714 912 646 873 397 159 155 136 463 363
199 113 941 933 375 651 414 891 129 938 862 572 698 128 363 478 214 841 314 437 792 874 926
918 481 797 621 743 827 377 916 966 426 657 246 423 277 685 533 937 223 582 946 323 626 519
335 662 875 282 617 274 635 379 287 791 334 139 117 963 448 957 451 585 821 829 267 512 638
477 776 339 818 251 916 581 232 372 374 799 461 276 486 274 791 369 774 795 681 458 938 171

653 489 538 216 446 849 914 337 993 459 325 614 771 244 429 874 557 119 122 417 882 714 769
749 824 721 967 287 556 628 843 725 731 553 253 183 653 988 431 788 426 875 838 457 927 475
522 967 259 532 618 624 396 562 134 563 932 441 834 787 231 958 232 537 439 956 531 345 352
475 172 986 859 925 932 282 924 842 642 797 565 399 896 596 282 441 784 258 684 625 662 291
894 333 612 728 869 487 741 259 476 127 286 736 257 168 847 316 969 692 786 549 949 559 526
116 218 464 191 132 218 573 786 258 296 471 372 618 935 353 747 123 863 644 161 793 196 847
381 641 393 375 354 193 165 615 587 384 119 187 965 572 112 695 615 941 361 375 376 871 633
968 755 847 643 773 765 439 478 611 978 868 898 546 319 775 169 896 275 513 222 114 233 184

742 421 226 286 522 618 471 218 397 745 461 477 478 535 957 674 132 228 442 225 444 171 151
859 878 392 311 659 772 935 447 834 117 658 161 754 654 176 883 855 195 637 751 586 948 513
964 593 137 574 288 994 582 961 746 336 983 782 611 988 833 265 969 584 564 683 197 214 326
177 636 674 897 167 157 856 524 662 598 145 926 362 777 415 931 313 317 195 137 959 536 985
228 755 915 955 946 233 647 653 425 674 719 543 549 826 669 429 576 773 756 392 632 725 879
591 214 851 669 394 349 299 192 179 261 332 294 896 299 782 397 791 659 921 569 811 683 762
636 167 789 438 413 565 118 889 253 452 577 859 125 141 241 746 444 841 313 446 225 362 248
415 982 543 743 835 826 364 776 988 923 224 615 283 462 328 512 228 466 278 874 373 499 437
383 349 468 122 771 481 723 335 511 889 896 338 937 313 594 158 687 932 889 918 768 857 694

975 973 235 811 761 226 637 382 741 767 894 371 128 972 161 911 427 164 461 991 792 256 194
257 752 667 227 813 488 598 198 979 388 921 926 715 349 644 846 879 242 695 222 633 595 526
723 395 174 453 276 732 323 866 583 826 562 817 397 556 786 358 755 996 249 676 461 614 485
448 524 951 982 455 999 451 434 695 693 788 493 951 231 259 667 318 655 374 559 577 873 747
539 881 529 664 594 555 779 629 168 442 377 685 449 128 532 232 241 418 536 733 348 162 919
661 469 312 748 942 671 284 777 354 939 116 158 583 615 977 525 193 871 833 818 154 449 333
394 647 493 599 628 317 846 255 416 174 449 269 276 883 828 193 984 529 758 164 215 938 272
882 216 786 376 187 864 912 941 837 551 233 744 634 464 313 474 536 333 927 345 889 387 658
116 138 848 135 339 143 165 513 222 215 655 532 862 797 495 789 662 787 112 487 926 721 861
```

From Cochran, W. G. and Cox, G. M., *Experimental Design,* John Wiley & Sons, New York, 1957. With permission.

TABLE T2
The Standard Normal Distribution

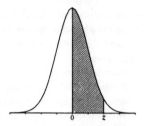

Instructions: See the Examples in Chapter 13.

z	0.00	0.01	0.02	0.03	0.04	0.05	0.06	0.07	0.08	0.09
0.0	0.0000	0.0040	0.0080	0.0120	0.0160	0.0199	0.0239	0.0279	0.0319	0.0359
0.1	0.0398	0.0438	0.0478	0.0517	0.0557	0.0596	0.0636	0.0675	0.0714	0.0753
0.2	0.0793	0.0832	0.0871	0.0910	0.0948	0.0987	0.1026	0.1064	0.1103	0.1141
0.3	0.1179	0.1217	0.1255	0.1293	0.1331	0.1368	0.1406	0.1443	0.1480	0.1517
0.4	0.1554	0.1591	0.1628	0.1664	0.1700	0.1736	0.1772	0.1808	0.1844	0.1879
0.5	0.1915	0.1950	0.1985	0.2019	0.2054	0.2088	0.2123	0.2157	0.2190	0.2224
0.6	0.2257	0.2291	0.2324	0.2357	0.2389	0.2422	0.2454	0.2486	0.2517	0.2549
0.7	0.2580	0.2611	0.2642	0.2673	0.2704	0.2734	0.2764	0.2794	0.2823	0.2852
0.8	0.2881	0.2910	0.2939	0.2967	0.2995	0.3023	0.3051	0.3078	0.3106	0.3133
0.9	0.3159	0.3186	0.3212	0.3238	0.3264	0.3289	0.3315	0.3340	0.3365	0.3389
1.0	0.3413	0.3438	0.3461	0.3485	0.3508	0.3531	0.3554	0.3577	0.3599	0.3621
1.1	0.3643	0.3665	0.3686	0.3708	0.3729	0.3749	0.3770	0.3790	0.3810	0.3830
1.2	0.3849	0.3869	0.3888	0.3907	0.3925	0.3944	0.3962	0.3980	0.3997	0.4015
1.3	0.4032	0.4049	0.4066	0.4082	0.4099	0.4115	0.4131	0.4147	0.4162	0.4177
1.4	0.4192	0.4207	0.4222	0.4236	0.4251	0.4265	0.4279	0.4292	0.4306	0.4319
1.5	0.4332	0.4345	0.4357	0.4370	0.4382	0.4394	0.4406	0.4418	0.4429	0.4441
1.6	0.4452	0.4463	0.4474	0.4484	0.4495	0.4505	0.4515	0.4525	0.4535	0.4545
1.7	0.4554	0.4564	0.4573	0.4582	0.4591	0.4599	0.4608	0.4616	0.4625	0.4633
1.8	0.4641	0.4649	0.4656	0.4664	0.4671	0.4678	0.4686	0.4693	0.4699	0.4706
1.9	0.4713	0.4719	0.4726	0.4732	0.4738	0.4744	0.4750	0.4756	0.4761	0.4767
2.0	0.4772	0.4778	0.4783	0.4788	0.4793	0.4798	0.4803	0.4808	0.4812	0.4817
2.1	0.4821	0.4826	0.4830	0.4834	0.4838	0.4842	0.4846	0.4850	0.4854	0.4857
2.2	0.4861	0.4864	0.4868	0.4871	0.4875	0.4878	0.4881	0.4884	0.4887	0.4890
2.3	0.4893	0.4896	0.4898	0.4901	0.4904	0.4906	0.4909	0.4911	0.4913	0.4916
2.4	0.4918	0.4920	0.4922	0.4925	0.4927	0.4929	0.4931	0.4932	0.4934	0.4936
2.5	0.4938	0.4940	0.4941	0.4943	0.4945	0.4946	0.4948	0.4949	0.4951	0.4952
2.6	0.4953	0.4955	0.4956	0.4957	0.4959	0.4960	0.4961	0.4962	0.4963	0.4964
2.7	0.4965	0.4966	0.4967	0.4968	0.4969	0.4970	0.4971	0.4972	0.4973	0.4974
2.8	0.4974	0.4975	0.4976	0.4977	0.4977	0.4978	0.4979	0.4979	0.4980	0.4981
2.9	0.4981	0.4982	0.4982	0.4983	0.4984	0.4984	0.4985	0.4985	0.4986	0.4986
3.0	0.4987	0.4987	0.4987	0.4988	0.4988	0.4989	0.4989	0.4989	0.4990	0.4990

TABLE T3
Upper-α Probability Points of Student's *t*-Distribution
(Entries Are $t_{\alpha:\nu}$)

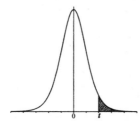

Instructions: (1) Enter the row of the table corresponding to the number of degrees of freedom (ν) for error.
(2) Pick the value of *t* in that row, from the column that corresponds to the predetermined α-level.

				α			
ν	0.25	0.10	0.05	0.025	0.01	0.005	0.0005
1	1.000	3.078	6.314	12.706	31.821	63.657	636.619
2	0.816	1.886	2.920	4.303	6.965	9.925	31.598
3	0.765	1.638	2.353	3.182	4.541	5.841	12.941
4	0.741	1.533	2.132	2.776	3.747	4.604	8.610
5	0.727	1.476	2.015	2.571	3.365	4.032	6.859
6	0.718	1.440	1.943	2.447	3.143	3.707	5.959
7	0.711	1.415	1.895	2.365	2.998	3.499	5.405
8	0.706	1.397	1.860	2.306	2.896	3.355	5.041
9	0.703	1.383	1.833	2.262	2.821	3.250	4.781
10	0.700	1.372	1.812	2.228	2.764	3.169	4.587
11	0.697	1.363	1.796	2.201	2.718	3.106	4.437
12	0.695	1.356	1.782	2.179	2.681	3.055	4.318
13	0.694	1.350	1.771	2.160	2.650	3.012	4.221
14	0.692	1.345	1.761	2.145	2.624	2.977	4.140
15	0.691	1.341	1.753	2.131	2.602	2.947	4.073
16	0.690	1.337	1.746	2.120	2.583	2.921	4.015
17	0.689	1.333	1.740	2.110	2.567	2.898	3.965
18	0.688	1.330	1.734	2.101	2.552	2.878	3.922
19	0.688	1.328	1.729	2.093	2.539	2.861	3.883
20	0.687	1.325	1.725	2.086	2.528	2.845	3.850
21	0.686	1.323	1.721	2.080	2.518	2.831	3.819
22	0.686	1.321	1.717	2.074	2.508	2.819	3.792
23	0.685	1.319	1.714	2.069	2.500	2.807	3.767
24	0.685	1.318	1.711	2.064	2.492	2.797	3.745
25	0.684	1.316	1.708	2.060	2.485	2.787	3.725
26	0.684	1.315	1.706	2.056	2.479	2.779	3.707
27	0.684	1.314	1.703	2.052	2.473	2.771	3.690
28	0.683	1.313	1.701	2.048	2.467	2.763	3.674
29	0.683	1.311	1.699	2.045	2.462	2.756	3.659
30	0.683	1.310	1.697	2.042	2.457	2.750	3.646
∞	0.674	1.282	1.645	1.960	2.326	2.576	3.291

TABLE T4
Percentage Points of the Studentized Range: Upper-α Critical Values for Tukey's HSD Multiple Comparison Procedure

Instructions:

(1) Enter the section of the table that corresponds to the predetermined α-level.
(2) Enter the row that corresponds to the degrees-of-freedom for error from the ANOVA.
(3) Pick the value of q in that row from the column that corresponds to the number of treatments being compared.

The Entries Are $q_{0.01}$ Where $p(q < q_{0.01}) = 0.99$

v	2	3	4	5	6	7	8	9	10	11	12	13	14	15	16	17	18	19	20
1	90.03	135.0	164.3	185.6	202.2	215.8	227.2	237.0	245.6	253.2	260.0	266.2	271.8	277.0	281.8	286.3	290.4	294.3	290.0
2	14.04	19.02	22.29	24.72	26.63	28.29	29.53	30.68	31.69	32.59	33.40	34.13	34.81	35.43	36.00	36.53	37.03	37.50	37.95
3	8.26	10.62	12.17	13.33	14.24	15.00	15.64	16.20	16.69	17.13	17.53	17.89	18.22	18.52	18.81	19.07	19.32	19.55	19.77
4	6.51	8.12	9.17	9.96	10.58	11.10	11.55	11.93	12.27	12.57	12.84	13.09	13.32	13.53	13.73	13.91	14.08	14.24	14.40
5	5.70	6.98	7.80	8.42	8.91	9.32	9.67	9.97	10.24	10.48	10.70	10.89	11.08	11.24	11.40	11.55	11.68	11.81	11.93
6	5.24	6.33	7.03	7.56	7.97	8.32	8.61	8.87	9.10	9.30	9.48	9.65	9.81	9.95	10.08	10.21	10.32	10.43	10.54
7	4.95	5.92	6.54	7.01	7.37	7.68	7.94	8.17	8.37	8.55	8.71	8.86	9.00	9.12	9.24	9.35	9.46	9.55	9.65
8	4.75	5.64	6.20	6.62	6.96	7.24	7.47	7.68	7.86	8.03	8.18	8.31	8.44	8.55	8.66	8.76	8.85	8.94	9.03
9	4.60	5.43	5.96	6.35	6.66	6.91	7.13	7.33	7.49	7.65	7.78	7.91	8.03	8.13	8.23	8.33	8.41	8.49	8.57
10	4.48	5.27	5.77	6.14	6.43	6.67	6.87	7.05	7.21	7.36	7.49	7.60	7.71	7.81	7.91	7.99	8.08	8.15	8.23
11	4.39	5.15	5.62	5.97	6.25	6.48	6.67	6.84	6.99	7.13	7.25	7.36	7.46	7.56	7.65	7.73	7.81	7.88	7.95
12	4.32	5.05	5.50	5.84	6.10	6.32	6.51	6.67	6.81	6.94	7.06	7.17	7.26	7.36	7.44	7.52	7.59	7.66	7.73
13	4.26	4.96	5.40	5.72	5.98	6.19	6.37	6.52	6.67	6.79	6.90	7.01	7.10	7.19	7.27	7.35	7.42	7.48	7.55
14	4.21	4.89	5.32	5.63	5.88	6.08	6.26	6.41	6.54	6.66	6.77	6.87	6.96	7.05	7.13	7.20	7.27	7.33	7.39
15	4.17	4.84	5.25	5.56	5.80	5.99	6.16	5.31	6.44	6.55	6.66	6.76	6.84	6.93	7.00	7.07	7.14	7.20	7.26
16	4.13	4.79	5.19	5.49	5.72	5.92	6.08	6.22	6.35	6.46	6.56	6.66	6.74	6.82	6.90	6.97	7.03	7.09	7.15
17	4.10	4.74	5.14	5.43	5.66	5.85	6.01	6.15	6.27	6.38	6.48	6.57	6.66	6.73	6.81	6.87	6.94	7.00	7.05
18	4.07	4.70	5.09	5.38	5.60	5.79	5.94	6.08	6.20	6.31	6.41	6.50	6.58	6.65	6.73	6.79	6.85	6.91	6.97
19	4.05	4.67	5.05	5.33	5.55	5.73	5.89	6.02	6.14	6.25	6.34	6.43	6.51	6.58	6.65	6.72	6.78	6.84	6.89
20	4.02	4.64	5.02	5.29	5.51	5.69	5.84	5.97	6.09	6.19	6.28	6.37	6.45	6.52	6.59	6.65	6.71	6.77	6.82

df	2	3	4	5	6	7	8	9	10	11	12	13	14	15	16	17	18	19	20
24	3.96	4.55	4.91	5.17	5.37	5.54	5.69	5.81	5.92	6.02	6.11	6.19	6.26	6.33	6.39	6.45	6.51	6.56	6.61
30	3.89	4.45	4.80	5.05	5.24	5.40	5.54	5.65	5.76	5.85	5.93	6.01	6.08	6.14	6.20	6.26	6.31	6.36	6.41
40	3.82	4.37	4.70	4.93	5.11	5.26	5.39	5.50	5.60	5.69	5.76	5.83	5.90	5.96	6.02	6.07	6.12	6.16	6.21
60	3.76	4.28	4.59	4.82	4.99	5.13	5.25	5.36	5.45	5.53	5.60	5.67	5.73	5.78	5.84	5.89	5.93	5.97	6.01
120	3.70	4.20	4.50	4.71	4.87	5.01	5.12	5.21	5.30	5.37	5.44	5.50	5.56	5.61	5.66	5.71	5.75	5.79	5.83
∞	3.64	4.12	4.40	4.60	4.76	4.88	4.99	5.08	5.16	5.23	5.29	5.35	5.40	5.45	5.49	5.54	5.57	5.61	5.65

The Entries Are $q_{0.05}$ Where $p(q < q_{0.05}) = 0.95$

df	2	3	4	5	6	7	8	9	10	11	12	13	14	15	16	17	18	19	20
1	17.97	26.98	32.82	37.08	40.41	43.12	45.40	47.36	49.07	50.59	51.96	53.20	54.33	55.36	56.32	57.22	58.04	58.83	59.56
2	6.08	8.33	9.80	10.88	11.74	12.44	13.03	13.54	13.99	14.39	14.75	15.08	15.38	15.65	15.91	16.14	16.37	16.57	16.77
3	4.50	5.91	6.82	7.50	8.04	8.48	8.85	9.18	9.46	9.72	9.95	10.15	10.35	10.53	10.69	10.84	10.98	11.11	11.24
4	3.93	5.04	5.76	6.29	6.71	7.05	7.35	7.60	7.83	8.03	8.21	8.37	8.52	8.66	8.79	8.91	9.03	9.13	9.23
5	3.64	4.60	5.22	5.67	6.03	6.33	6.58	6.80	6.99	7.17	7.32	7.47	7.60	7.72	7.83	7.93	8.03	8.12	8.21
6	3.46	4.34	4.90	5.30	5.63	5.90	6.12	6.32	6.49	6.65	6.79	6.92	7.03	7.14	7.24	7.34	7.43	7.51	7.59
7	3.34	4.16	4.68	5.06	5.36	5.61	5.82	6.00	6.16	6.30	6.43	6.55	6.66	6.76	6.85	6.94	7.02	7.10	7.17
8	3.26	4.04	4.53	4.89	5.17	5.40	5.60	5.77	5.92	6.05	6.18	6.29	6.39	6.48	6.57	6.65	6.73	6.80	6.87
9	3.20	3.95	4.41	4.76	5.02	5.24	5.43	5.59	5.74	5.87	5.98	6.09	6.19	6.28	6.36	6.44	6.51	6.58	6.64
10	3.15	3.88	4.33	4.65	4.91	5.12	5.30	5.46	5.60	5.72	5.83	5.93	6.03	6.11	6.19	6.27	6.34	6.40	6.47
11	3.11	3.82	4.26	4.57	4.82	5.03	5.20	5.35	5.49	5.61	5.71	5.81	5.90	5.98	6.06	6.13	6.20	6.27	6.33
12	3.08	3.77	4.20	4.51	4.75	4.95	5.12	5.27	5.39	5.51	5.61	5.71	5.80	5.88	5.95	6.02	6.09	6.15	6.21
13	3.06	3.73	4.15	4.45	4.69	4.88	5.05	5.19	5.32	5.43	5.53	5.63	5.71	5.79	5.86	5.93	5.99	6.05	6.11
14	3.03	3.70	4.11	4.41	4.64	4.83	4.99	5.13	5.25	5.36	5.46	5.55	5.64	5.71	5.79	5.85	5.91	5.97	6.03
15	3.01	3.67	4.08	4.37	4.59	4.78	4.94	5.08	5.20	5.31	5.40	5.49	5.57	5.65	5.72	5.78	5.85	5.90	5.96
16	3.00	3.65	4.05	4.33	4.56	4.74	4.90	5.03	5.15	5.26	5.35	5.44	5.52	5.59	5.66	5.73	5.79	5.84	5.90
17	2.98	3.63	4.02	4.30	4.52	4.70	4.86	4.99	5.11	5.21	5.31	5.39	5.47	5.54	5.61	5.67	5.73	5.79	5.84
18	2.97	3.61	4.00	4.28	4.49	4.67	4.82	4.96	5.07	5.17	5.27	5.35	5.43	5.50	5.57	5.63	5.69	5.74	5.79
19	2.96	3.59	3.98	4.25	4.47	4.65	4.79	4.92	5.04	5.14	5.23	5.31	5.39	5.46	5.53	5.59	5.65	5.70	5.75
20	2.95	3.58	3.96	4.23	4.45	4.62	4.77	4.90	5.01	5.11	5.20	5.28	5.36	5.43	5.49	5.55	5.61	5.66	5.71

TABLE T4 (continued)
Percentage Points of the Studentized Range: Upper-α Critical Values for Tukey's HSD Multiple Comparison Procedure

v	2	3	4	5	6	7	8	9	10	11	12	13	14	15	16	17	18	19	20
24	2.92	3.53	3.90	4.17	4.37	4.54	4.68	4.81	4.92	5.01	5.10	5.18	5.25	5.32	5.38	5.44	5.49	5.55	5.59
30	2.89	3.49	3.85	4.10	4.30	4.46	4.60	4.72	4.82	4.92	5.00	5.08	5.15	5.21	5.27	5.33	5.38	5.43	5.47
40	2.86	3.44	3.79	4.04	4.23	4.39	4.52	4.63	4.73	4.82	4.90	4.98	5.04	4.11	5.16	5.22	5.27	5.31	5.36
60	2.83	3.40	3.74	3.98	4.16	4.31	4.44	4.55	4.65	4.73	4.81	4.88	4.94	5.00	5.06	5.11	5.15	5.20	5.24
120	2.80	3.36	3.68	3.92	4.10	4.24	4.36	4.47	4.56	4.64	4.71	4.78	4.84	4.90	4.95	5.00	5.04	5.09	5.13
∞	2.77	3.31	3.63	3.86	4.03	4.17	4.29	4.39	4.47	4.55	4.62	4.68	4.74	4.80	4.85	4.89	4.93	4.97	5.01

The Entries Are $q_{0.10}$ Where $p(q < q_{0.10}) = 0.90$

v	2	3	4	5	6	7	8	9	10	11	12	13	14	15	16	17	18	19	20
1	8.93	13.44	16.36	18.49	20.15	21.51	22.64	23.62	24.48	25.24	25.92	26.54	27.10	27.62	28.10	28.54	28.96	29.35	29.71
2	4.13	5.73	6.77	7.54	8.14	8.63	9.05	9.41	9.72	10.01	10.26	10.49	10.70	10.89	11.07	11.24	11.39	11.54	11.68
3	3.33	4.47	5.20	5.74	6.16	6.51	6.81	7.06	7.29	7.49	7.67	7.83	7.98	8.12	8.25	8.27	8.48	8.58	8.68
4	3.01	3.98	4.59	5.03	5.39	5.68	5.93	6.14	6.33	6.49	6.65	6.78	6.91	7.02	7.13	7.23	7.33	7.41	7.50
5	2.85	3.72	4.26	4.66	4.98	5.24	5.46	5.65	5.82	5.97	6.10	6.22	6.34	6.44	6.54	6.63	6.71	6.79	6.86
6	2.75	3.56	4.07	4.44	4.73	4.97	5.17	5.34	5.50	5.64	5.76	5.87	5.98	6.07	6.16	6.25	6.32	6.40	6.47
7	2.68	3.45	3.93	4.28	4.55	4.78	4.97	5.14	5.28	5.41	5.53	5.64	5.74	5.83	5.91	5.99	6.06	6.13	6.19
8	2.63	3.37	3.83	4.17	4.43	4.65	4.83	4.99	5.13	5.25	5.36	5.46	5.56	5.64	5.72	5.80	5.87	5.93	6.00
9	2.59	3.32	3.76	4.08	4.34	4.54	4.72	4.87	5.01	5.13	5.23	5.33	5.42	5.51	5.58	5.66	5.72	5.79	5.85
10	2.56	3.27	3.70	4.02	4.26	4.47	4.64	4.78	4.91	5.03	5.13	5.23	5.32	5.40	5.47	5.54	5.61	5.67	5.73
11	2.54	3.23	3.66	3.96	4.20	4.40	4.57	4.71	4.84	4.95	5.05	5.15	5.23	5.31	5.38	5.45	5.51	5.57	5.63
12	2.52	3.20	3.62	3.92	4.16	4.35	4.51	4.65	4.78	4.89	4.99	5.08	5.16	5.24	5.31	5.37	5.44	5.49	5.55
13	2.50	3.18	3.59	3.88	4.12	4.20	4.46	4.60	4.72	4.83	4.93	5.02	5.10	5.28	5.25	5.31	5.37	5.43	5.48
14	2.49	3.16	3.56	3.85	4.08	4.27	4.42	4.56	4.68	4.79	4.88	4.97	5.05	5.12	5.19	5.26	5.32	5.37	5.43
15	2.48	3.14	3.54	3.83	4.05	4.23	4.39	4.52	4.64	4.75	4.84	4.93	5.01	5.08	5.15	5.21	5.27	5.32	5.38

16	2.47	3.12	3.52	3.80	4.03	4.21	4.36	4.49	4.61	4.71	4.81	4.89	4.97	5.04	5.11	5.17	5.23	5.28	5.33
17	2.46	3.11	3.50	3.78	4.00	4.18	4.33	4.46	4.58	4.68	4.77	4.86	4.93	5.01	5.07	5.13	5.19	5.24	5.30
18	2.45	3.10	3.49	3.77	3.98	4.16	4.31	4.44	4.55	4.65	4.75	4.83	4.90	4.98	5.04	5.10	5.16	5.21	5.26
19	2.45	3.09	3.47	3.75	3.97	4.14	4.29	4.42	4.53	4.63	4.72	4.80	4.88	4.95	5.01	5.07	5.13	5.18	5.23
20	2.44	3.08	3.46	3.74	3.95	4.12	4.27	4.40	4.51	4.61	4.70	4.78	4.85	4.92	4.99	5.05	5.10	5.16	5.20
24	2.42	3.05	3.42	3.69	3.90	4.07	4.21	4.34	4.44	4.54	4.63	4.71	4.78	4.85	4.91	4.97	5.02	5.07	5.12
30	2.40	3.02	3.39	3.65	3.85	4.02	4.16	4.28	4.38	4.47	4.56	4.64	4.71	4.77	4.83	4.89	4.94	4.99	5.03
40	2.38	2.99	3.35	3.60	3.80	3.96	4.10	4.21	4.32	4.41	4.49	4.56	4.63	4.69	4.75	4.81	4.86	4.90	4.95
60	2.36	2.96	3.31	3.56	3.75	3.91	4.04	4.16	4.25	4.34	4.42	4.49	4.56	4.62	4.67	4.73	4.78	4.82	4.86
120	2.34	2.93	3.28	3.52	3.71	3.86	3.99	4.10	4.19	4.28	4.35	4.42	4.48	4.54	4.60	4.65	4.69	4.74	4.78
∞	2.33	2.90	3.24	3.48	3.66	3.81	3.93	4.04	4.13	4.21	4.28	4.35	4.41	4.47	4.52	4.57	4.61	4.65	4.69

TABLE T5
Upper-α Probability Points of χ^2-Distribution (Entries Are $\chi^2_{\alpha:\nu}$)

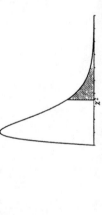

Instructions: (1) Enter the row of the table corresponding to the number of degrees of freedom (ν) for χ^2.
(2) Pick the value of χ^2 in that row from the column that corresponds to the predetermined α-level.

ν	0.995	0.990	0.975	0.950	0.900	0.750	0.500	0.250	0.100	0.050	0.025	0.010	0.005
1	0.0000393	0.000157	0.000982	0.00393	0.0158	0.102	0.455	1.32	2.71	3.84	5.02	6.63	7.88
2	0.0100	0.0201	0.0506	0.103	0.211	0.575	1.39	2.77	4.61	5.99	7.38	9.21	10.6
3	0.0717	0.115	0.216	0.352	0.584	1.21	2.37	4.11	6.25	7.81	9.35	11.3	12.8
4	0.207	0.297	0.484	0.711	1.06	1.92	3.36	5.39	7.78	9.49	11.1	13.3	14.9
5	0.412	0.554	0.831	1.15	1.61	2.67	4.35	6.63	9.24	11.1	12.8	15.1	16.7
6	0.676	.872	1.24	1.64	2.20	3.45	5.35	7.84	10.6	12.6	14.4	16.8	18.5
7	0.989	1.24	1.69	2.17	2.83	4.25	6.35	9.04	12.0	14.1	16.0	18.5	20.3
8	1.34	1.65	2.18	2.73	3.49	5.07	7.34	10.2	13.4	15.5	17.5	20.1	22.0
9	1.73	2.09	2.70	3.33	4.17	5.90	8.34	11.4	14.7	16.9	19.0	21.7	23.6
10	2.16	2.56	3.25	3.94	4.87	6.74	9.34	12.5	16.0	18.3	20.5	23.2	25.2

df													
11	2.60	3.05	3.82	4.57	5.58	7.58	10.3	13.7	17.3	19.7	21.9	24.7	26.8
12	3.07	3.57	4.40	5.23	6.30	8.44	11.3	14.8	18.5	21.0	23.3	26.2	28.3
13	3.57	4.11	5.01	5.89	7.04	9.30	12.3	16.0	19.8	22.4	24.7	27.7	29.8
14	4.07	4.66	5.63	6.57	7.79	10.2	13.3	17.1	21.1	23.7	26.1	29.1	31.3
15	4.60	5.23	6.26	7.26	8.55	11.0	14.3	18.2	22.3	25.0	27.5	30.6	32.8
16	5.14	5.81	6.91	7.96	9.31	11.9	15.3	19.4	23.5	26.3	28.8	32.0	34.3
17	5.70	6.41	7.56	8.67	10.1	12.8	16.3	20.5	24.8	27.6	30.2	33.4	35.7
18	6.26	7.01	8.23	9.39	10.9	13.7	17.3	21.6	26.0	28.9	31.5	34.8	37.2
19	6.84	7.63	8.91	10.1	11.7	14.6	18.3	22.7	27.2	30.1	32.9	36.2	38.6
20	7.43	8.26	9.59	10.9	12.4	15.5	19.3	23.8	28.4	31.4	34.2	37.6	40.0
21	8.03	8.90	10.3	11.6	13.2	16.3	20.3	24.9	29.6	32.7	35.5	38.9	41.4
22	8.64	9.54	11.0	12.3	14.0	17.2	21.3	26.0	30.8	33.9	36.8	40.3	42.8
23	9.26	10.2	11.7	13.1	14.8	18.1	22.3	27.1	32.0	35.2	38.1	41.6	44.2
24	9.89	10.9	12.4	13.8	15.7	19.0	23.3	28.2	33.2	36.4	39.4	43.0	45.6
25	10.5	11.5	13.1	14.6	16.5	19.9	24.3	29.3	34.4	37.7	40.6	44.3	46.9
26	11.2	12.2	13.8	15.4	17.3	20.8	25.3	30.4	35.6	38.9	41.9	45.6	48.3
27	11.8	12.9	14.6	16.2	18.1	21.7	26.3	31.5	36.7	40.1	43.2	47.0	49.6
28	12.5	13.6	15.3	16.9	18.9	22.7	27.3	32.6	37.9	41.3	44.5	48.3	51.0
29	13.1	14.3	16.0	17.7	19.8	23.6	28.3	33.7	39.1	42.6	45.7	49.6	52.3
30	13.8	15.0	16.8	18.5	20.6	24.5	29.3	34.8	40.3	43.8	47.0	50.9	53.7

TABLE T6
Upper-α Probability Points of F-Distribution (Entries Are $F_{\alpha:v_1,v_2}$)

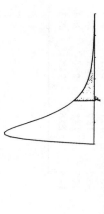

Instructions: (1) Enter the section of the table corresponding to the predetermined α-level.
(2) Enter the row that corresponds to the denominator degrees of freedom (v_2)
(3) Pick the value of F in that row from the column that corresponds to the numerator degrees of freedom (v_1).

v_1

α = 0.10

v_2	1	2	3	4	5	6	7	8	9	10	12	15	20	24	30	40	60	120	∞
1	39.86	49.50	53.59	55.83	57.24	58.20	58.91	59.44	59.86	60.19	60.71	61.22	61.74	62.00	62.26	62.53	62.79	63.06	63.33
2	8.53	9.00	9.16	9.24	9.29	9.33	9.35	9.37	9.38	9.39	9.41	9.42	9.44	9.45	9.46	9.47	9.47	9.48	9.49
3	5.54	5.46	5.39	5.34	5.31	5.28	5.27	5.25	5.24	5.23	5.22	5.20	5.18	5.18	5.17	5.16	5.15	5.14	5.13
4	4.54	4.32	4.19	4.11	4.05	4.01	3.98	3.95	3.94	3.92	3.90	3.87	3.84	3.83	3.82	3.80	3.79	3.78	3.76
5	4.06	3.78	3.62	3.52	3.45	3.40	3.37	3.34	3.32	3.30	3.27	3.24	3.21	3.19	3.17	3.16	3.14	3.12	3.10
6	3.78	3.46	3.29	3.18	3.11	3.05	3.01	2.98	2.96	2.94	2.90	2.87	2.84	2.82	2.80	2.78	2.76	2.74	2.72
7	3.59	3.26	3.07	2.96	2.88	2.83	2.78	2.75	2.72	2.70	2.67	2.63	2.59	2.58	2.56	2.54	2.51	2.49	2.47
8	3.46	3.11	2.92	2.81	2.73	2.67	2.62	2.59	2.56	2.54	2.50	2.46	2.42	2.40	2.38	2.36	2.34	2.32	2.29
9	3.36	3.01	2.81	2.69	2.61	2.55	2.51	2.47	2.44	2.42	2.38	2.34	2.30	2.28	2.25	2.23	2.21	2.18	2.16
10	3.29	2.92	2.73	2.61	2.52	2.46	2.41	2.38	2.35	2.32	2.28	2.24	2.20	2.18	2.16	2.13	2.11	2.08	2.06

11	3.23	2.86	2.66	2.54	2.45	2.39	2.34	2.30	2.27	2.25	2.21	2.17	2.12	2.10	2.08	2.05	2.03	2.00	1.97
12	3.18	2.81	2.61	2.48	2.39	2.33	2.28	2.24	2.21	2.19	2.15	2.10	2.06	2.04	2.01	1.99	1.96	1.93	1.90
13	3.14	2.76	2.56	2.43	2.35	2.28	2.23	2.20	2.16	2.14	2.10	2.05	2.01	1.98	1.96	1.93	1.90	1.88	1.85
14	3.10	2.73	2.52	2.39	2.31	2.24	2.19	2.15	2.12	2.10	2.05	2.01	1.96	1.94	1.91	1.89	1.86	1.83	1.80
15	3.07	2.70	2.49	2.36	2.27	2.21	2.16	2.12	2.09	2.06	2.02	1.97	1.92	1.90	1.87	1.85	1.82	1.79	1.76
16	3.05	2.67	2.46	2.33	2.24	2.18	2.13	2.09	2.06	2.03	1.99	1.94	1.89	1.87	1.84	1.81	1.78	1.75	1.72
17	3.03	2.64	2.44	2.31	2.22	2.15	2.10	2.06	2.03	2.00	1.96	1.91	1.86	1.84	1.81	1.78	1.75	1.72	1.69
18	3.01	2.62	2.42	2.29	2.20	2.13	2.08	2.04	2.00	1.98	1.93	1.89	1.84	1.81	1.78	1.75	1.72	1.69	1.66
19	2.99	2.61	2.40	2.27	2.18	2.11	2.06	2.02	1.98	1.96	1.91	1.86	1.81	1.79	1.76	1.73	1.70	1.67	1.63
20	2.97	2.59	2.38	2.25	2.16	2.09	2.04	2.00	1.96	1.94	1.89	1.84	1.79	1.77	1.74	1.71	1.68	1.64	1.61
21	2.96	2.57	2.36	2.23	2.14	2.08	2.02	1.98	1.95	1.92	1.87	1.83	1.78	1.75	1.72	1.69	1.66	1.62	1.59
22	2.95	2.56	2.35	2.22	2.13	2.06	2.01	1.97	1.93	1.90	1.86	1.81	1.76	1.73	1.70	1.67	1.64	1.60	1.57
23	2.94	2.55	2.34	2.21	2.11	2.05	1.99	1.95	1.92	1.89	1.84	1.80	1.74	1.72	1.69	1.66	1.62	1.59	1.55
24	2.93	2.54	2.33	2.19	2.10	2.04	1.98	1.94	1.91	1.88	1.83	1.78	1.73	1.70	1.67	1.64	1.61	1.57	1.53
25	2.92	2.53	2.32	2.18	2.09	2.02	1.97	1.93	1.89	1.87	1.82	1.77	1.72	1.69	1.66	1.63	1.59	1.56	1.52
26	2.91	2.52	2.31	2.17	2.08	2.01	1.96	1.92	1.88	1.86	1.81	1.76	1.71	1.68	1.65	1.61	1.58	1.54	1.50
27	2.90	2.51	2.30	2.17	2.07	2.00	1.95	1.91	1.87	1.85	1.80	1.75	1.70	1.67	1.64	1.60	1.57	1.53	1.49
28	2.89	2.50	2.29	2.16	2.06	2.00	1.94	1.90	1.87	1.84	1.79	1.74	1.69	1.66	1.63	1.59	1.56	1.52	1.48
29	2.89	2.50	2.28	2.15	2.06	1.99	1.93	1.89	1.86	1.83	1.78	1.73	1.68	1.65	1.62	1.58	1.55	1.51	1.47
30	2.88	2.49	2.28	2.14	2.05	1.98	1.93	1.88	1.85	1.82	1.77	1.72	1.67	1.64	1.61	1.57	1.54	1.50	1.46
40	2.84	2.44	2.23	2.09	2.00	1.93	1.87	1.83	1.79	1.76	1.71	1.66	1.61	1.57	1.54	1.51	1.47	1.42	1.38
60	2.79	2.39	2.18	2.04	1.95	1.87	1.82	1.77	1.74	1.71	1.66	1.60	1.54	1.51	1.48	1.44	1.40	1.35	1.29
120	2.75	2.35	2.13	1.99	1.90	1.82	1.77	1.72	1.68	1.65	1.60	1.55	1.48	1.45	1.41	1.37	1.32	1.26	1.19
∞	2.71	2.30	2.08	1.94	1.85	1.77	1.72	1.67	1.63	1.60	1.55	1.49	1.42	1.38	1.34	1.30	1.24	1.17	1.00

TABLE T6 (continued)
Upper-α Probability Points of F-Distribution (Entries Are $F_{\alpha:\nu_1,\nu_2}$)

$\alpha = 0.05$

ν_2	ν_1 1	2	3	4	5	6	7	8	9	10	12	15	20	24	30	40	60	120	∞
1	161.4	199.5	215.7	224.6	230.2	234.0	236.8	238.9	240.5	241.9	243.9	245.9	248.0	249.1	250.1	251.1	252.2	253.3	254.3
2	18.51	19.00	19.16	19.25	19.30	19.33	19.35	19.37	19.38	19.40	19.41	19.43	19.45	19.45	19.46	19.47	19.48	19.49	19.50
3	10.13	9.55	9.28	9.12	9.01	8.94	8.89	8.85	8.81	8.79	8.74	8.70	8.66	8.64	8.62	8.59	8.57	8.55	8.53
4	7.71	6.94	6.59	6.39	6.26	6.16	6.09	6.04	6.00	5.96	5.91	5.86	5.80	5.77	5.75	5.72	5.69	5.66	5.63
5	6.61	5.79	5.41	5.19	5.05	4.95	4.88	4.82	4.77	4.74	4.68	4.62	4.56	4.53	4.50	4.46	4.43	4.40	4.36
6	5.99	5.14	4.76	4.53	4.39	4.28	4.21	4.15	4.10	4.06	4.00	3.94	3.87	3.84	3.81	3.77	3.74	3.70	3.67
7	5.59	4.74	4.35	4.12	3.97	3.87	3.79	3.73	3.68	3.64	3.57	3.51	3.44	3.41	3.38	3.34	3.30	3.27	3.23
8	5.32	4.46	4.07	3.84	3.69	3.58	3.50	3.44	3.39	3.35	3.28	3.22	3.15	3.12	3.08	3.04	3.01	2.97	2.93
9	5.12	4.26	3.86	3.63	3.48	3.37	3.29	3.23	3.18	3.14	3.07	3.01	2.94	2.90	2.86	2.83	2.79	2.75	2.71
10	4.96	4.10	3.71	3.48	3.33	3.22	3.14	3.07	3.02	2.98	2.91	2.85	2.77	2.74	2.70	2.66	2.62	2.58	2.54
11	4.84	3.98	3.59	3.36	3.20	3.09	3.01	2.95	2.90	2.85	2.79	2.72	2.65	2.61	2.57	2.53	2.49	2.45	2.40
12	4.75	3.89	3.49	3.26	3.11	3.00	2.91	2.85	2.80	2.75	2.69	2.62	2.54	2.51	2.47	2.43	2.38	2.34	2.30
13	4.67	3.81	3.41	3.18	3.03	2.92	2.83	2.77	2.71	2.67	2.60	2.53	2.46	2.42	2.38	2.34	2.30	2.25	2.21
14	4.60	3.74	3.34	3.11	2.96	2.85	2.76	2.70	2.65	2.60	2.53	2.46	2.39	2.35	2.31	2.27	2.22	2.18	2.13
15	4.54	3.68	3.29	3.06	2.90	2.79	2.71	2.64	2.59	2.54	2.48	2.40	2.33	2.29	2.25	2.20	2.16	2.11	2.07

16	4.49	3.63	3.24	3.01	2.85	2.74	2.66	2.59	2.54	2.49	2.42	2.35	2.28	2.24	2.19	2.15	2.11	2.06	2.01
17	4.45	3.59	3.20	2.96	2.81	2.70	2.61	2.55	2.49	2.45	2.38	2.31	2.23	2.19	2.15	2.10	2.06	2.01	1.96
18	4.41	3.55	3.16	2.93	2.77	2.66	2.58	2.51	2.46	2.41	2.34	2.27	2.19	2.15	2.11	2.06	2.02	1.97	1.92
19	4.38	3.52	3.13	2.90	2.74	2.63	2.54	2.48	2.42	2.38	2.31	2.23	2.16	2.11	2.07	2.03	1.98	1.93	1.88
20	4.35	3.49	3.10	2.87	2.71	2.60	2.51	2.45	2.39	2.35	2.28	2.20	2.12	2.08	2.04	1.99	1.95	1.90	1.84
21	4.32	3.47	3.07	2.84	2.68	2.57	2.49	2.42	2.37	2.32	2.25	2.18	2.10	2.05	2.01	1.96	1.92	1.87	1.81
22	4.30	3.44	3.05	2.82	2.66	2.55	2.46	2.40	2.34	2.30	2.23	2.15	2.07	2.03	1.98	1.94	1.89	1.84	1.78
23	4.28	3.42	3.03	2.80	2.64	2.53	2.44	2.37	2.32	2.27	2.20	2.13	2.05	2.01	1.96	1.91	1.86	1.81	1.76
24	4.26	3.40	3.01	2.78	2.62	2.51	2.42	2.36	2.30	2.25	2.18	2.11	2.03	1.98	1.94	1.89	1.84	1.79	1.73
25	4.24	3.39	2.99	2.76	2.60	2.49	2.40	2.34	2.28	2.24	2.16	2.09	2.01	1.96	1.92	1.87	1.82	1.77	1.71
26	4.23	3.37	2.98	2.74	2.59	2.47	2.39	2.32	2.27	2.22	2.15	2.07	1.99	1.95	1.90	1.85	1.80	1.75	1.69
27	4.21	3.35	2.96	2.73	2.57	2.46	2.37	2.31	2.25	2.20	2.13	2.06	1.97	1.93	1.88	1.84	1.79	1.73	1.67
28	4.20	3.34	2.95	2.71	2.56	2.45	2.36	2.29	2.24	2.19	2.12	2.04	1.96	1.91	1.87	1.82	1.77	1.71	1.65
29	4.18	3.33	2.93	2.70	2.55	2.43	2.35	2.28	2.22	2.18	2.10	2.03	1.94	1.90	1.85	1.81	1.75	1.70	1.64
30	4.17	3.32	2.92	2.69	2.53	2.42	2.33	2.27	2.21	2.16	2.09	2.01	1.93	1.89	1.84	1.79	1.74	1.68	1.62
40	4.08	3.23	2.84	2.61	2.45	2.34	2.25	2.18	2.12	2.08	2.00	1.92	1.84	1.79	1.74	1.69	1.64	1.58	1.51
60	4.00	3.15	2.76	2.53	2.37	2.25	2.17	2.10	2.04	1.99	1.92	1.84	1.75	1.70	1.65	1.59	1.53	1.47	1.39
120	3.92	3.07	2.68	2.45	2.29	2.17	2.09	2.02	1.96	1.91	1.83	1.75	1.66	1.61	1.55	1.50	1.43	1.35	1.25
∞	3.84	3.00	2.60	2.37	2.21	2.10	2.01	1.94	1.88	1.83	1.75	1.67	1.57	1.52	1.46	1.39	1.32	1.22	1.00

TABLE T6 (continued)
Upper-α Probability Points of F-Distribution (Entries Are $F_{\alpha:v_1,v_2}$)

$\alpha = 0.01$

v_2	v_1																		
	1	2	3	4	5	6	7	8	9	10	12	15	20	24	30	40	60	120	∞
1	4052	4999.5	5403	5625	5764	5859	5928	5982	6022	6056	6106	6157	6209	6235	6261	6287	6313	6339	6366
2	98.50	99.00	99.17	99.25	99.30	99.33	99.36	99.37	99.39	99.40	99.42	99.43	99.45	99.46	99.47	99.47	99.48	99.49	99.50
3	34.12	30.82	29.46	28.71	28.24	27.91	27.67	27.49	27.35	27.23	27.05	26.87	26.69	26.60	26.50	26.41	26.32	26.22	26.13
4	21.20	18.00	16.69	15.98	15.52	15.21	14.98	14.80	14.66	14.55	14.37	14.20	14.02	13.93	13.84	13.75	13.65	13.56	13.46
5	16.26	13.27	12.06	11.39	10.97	10.67	10.46	10.29	10.16	10.05	9.89	9.72	9.55	9.47	9.38	9.29	9.20	9.11	9.02
6	13.75	10.92	9.78	9.15	8.75	8.47	8.26	8.10	7.98	7.87	7.72	7.56	7.40	7.31	7.23	7.14	7.06	6.97	6.88
7	12.25	9.55	8.45	7.85	7.46	7.19	6.99	6.84	6.72	6.62	6.47	6.31	6.16	6.07	5.99	5.91	5.82	5.74	5.65
8	11.26	8.65	7.59	7.01	6.63	6.37	6.18	6.03	5.91	5.81	5.67	5.52	5.36	5.28	5.20	5.12	5.03	4.95	4.86
9	10.56	8.02	6.99	6.42	6.06	5.80	5.61	5.47	5.35	5.26	5.11	4.96	4.81	4.73	4.65	4.57	4.48	4.40	4.31
10	10.04	7.56	6.55	5.99	5.64	5.39	5.20	5.06	4.94	4.85	4.71	4.56	4.41	4.33	4.25	4.17	4.08	4.00	3.91
11	9.65	7.21	6.22	5.67	5.32	5.07	4.89	4.74	4.63	4.54	4.40	4.25	4.10	4.02	3.94	3.86	3.78	3.69	3.60
12	9.33	6.93	5.95	5.41	5.06	4.82	4.64	4.50	4.39	4.30	4.16	4.01	3.86	3.78	3.70	3.62	3.54	3.45	3.36
13	9.07	6.70	5.74	5.21	4.86	4.62	4.44	4.30	4.19	4.10	3.96	3.82	3.66	3.59	3.51	3.43	3.34	3.25	3.17
14	8.86	6.51	5.56	5.04	4.69	4.46	4.28	4.14	4.03	3.94	3.80	3.66	3.51	3.43	3.35	3.27	3.18	3.09	3.00
15	8.68	6.36	5.42	4.89	4.56	4.32	4.14	4.00	3.89	3.80	3.67	3.52	3.37	3.29	3.21	3.13	3.05	2.96	2.87

Table T-6. *F*-Distribution

df																			
16	8.53	6.23	5.29	4.77	4.44	4.20	4.03	3.89	3.78	3.69	3.55	3.41	3.26	3.18	3.10	3.02	2.93	2.84	2.75
17	8.40	6.11	5.18	4.67	4.34	4.10	3.93	3.79	3.68	3.59	3.46	3.31	3.16	3.08	3.00	2.92	2.83	2.75	2.65
18	8.29	6.01	5.09	4.58	4.25	4.01	3.84	3.71	3.60	3.51	3.37	3.23	3.08	3.00	2.92	2.84	2.75	2.66	2.57
19	8.18	5.93	5.01	4.50	4.17	3.94	3.77	3.63	3.52	3.43	3.30	3.15	3.00	2.92	2.84	2.76	2.67	2.58	2.49
20	8.10	5.85	4.94	4.43	4.10	3.87	3.70	3.56	3.46	3.37	3.23	3.09	2.94	2.86	2.78	2.69	2.61	2.52	2.42
21	8.02	5.78	4.87	4.37	4.04	3.81	3.64	3.51	3.40	3.31	3.17	3.03	2.88	2.80	2.72	2.64	2.55	2.46	2.36
22	7.95	5.72	4.82	4.31	3.99	3.76	3.59	3.45	3.35	3.26	3.12	2.98	2.83	2.75	2.67	2.58	2.50	2.40	2.31
23	7.88	5.66	4.76	4.26	3.94	3.71	3.54	3.41	3.30	3.21	3.07	2.93	2.78	2.70	2.62	2.54	2.45	2.35	2.26
24	7.82	5.61	4.72	4.22	3.90	3.67	3.50	3.36	3.26	3.17	3.03	2.89	2.74	2.66	2.58	2.49	2.40	2.31	2.21
25	7.77	5.57	4.68	4.18	3.85	3.63	3.46	3.32	3.22	3.13	2.99	2.85	2.70	2.62	2.54	2.45	2.36	2.27	2.17
26	7.72	5.53	4.64	4.14	3.82	3.59	3.42	3.29	3.18	3.09	2.96	2.81	2.66	2.58	2.50	2.42	2.33	2.23	2.13
27	7.68	5.49	4.60	4.11	3.78	3.56	3.39	3.26	3.15	3.06	2.93	2.78	2.63	2.55	2.47	2.38	2.29	2.20	2.10
28	7.64	5.45	4.57	4.07	3.75	3.53	3.36	3.23	3.12	3.03	2.90	2.75	2.60	2.52	2.44	2.35	2.26	2.17	2.06
29	7.60	5.42	4.54	4.04	3.73	3.50	3.33	3.20	3.09	3.00	2.87	2.73	2.57	2.49	2.41	2.33	2.23	2.14	2.03
30	7.56	5.39	4.51	4.02	3.70	3.47	3.30	3.17	3.07	2.98	2.84	2.70	2.55	2.47	2.39	2.30	2.21	2.11	2.01
40	7.31	5.18	4.31	3.83	3.51	3.29	3.12	2.99	2.89	2.80	2.66	2.52	2.37	2.29	2.20	2.11	2.02	1.92	1.80
60	7.08	4.98	4.13	3.65	3.34	3.12	2.95	2.82	2.72	2.63	2.50	2.35	2.20	2.12	2.03	1.94	1.84	1.73	1.60
120	6.85	4.79	3.95	3.48	3.17	2.96	2.79	2.66	2.56	2.47	2.34	2.19	2.03	1.95	1.86	1.76	1.66	1.53	1.38
∞	6.63	4.61	3.78	3.32	3.02	2.80	2.64	2.51	2.41	2.32	2.18	2.04	1.88	1.79	1.70	1.59	1.47	1.32	1.00

TABLE T7
Minimum Number of Assessments in a Triangle Test
(Entries are n_{α,β,p_d})

Entries are the sample sizes (n) required in a Triangle test to deliver sensitivity defined by the values chosen for α, β, and p_d. Enter the table in the section corresponding to the chosen value of p_d and the row corresponding to the chosen value of α. Read the required sample size, n, from the column corresponding to the chosen value of β.

α	β 0.50	0.40	0.30	0.20	0.10	0.05	0.01	0.001
$p_d = 50\%$								
0.40	3	3	3	6	8	9	15	26
0.30	3	3	3	7	8	11	19	30
0.20	4	6	7	7	12	16	25	36
0.10	7	8	8	12	15	20	30	43
0.05	7	9	11	16	20	23	35	48
0.01	13	15	19	25	30	35	47	62
0.001	22	26	30	36	43	48	62	81
$p_d = 40\%$								
0.40	3	3	6	6	9	15	26	41
0.30	3	3	7	8	11	19	30	47
0.20	6	7	7	12	17	25	36	55
0.10	8	10	15	17	25	30	46	67
0.05	11	15	16	23	30	40	57	79
0.01	21	26	30	35	47	56	76	102
0.001	36	39	48	55	68	76	102	130
$p_d = 30\%$								
0.40	3	6	6	9	15	26	44	73
0.30	3	8	8	16	22	30	53	84
0.20	7	12	17	20	28	39	64	97
0.10	15	15	20	30	43	54	81	119
0.05	16	23	30	40	53	66	98	136
0.01	33	40	52	62	82	97	131	181
0.001	61	69	81	93	120	138	181	233
$p_d = 20\%$								
0.40	6	9	12	18	35	50	94	153
0.30	8	11	19	30	47	67	116	183
0.20	12	20	28	39	64	86	140	212
0.10	25	33	46	62	89	119	178	260
0.05	40	48	66	87	117	147	213	305
0.01	72	92	110	136	176	211	292	397
0.001	130	148	176	207	257	302	396	513
$p_d = 10\%$								
0.40	9	18	38	70	132	197	360	598
0.30	19	36	64	102	180	256	430	690
0.20	39	64	103	149	238	325	539	819
0.10	89	125	175	240	348	457	683	1011
0.05	144	191	249	325	447	572	828	1178
0.01	284	350	425	525	680	824	1132	1539
0.001	494	579	681	803	996	1165	1530	1992

43 × 2

TABLE T8
Critical Number of Correct Responses in a Triangle Test
(Entries are $x_{\alpha,n}$)

Entries are the minimum number of correct responses required for significance at the stated α-level (i.e., column) for the corresponding number of respondents, n (i.e., row). Reject the assumption of "no difference" if the number of correct responses is greater than or equal to the tabled value.

n	\multicolumn{7}{c}{α}	n	\multicolumn{7}{c}{α}

n	0.40	0.30	0.20	0.10	0.05	0.01	0.001	n	0.40	0.30	0.20	0.10	0.05	0.01	0.001
								31	12	13	14	15	16	18	20
								32	12	13	14	15	16	18	20
3	2	2	3	3	3	—	—	33	13	13	14	15	17	18	21
4	3	3	3	4	4	—	—	34	13	14	15	16	17	19	21
5	3	3	4	4	4	5	—	35	13	14	15	16	17	19	22
6	3	4	4	5	5	6	—	36	14	14	15	17	18	20	22
7	4	4	4	5	5	6	7	42	16	17	18	19	20	22	25
8	4	4	5	5	6	7	8	48	18	19	20	21	22	25	27
9	4	5	5	6	6	7	8	54	20	21	22	23	25	27	30
10	5	5	6	6	7	8	9	60	22	23	24	26	27	30	33
11	5	5	6	7	7	8	10	66	24	25	26	28	29	32	35
12	5	6	6	7	8	9	10	72	26	27	28	30	32	34	38
13	6	6	7	8	8	9	11	78	28	29	30	32	34	37	40
14	6	7	7	8	9	10	11	84	30	31	33	35	36	39	43
15	6	7	8	8	9	10	12	90	32	33	35	37	38	42	45
16	7	7	8	9	9	11	12	96	34	35	37	39	41	44	48
17	7	8	8	9	10	11	13	102	36	37	39	41	43	46	50
18	7	8	9	10	10	12	13	108	38	40	41	43	45	49	53
19	8	8	9	10	11	12	14	114	40	42	43	45	47	51	55
20	8	9	9	10	11	13	14	120	42	44	45	48	50	53	57
21	8	9	10	11	12	13	15	126	44	46	47	50	52	56	60
22	9	9	10	11	12	14	15	132	46	48	50	52	54	58	62
23	9	10	11	12	12	14	16	138	48	50	52	54	56	60	64
24	10	10	11	12	13	15	16	144	50	52	54	56	58	62	67
25	10	11	11	12	13	15	17	150	52	54	56	58	61	65	69
26	10	11	12	13	14	15	17	156	54	56	58	61	63	67	72
27	11	11	12	13	14	16	18	162	56	58	60	63	65	69	74
28	11	12	12	14	15	16	18	168	58	60	62	65	67	71	76
29	11	12	13	14	15	17	19	174	61	62	64	67	69	74	79
30	12	12	13	14	15	17	19	180	63	64	66	69	71	76	81

Note: For values of n not in the table, compute $z = (k - 1(1/3)n)/\sqrt{(2/9)n}$, where k is the number of correct responses. Compare the value of z to the α-critical value of a standard normal variable, i.e., the values in the last row of Table T3 $(z_a = t_{\alpha,\infty})$.

TABLE T9
Minimum Number of Assessments in a Duo-Trio or One-Sided Directional Difference Test (Entries are n_{α,β,p_d})

Entries are the sample sizes (n) required in a Duo-trio or One-Sided Directional Difference test to deliver the sensitivity defined by the values chosen for α, β, and p_d. Enter the table in the section corresponding to the chosen value of p_d for a Duo-trio test or p_{max} for a Directional Difference test and the row corresponding to the chosen value of α. Read the required sample size, n, from the column corresponding to the chosen value of β.

α		β 0.50	0.40	0.30	0.20	0.10	0.05	0.01	0.001
	$p_d = 50\%$								
0.40	$p_{max} = 75\%$	2	4	4	6	10	14	27	41
0.30		2	5	7	9	13	20	30	47
0.20		5	5	10	12	19	26	39	58
0.10		9	9	14	19	26	33	48	70
0.05		13	16	18	23	33	42	58	82
0.01		22	27	33	40	50	59	80	107
0.001		38	43	51	61	71	83	107	140
	$p_d = 40\%$								
0.40	$p_{max} = 70\%$	4	4	6	8	14	25	41	70
0.30		5	7	9	13	22	28	49	78
0.20		5	10	12	19	30	39	60	94
0.10		14	19	21	28	39	53	79	113
0.05		18	23	30	37	53	67	93	132
0.01		35	42	52	64	80	96	130	174
0.001		61	71	81	95	117	135	176	228
	$p_d = 30\%$								
0.40	$p_{max} = 65\%$	4	6	8	14	29	41	76	120
0.30		7	9	13	24	39	53	88	144
0.20		10	17	21	32	49	68	110	166
0.10		21	28	37	53	72	96	145	208
0.05		30	42	53	69	93	119	173	243
0.01		64	78	89	112	143	174	235	319
0.001		107	126	144	172	210	246	318	412
	$p_d = 20\%$								
0.40	$p_{max} = 60\%$	6	10	23	35	59	94	171	282
0.30		11	22	30	49	84	119	205	327
0.20		21	32	49	77	112	158	253	384
0.10		46	66	85	115	168	214	322	471
0.05		71	93	119	158	213	268	392	554
0.01		141	167	207	252	325	391	535	726
0.001		241	281	327	386	479	556	731	944
	$p_d = 10\%$								
0.40	$p_{max} = 55\%$	10	35	61	124	237	362	672	1124
0.30		30	72	117	199	333	479	810	1302
0.20		81	129	193	294	451	618	1006	1555
0.10		170	239	337	461	658	861	1310	1905
0.05		281	369	475	620	866	1092	1583	2237
0.01		550	665	820	1007	1301	1582	2170	2927
0.001		961	1125	1309	1551	1908	2248	2937	3812

TABLE T10
Critical Number of Correct Responses in a Duo-Trio or One-Sided Directional Difference Test
(Entries are $x_{\alpha,n}$)

Entries are the minimum number of correct responses required for significance at the stated α-level (i.e., column) for the corresponding number of respondents, n (i.e., row). Reject the assumption of "no difference" if the number of correct responses is greater than or equal to the tabled value.

n	0.40	0.30	0.20	0.10	0.05	0.01	0.001	n	0.40	0.30	0.20	0.10	0.05	0.01	0.001
								31	17	18	19	20	21	23	25
2	2	2	—	—	—	—	—	32	18	18	19	21	22	24	26
3	3	3	3	—	—	—	—	33	18	19	20	21	22	24	26
4	3	4	4	4	—	—	—	34	19	20	20	22	23	25	27
5	4	4	4	5	5	—	—	35	19	20	21	22	23	25	27
6	4	5	5	6	6	—	—	36	20	21	22	23	24	26	28
7	5	5	6	6	7	7	—	40	22	23	24	25	26	28	31
8	5	6	6	7	7	8	—	44	24	25	26	27	28	31	33
9	6	6	7	7	8	9	—	48	26	27	28	29	31	33	36
10	6	7	7	8	9	10	10	52	28	29	30	32	33	35	38
11	7	7	8	9	9	10	11	56	30	31	32	34	35	38	40
12	7	8	8	9	10	11	12	60	32	33	34	36	37	40	43
13	8	8	9	10	10	12	13	64	34	35	36	38	40	42	45
14	8	9	10	10	11	12	13	68	36	37	38	40	42	45	48
15	9	10	10	11	12	13	14	72	38	39	41	42	44	47	50
16	10	10	11	12	12	14	15	76	40	41	43	45	46	49	52
17	10	11	11	12	13	14	16	80	42	43	45	47	48	51	55
18	11	11	12	13	13	15	16	84	44	45	47	49	51	54	57
19	11	12	12	13	14	15	17	88	46	47	49	51	53	56	59
20	12	12	13	14	15	16	18	92	48	50	51	53	55	58	62
21	12	13	13	14	15	17	18	96	50	52	53	55	57	60	64
22	13	13	14	15	16	17	19	100	52	54	55	57	59	63	66
23	13	14	15	16	16	18	20	104	54	56	57	60	61	65	69
24	14	14	15	16	17	19	20	108	56	58	59	62	64	67	71
25	14	15	16	17	18	19	21	112	58	60	61	64	66	69	73
26	15	15	16	17	18	20	22	116	60	62	64	66	68	71	76
27	15	16	17	18	19	20	22	122	63	65	67	69	71	75	79
28	16	16	17	18	19	21	23	128	66	68	70	72	74	78	82
29	16	17	18	19	20	22	24	134	69	71	73	75	78	81	86
30	17	17	18	20	20	22	24	140	72	74	76	79	81	85	89

Note: For values of n not in the table, compute $z = (k - 0.5n)/\sqrt{0.25n}$, where k is the number of correct responses. Compare the value of z to the α-critical value of a standard normal variable, i.e., the values in the last row of Table T3 ($z_\alpha = t_{\alpha,\infty}$).

TABLE T11

Minimum Number of Assessments in a Two-Sided Directional Difference Test (Entries are $n_{\alpha,\beta,p_{max}}$)

Entries are the sample sizes (n) required in a Two-Sided Directional Difference test to deliver sensitivity defined by the values chosen for α, β, and p_{max}. Enter the table in the section corresponding to the chosen value of p_{max} and the row corresponding to the chosen value of α. Read the required sample size, n, from the column corresponding to the chosen value of β.

α	β							
	0.50	0.40	0.30	0.20	0.10	0.05	0.01	0.001
$p_{max} = 75\%$								
0.40	5	5	10	12	19	26	39	58
0.30	6	8	11	16	22	29	42	64
0.20	9	9	14	19	26	33	48	70
0.10	13	16	18	23	33	42	58	82
0.05	17	20	25	30	42	49	67	92
0.01	26	34	39	44	57	66	87	117
0.001	42	50	58	66	78	90	117	149
$p_{max} = 70\%$								
0.40	5	10	12	19	30	39	60	94
0.30	8	13	18	22	33	44	68	102
0.20	14	19	21	28	39	53	79	113
0.10	18	23	30	37	53	67	93	132
0.05	25	35	40	49	65	79	110	149
0.01	44	49	59	73	92	108	144	191
0.001	68	78	90	102	126	147	188	240
$p_{max} = 65\%$								
0.40	10	17	21	32	49	68	110	166
0.30	13	20	29	42	59	81	125	188
0.20	21	28	37	53	72	96	145	208
0.10	30	42	53	69	93	119	173	243
0.05	44	56	67	90	114	145	199	176
0.01	73	92	108	131	164	195	261	345
0.001	121	140	161	188	229	267	342	440
$p_{max} = 60\%$								
0.40	21	32	49	77	112	158	253	384
0.30	31	44	66	89	133	179	283	425
0.20	46	66	85	115	168	214	322	471
0.10	71	93	119	158	213	268	392	554
0.05	101	125	158	199	263	327	455	635
0.01	171	204	241	291	373	446	596	796
0.001	276	318	364	425	520	604	781	1010
$p_{max} = 55\%$								
0.40	81	129	193	294	451	618	1006	1555
0.30	110	173	254	359	550	721	1130	1702
0.20	170	239	337	461	658	861	1310	1905
0.10	281	369	475	620	866	1092	1583	2237
0.05	390	497	620	786	1055	1302	1833	2544
0.01	670	802	963	1167	1493	1782	2408	3203
0.001	1090	1260	1461	1707	2094	2440	3152	4063

TABLE T12
Critical Number of Correct Responses in a Two-Sided Directional Difference Test (Entries are $x_{\alpha,n}$)

Entries are the minimum number of correct responses required for significance at the stated α-level (i.e., column) for the corresponding number of respondents, n (i.e., row). Reject the assumption of "no difference" if the number of correct responses is greater than or equal to the tabled value.

n	0.40	0.30	0.20	0.10	0.05	0.01	0.001	n	0.40	0.30	0.20	0.10	0.05	0.01	0.001
								31	19	19	20	21	22	24	25
2	—	—	—	—	—	—	—	32	19	20	21	22	23	24	26
3	3	3	—	—	—	—	—	33	20	20	21	22	23	25	27
4	4	4	4	—	—	—	—	34	20	21	22	23	24	25	27
5	4	5	5	5	—	—	—	35	21	22	22	23	24	26	28
6	5	5	6	6	6	—	—	36	22	22	23	24	25	27	29
7	6	6	6	7	7	—	—	40	24	24	25	26	27	29	31
8	6	6	7	7	8	8	—	44	26	26	27	28	29	31	34
9	7	7	7	8	8	9	—	48	28	29	29	31	32	34	36
10	7	8	8	9	9	10	—	52	30	31	32	33	34	36	39
11	8	8	9	9	10	11	11	56	32	33	34	35	36	39	41
12	8	9	9	10	10	11	12	60	34	35	36	37	39	41	44
13	9	9	10	10	11	12	13	64	36	37	38	40	41	43	46
14	10	10	10	11	12	13	14	68	38	39	40	42	43	46	48
15	10	11	11	12	12	13	14	72	41	41	42	44	45	48	51
16	11	11	12	12	13	14	15	76	43	44	45	46	48	50	53
17	11	12	12	13	13	15	16	80	45	46	47	48	50	52	56
18	12	12	13	13	14	15	17	84	47	48	49	51	52	55	58
19	12	13	13	14	15	16	17	88	49	50	51	53	54	57	60
20	13	13	14	15	15	17	18	92	51	52	53	55	56	59	63
21	13	14	14	15	16	17	19	96	53	54	55	57	59	62	65
22	14	14	15	16	17	18	19	100	55	56	57	59	61	64	67
23	15	15	16	16	17	19	20	104	57	58	60	61	63	66	70
24	15	16	16	17	18	19	21	108	59	60	62	64	65	68	72
25	16	16	17	18	18	20	21	112	61	62	64	66	67	71	74
26	16	17	17	18	19	20	22	116	64	65	66	68	70	73	77
27	17	17	18	19	20	21	23	122	67	68	69	71	73	76	80
28	17	18	18	19	20	22	23	128	70	71	72	74	76	80	83
29	18	18	19	20	21	22	24	134	73	74	75	78	79	83	87
30	18	19	20	20	21	23	25	140	76	77	79	81	83	86	90

Note: For values of n not in the table, compute $z = (k - 0.5n)/\sqrt{0.25n}$, where k is the number of correct responses. Compare the value of z to the $\alpha/2$-critical value of a standard normal variable, i.e., the values in the last row of Table T3 ($z_{\alpha/2} = t_{\alpha/2,\infty}$).

TABLE T13
Minimum Number of Assessments in a Two-out-of-Five Test
(Entries are n_{α,β,p_d})

Entries are the sample sizes (n) required in a Two-out-of-five test to deliver the sensitivity defined by the values chosen for α, β, and p_d. Enter the table in the section corresponding to the chosen value of p_d and the row corresponding to the chosen value of α. Read the required sample size, n, from the column corresponding to the chosen value of β.

	β							
α	0.50	0.40	0.30	0.20	0.10	0.05	0.01	0.001
$p_d = 50\%$								
0.40	3	4	4	5	6	7	9	13
0.30	3	4	4	5	6	7	9	16
0.20	3	4	4	5	6	7	12	18
0.10	3	4	4	5	8	9	15	18
0.05	3	6	6	7	8	12	17	24
0.01	5	7	8	9	13	14	22	29
0.001	9	9	12	13	17	21	27	36
$p_d = 40\%$								
0.40	4	4	5	6	7	9	12	20
0.30	4	4	5	6	7	9	15	23
0.20	4	4	5	6	7	12	15	23
0.10	4	4	5	9	10	15	18	30
0.05	6	7	7	11	13	18	24	33
0.01	8	9	12	14	18	23	30	42
0.001	12	13	17	21	26	31	41	54
$p_d = 30\%$								
0.40	5	5	6	8	9	11	20	30
0.30	5	5	6	8	9	15	24	35
0.20	5	5	6	8	13	15	28	39
0.10	5	5	9	11	17	22	32	47
0.05	7	8	12	14	20	26	39	54
0.01	13	14	18	23	30	36	49	69
0.001	21	22	27	32	42	49	66	87
$p_d = 20\%$								
0.40	6	7	8	10	13	21	38	59
0.30	6	7	8	10	18	26	43	69
0.20	6	7	8	15	22	30	53	79
0.10	10	11	17	23	31	40	62	94
0.05	13	19	24	27	40	53	76	108
0.01	24	30	36	43	57	70	99	136
0.001	38	48	55	67	81	99	129	172
$p_d = 10\%$								
0.40	9	11	13	22	40	60	108	184
0.30	9	16	19	34	54	80	128	212
0.20	14	22	31	47	73	99	161	245
0.10	25	38	54	70	103	130	206	297
0.05	41	55	70	94	127	167	244	349
0.01	77	98	121	145	192	233	330	449
0.001	135	158	187	224	278	332	438	572

TABLE T14
Critical Number of Correct Responses in a Two-out-of-Five Test
(Entries are $x_{\alpha,n}$)

Entries are the minimum number of correct responses required for significance at the stated α-level (i.e., column) for the corresponding number of respondents, n (i.e., row). Reject the assumption of "no difference" if the number of correct responses is greater than or equal to the tabled value.

n	0.40	0.30	0.20	0.10	0.05	0.01	0.001		n	0.40	0.30	0.20	0.10	0.05	0.01	0.001
									31	4	5	5	6	7	8	10
									32	4	5	6	6	7	9	10
3	1	1	2	2	2	3	3		33	5	5	6	7	7	9	11
4	1	2	2	2	3	3	4		34	5	5	6	7	7	9	11
5	2	2	2	2	3	3	4		35	5	5	6	7	8	9	11
6	2	2	2	3	3	4	5		36	5	5	6	7	8	9	11
7	2	2	2	3	3	4	5		37	5	6	6	7	8	9	11
8	2	2	2	3	3	4	5		38	5	6	6	7	8	10	11
9	2	2	3	3	4	4	5		39	5	6	6	7	8	10	12
10	2	2	3	3	4	5	6		40	5	6	7	7	8	10	12
11	2	3	3	3	4	5	6		41	5	6	7	8	8	10	12
12	2	3	3	4	4	5	6		42	6	6	7	8	9	10	12
13	2	3	3	4	4	5	6		43	6	6	7	8	9	10	12
14	3	3	3	4	4	5	7		44	6	6	7	8	9	11	12
15	3	3	3	4	5	6	7		45	6	6	7	8	9	11	13
16	3	3	4	4	5	6	7		46	6	7	7	8	9	11	13
17	3	3	4	4	5	6	7		47	6	7	7	8	9	11	13
18	3	3	4	4	5	6	8		48	6	7	8	9	9	11	13
19	3	3	4	5	5	6	8		49	6	7	8	9	10	11	13
20	3	4	4	5	5	7	8		50	6	7	8	9	10	11	14
21	3	4	4	5	6	7	8		51	7	7	8	9	10	12	14
22	3	4	4	5	6	7	8		52	7	7	8	9	10	12	14
23	4	4	4	5	6	7	9		53	7	7	8	9	10	12	14
24	4	4	5	5	6	7	9		54	7	7	8	9	10	12	14
25	4	4	5	5	6	7	9		55	7	8	8	9	10	12	14
26	4	4	5	6	6	8	9		56	7	8	8	10	10	12	14
27	4	4	5	6	6	8	9		57	7	8	9	10	11	12	15
28	4	5	5	6	7	8	10		58	7	8	9	10	11	13	15
29	4	5	5	6	7	8	10		59	7	8	9	10	11	13	15
30	4	5	5	6	7	8	10		60	7	8	9	10	11	13	15

Note: For values of n not in the table, compute $z = (k - 0.1n)/\sqrt{0.09n}$, where k is the number of correct responses. Compare the value of z to the α-critical value of a standard normal variable, i.e., the values in the last row of Table T3 $(z_\alpha = t_{\alpha\infty})$.

Index

A

B